AN INTRODUCTION TO
METABOLIC AND
CELLULAR ENGINEERING

Second Edition

AN INTRODUCTION TO
METABOLIC AND
CELLULAR ENGINEERING

Second Edition

S Cortassa
Johns Hopkins University, USA

M A Aon
Johns Hopkins University, USA

A A Iglesias
Universidad del Litoral, Argentina

J C Aon
GlaxoSmithKline, USA

D Lloyd
Cardiff University, UK

World Scientific

NEW JERSEY · LONDON · SINGAPORE · BEIJING · SHANGHAI · HONG KONG · TAIPEI · CHENNAI

Published by

World Scientific Publishing Co. Pte. Ltd.

5 Toh Tuck Link, Singapore 596224

USA office: 27 Warren Street, Suite 401-402, Hackensack, NJ 07601

UK office: 57 Shelton Street, Covent Garden, London WC2H 9HE

British Library Cataloguing-in-Publication Data
A catalogue record for this book is available from the British Library.

AN INTRODUCTION TO METABOLIC AND CELLULAR ENGINEERING
2nd Edition

ISBN-13 978-981-4365-71-0
ISBN-10 981-4365-71-8
ISBN-13 978-981-4365-72-7 (pbk)
ISBN-10 981-4365-72-6 (pbk)

Typeset by Stallion Press
Email: enquiries@stallionpress.com

Printed in Singapore.

To Juan Ernesto, Nehuen Quimey, Raul and Pedro.
In memoriam to Gladys.
To Francisca and Maris. In memoriam to Miguel.
To Silvia. In memoriam to Ruben, Amalia, Indalecio, and Norberto.
To Maria Elena and Benjamin

Contents

Preface

In the opening sentence of the Preface to the First Edition of *An Introduction to Metabolic and Cellular Engineering*, almost ten years ago, we stated: "Metabolic and Cellular Engineering, although as yet only at the beginning, promises huge advances in all fields of the life sciences." Needless to say, this statement passed the test of time; Metabolic and Cellular Engineering (MCE) still promises huge advances, new discoveries, and new avenues of research. MCE also produces new expectations by becoming the vibrant and exciting reality of a quickly diversifying and multidisciplinary research field.

We also wanted our book to promote interest in the MCE endeavor by attracting researchers and the attention of the scientific and technological communities at large from academia and industry. Hopefully, the First Edition contributed to the fulfillment of that aim. It is with the same inspirational theme, and thanks to the initiative of World Scientific, that we updated and enlarged the Second Edition of *An Introduction to Metabolic and Cellular Engineering*.

Today, MCE is more than an exciting scientific enterprise; it has become the cornerstone for coping with the challenges ahead for mankind. Continuous developments, conceptual and technological innovation, deal with new challenges, as well as those present ten years ago. Challenges are still everywhere — from unraveling fundamental aspects of cellular function to the increase in energy and food demand that rises in parallel with population growth. Then there is the ongoing shift between a fossil fuel and a clean energy-based world economy as a reflection of the mounting recognition of global climate change. These topics are treated in Chapters 1 and 10.

In charting the progress of MCE during the last decade, we could but feel in awe of the enormous strides of progress made from nascent Metabolic Engineering to the Systems Bioengineering of today. The burgeoning availability of genomic sequences from diverse species has been spectacular. It has become the engine that drives genetic means for modification of existing organisms and in the generation of synthetic, manmade, ones. From the initial attempts at purposeful genetic modification of a cell for the production of valuable compounds, we have now moved on to change microbes genetically or metabolically. This progression is treated through specific examples from bacteria to plant cells in Chapters 3, 4, and 7. Additionally, Chapter 7 details the actual potential of MCE as applied to microalgae and plant cells, highlighting their photosynthetic capacity and metabolic versatility.

The arsenal of experimental and theoretical tools available for MCE has expanded enormously, driven by the re-emergence of Physiology as Systems Biology. The revival of the concept of networks fueled by new developments has become central to Systems Biology. This is now focused on the study of their topological and dynamic properties. Networks represent an integrative vision of how processes of disparate nature relate to each other and as such is becoming a key analytical and conceptual tool for MCE. The combination of high throughput technologies with computational modeling and data mining are among the tools being utilized for handling and interpreting massive amounts of data. These developments and improvements are reflected mainly in Chapters 5 and 6 on Modeling Networks and Dynamic Process Behavior, respectively, which together with Chapters 2 to 4 give the essentials on concepts and analytical tools underlying the whole book.

Our initial focus was the organism, and its performance in a specific biotransformation and this continues to be so in this Second Edition. During industrial scale-up, when most microbial modifications become necessary, we stress that process robustness is a dominant feature (Chapter 9).

At the time of this Second Edition, researchers from independent laboratories around the world have been able to induce pluripotency in somatic cells to obtain inducible Pluripotent Stem cells (iPSCs). These findings open enormous possibilities for organ regeneration and their engineering. Chapter 10 briefly reviews this nascent field, and Chapter 8 contains a brief introduction to cell culture techniques.

For the Second Edition of this book we would like to gratefully acknowledge the following people, whose work and expertise have translated into fruitful and long-term collaborations, enlightening and enriching the authors: Brian O'Rourke (Johns Hopkins University, MD, USA), Raimond Winslow (Johns Hopkins University, MD, USA), Gordon Tomaselli (Johns Hopkins University, MD, USA), Eduardo Marbán (Cedars-Sinai Heart Institute, CA, USA), Nazareno Paolocci (Johns Hopkins University, MD, USA), Fadi G. Akar (Mount Sinai Hospital, NY, USA), Christoph Maack (Universitätsklinikum des Saarlandes, Homburg, Germany), Katey Lemar (Treorchy Comprehensive School, Wales, UK), Douglas Murray (Keio University, Tsuruoka, Japan), Marc Roussel (Lethbridge University, Alberta, Canada), Lufang Zhou (Johns Hopkins University, MD, USA), Ting Liu (Johns Hopkins University, MD, USA), Gabriele Tocchetti (National Cancer Institute, Naples, Italy), Brian A. Stanley and Vidhya Sivakumaran (Johns Hopkins University, MD, USA), Hannah Sivak (Arizona, USA), Jack Preiss (Michigan State University, MI, USA), Miguel A. Ballicora (Loyola University Chicago, IL, USA), Florencio E. Podestá (CEFOBI, CONICET, Argentina), Diego F. Gomez Casati (CEFOBI, CONICET, Argentina), Diego M. Bustos (INTECH, CONICET, Argentina), Sergio A. Guerrero (IAL, UNL-CONICET, Argentina), Alejandro J. Beccaria (IAL, UNL-CONICET, Argentina), Mabel Aleanzi (IAL, UNL-CONICET, Argentina), Claudia V. Piattoni (IAL, UNL-CONICET, Argentina), Raúl Comelli (INCAPE, UNL-CONICET, Argentina).

Sonia Cortassa
Miguel Antonio Aon
Alberto Alvaro Iglesias
Juan Carlos Aon
David Lloyd

Chapter 1

Introduction

Since the First Edition of this book in 2002, we have witnessed remarkable advance in our potential to address and solve increasingly complex problems in fundamental and applied research, expanding our horizons in many aspects of human endeavor. Major driving forces underlying these developments are:

(1) Completion of genome sequencing of more than 350 species (December 2010), including that of *Homo sapiens*.
(2) Powerful methods for collecting massive amounts of data by way of high throughput automated technologies.
(3) Increased computer power and feasibility of developing and applying mathematical modeling.
(4) Automation and accessibility of data bases by newly developing methods of bioinformatics.
(5) The successful development of new theoretical tools related to networks, which has become the predominant paradigm for assembly and interpretation of huge amounts of information.
(6) The ever-increasing connectivity among researchers throughout the world, which has augmented the scale of the projects undertaken.
(7) The generation of synthetic organisms; the first one is a bacterium whose whole genome has been chemically synthesized.

In parallel to all these advances, the magnitude of the challenges facing mankind has arisen steeply. Increasing global food demands and food security, industrial units, environmental pollution, the self-reinforced (potentially exponential) loop of global warming, and energy crises are leading the need for energy derived from renewable sources and, in a

1

wider scope, the conversion of the industry producing chemicals and other marketable products into "biorefineries." All these innovations together drive continuing developments.

The re-emerging field of physiology, now widely referred to as *Systems Biology*, reflects the major shift between the *analytical* and *integrative* philosophies in biology, which took place over the past decade. Major transitions, historically occurring at the *"les fins des siècles"* phenomena represent hinges in the accelerating development of the scientific enterprise.

Metabolic Engineering (ME), defined as the purposeful modification of cells by means of DNA recombinant technologies, represented an early development in this nascent period. The first edition of this book echoed the major initial stages of the implementation of ME in bioprocesses, focusing on the cellular level, and on microbes, in particular. Our interdisciplinary approach made reference to the integrative nature of the scientific endeavor as practiced today. The major disciplines involved — thermodynamics, enzyme kinetics, biomathematics, biochemistry, genetics, molecular biology, microbial physiology, and bioenergetics — still prevail, while new ones *Bioinformatics* and *Network Theory* join the scene. In this integrative framework, ME has driven major improvements of microbial performance (in rates and yields) for production of fine chemicals, amino acids, biodegradable polymers, biofuels, bulk-chemicals, and biomaterials from renewable nonfood biomass. This long list adds to items listed in our first edition (i.e. production of metabolites and heterologous proteins, introduction of heterologous metabolic pathways into microorganisms to give them the ability to degrade xenobiotics, modification of enzymatic activities for metabolite production especially of high value pharmaceuticals and biofuels).

These advances further improve and expand the potential of ME. At this stage, however, more attention is being paid to the global function of metabolic, transport, and signaling networks rather than selected pathways. This development, already anticipated in our First Edition, was inherent in our title: *Metabolic and Cellular Engineering* (MCE). The current achievements of MCE, are to improve the overall physiological behavior of cells (i.e. yields, growth rates, capabilities for proliferation, strain adaptability and robustness to harsh environments, and stress tolerance) throughout the entire production process encompassing not only the mass–energy transducing networks of the organism but also the information (signaling) process. The mass and energy transduction networks intertwine with transcription factor and second messenger functions, as well as the production or remodeling of supramolecular structures, thereby forming networks of their own. Cellular functions such as growth (macromolecular syntheses), division (proliferation), differentiation, transport, electrical conduction, and rhythms, are the emergent outcomes resulting from the dynamic interplay between all these networks. With all these tools at hand, the major challenge now is to find how (supra)cellular and organ networks functionally coordinate at a global level. The anticipated pay-offs of achievements in fundamental research in this area are:

(1) Introduction of physiological functions into cells, tissues, and organs.
(2) Control of proliferation and differentiation of stem cells for regenerating tissues or whole organs.
(3) Redirection of malign, uncontrolled, proliferation of cancer cells into controlled, sociable patterns of behavior.
(4) Targeted pharmacology with attenuated or negligible side effects.

Some discoveries and undergoing developments in the research areas of network properties encourage hope for the understanding of the global behavior of cells. These achievements culminate in the recent advances (during the last 15 years) referred to above, as well as in the improved implementation of earlier ones. Worthy of mention among the latter are the extension and refinement of experimental and theoretical tools such as the combination of fermentation and DNA recombinant technologies, Metabolic Control Analysis (MCA), Flux Balance or Metabolic Flux Analysis (MFA), and Biochemical Systems Theory. The classical studies emerging from the so called black-box approach, microbial biochemistry and physiology and Mosaic Non Equilibrium Thermodynamics, are also major achievements.

The powerful tools of nonlinear kinetics and stability analysis, essential for describing the self-organized dynamic behavior of networks in general, have been revived in a biological context under a flurry of new applications. All new advances and previously developed tools and approaches allow us to analyze, quantify, and predict the behavior of networks of disparate nature.

In this overall new landscape, the emerging challenge is represented by the interdisciplinary effort directed to redesign complex biosystems, or the attempt to design altogether completely new ones ("artificial life") as recent developments testify. Notwithstanding, as successes and new findings accumulate, many hitherto unknown or unrecognized aspects arise, such as the overall functional control and regulation of networks in cells, organs, and organisms, also including their spatiotemporal organization.

MCE is at the interphase between fundamental and biotechnological research as an iterative, self-correcting and self-optimizing process for facing the multiple challenges posed by continuing elaboration of our understanding of life processes.

Metabolic and Cellular Engineering in the Context of Bioprocess Engineering

Traditionally, bioprocesses underpin the food and pharmaceutical industries. Thus, microbial, plant cells, or their components are used for product manufacture, or for the degradation of toxic wastes. The use of microorganisms for making fermented foods has a very long history. Since Sumerian times, many different bioprocesses have been developed to give an enormous variety of commercial product commodities, e.g. ethanol, from cheap ones to more recently very expensive ones (e.g. antibiotics, therapeutic proteins, or vaccines). Enzymes and microorganisms such as bakers' yeast are also commercially obtainable by processes from malted barley.

The bases of the broad and highly multidisciplinary field of ME have been established (see Aon and Cortassa, 2005; Bailey, 1991; Cameron and Chaplen, 1997; Cameron and Tong, 1993; Farmer and Liao, 1996; Stephanopoulos, 2008 for reviews). ME within the context of bioprocess engineering constitutes a thoroughly transdisciplinary effort toward the development of rationally designed cells with specific biotransformation capabilities. This transdisciplinary effort requires the participation of scientists from different disciplines.

As a new approach for rationally designing biological systems, the introduction of specific modifications to metabolic networks for the purpose of improving cellular properties is becoming ever more important for biotechnological production processes and medicine. Because the challenge of this interdisciplinary effort is to redesign complex biosystems, a rigorous understanding of the interactions between metabolic and regulatory networks is critical. At this point we will adopt the notation MCE throughout the book, since it defines more accurately the panoply of activities being undertaken in the field. As such, an important component of MCE is the emphasis of the regulation of metabolic reactions *in vivo*. This goal differentiates the field from those related areas of life science that adopt the reductionistic approach as a primary objective, although it is inevitable that fundamental understanding based on detailed information gleaned from purified constituents separated from living organisms may be necessary. Concepts and methodologies of MCE have potential value in the application of metabolic cellular systems for the design of biotechnology production processes. These include cell-based processes as well as gene therapies, and degradation of recalcitrant pollutants. They promise a great impact on many areas of medicine and plant biotechnology (Beer *et al.*, 2009; Capell and Christou, 2004; Cascante *et al.*, 2010; Wong and Chiu, 2010; Yarmush and Berthiaume, 1997).

Some examples of research areas in MCE are

 (1) Experimental and computational tools.
 (2) Production of pharmaceuticals.
 (3) Manufacture of fuels and chemicals.
 (4) Biomaterials.
 (5) Modification of microbes, plant, and animal cells.
 (6) Large-scale production of proteins.
 (7) Strategies for strain improvement via ME.
 (8) Higher level ME through regulatory genes.
 (9) Medical applications and gene therapy.
(10) Environmental applications.
(11) Nanotechnology.
(12) Stem cells for repair of damaged tissues or organs (Atala, 2009; Cameron and Tong, 1993; Kruse and Hankamer, 2010; Lee and Papoutsakis, 1999; Stephanopoulos, 2008; Wackett, 2010) (see also Table 1.1).

Table 1.1   Examples of ME developments in main areas of biotechnological interest

Chemical	Host Organism	Notes (pathway engineered or modification method applied)	Reference
Improved production of chemicals and proteins produced by the host organism				
Xylanase	*Streptomyces lividans*	The signal peptide of xylanase A (XlnA) was replaced with signal peptides of mannanase A and cellulase A increasing the expression level 1.5–2.5 fold, according to the length of the signal peptide sequence	(Page *et al.*, 1996)
Aromatic compounds	*Escherichia coli* non-PTS glucose transporting mutants	*E. coli* with an inactivated PTS system with Gal permease, glucokinase, to internalize and phosphorylate glucose, plus the tklA coding for transketolase allowed higher yields of aromatic compounds	(Flores *et al.*, 1996)
Riboflavin, folic acid, and purine nucleosides	*Bacillus subtilis*	A diagnosis of the controlling steps driving carbon to the synthesis of purine-related compounds revealed that the central amphibolic pathways supplying carbon intermediates and reduced cofactors, are not limiting	(Sauer *et al.*, 1997; Sauer *et al.*, 1998)
Organic acids	*Saccharomyces cerevisiae*	Overexpression of malate dehydrogenase results in accumulation of malic, citric, and fumaric acids thereby revealing pyruvate carboxylase as a limiting factor for malic acid production	(Pines *et al.*, 1997)
Diacetyl and/or acetoin	*Lactococcus lactis*	Expression of *Streptococcus* mutans NADH oxidase that decreased the NADH/NAD ratio resulted in a shift from homolactic to mixed acid fermentation by activation of the acetolactate synthase and acetoin or diacetyl production	(Lopez de Felipe *et al.*, 1998)

(Continued)

Table 1.1   (*Continued*)

Chemical	Host Organism	Notes (pathway engineered or modification method applied)	Reference
Cellulose accumulation and decreased lignin synthesis	Aspen (*Populus tremuloides*)	The lignin biosynthetic pathway was down-regulated by antisense mRNA expression of 4-coumarate coenzyme A ligase. Trees exhibited 45% decrease in lignin and a compensatory 15% increase in cellulose	(Hu *et al.*, 1999)
Lactic acid	*Escherichia coli*	Overexpression of homologous and heterologous phosphofructokinase and pyruvate kinase resulted in altered fermentation patterns: increase in lactic acid with decreased ethanol production	(Emmerling *et al.*, 1999)
Cephalosporin	*Streptomyces clavuligerus*	Kinetic analysis and modeling indicated that the rate controlling step was the lysine-epsilon-aminotransferase that when overexpressed exhibited higher yield of the β-lactam antibiotic	(Khetan *et al.*, 1996)
Carbon dioxide	*Saccharomyces cerevisiae*	Overexpression of the enzymes catalyzing the lower glycolytic pathway led to increased rates of CO_2 production but to increased ethanol production only during glucose pulses	(Smits *et al.*, 2000)
Pyruvate	*Torulopsis glabrata*	MFA was used to point the most appropriate culture conditions regarding glucose, dissolved oxygen, and vitamin concentrations to optimize the production of pyruvate under fed-batch cultivation	(Hua *et al.*, 2001)
L-threonine	*Escherichia coli, Corynebacterium glutamicum*	Engineering of L-threonine biosynthesis, intracellular consumption, and export from the cell	Reviewed in (Dong *et al.*, 2011)

(*Continued*)

Table 1.1   (*Continued*)

Chemical	Host Organism	Notes (pathway engineered or modification method applied)	Reference
Hydrogen	*Chlamydomonas Reindhardtii*	Heterologous expression of the hexose/H^+ symporter (HUP1) further enhancing H_2 production by enabling the organism to couple glucose oxidation to hydrogenase activity	(Doebbe *et al.*, 2007)
L-tyrosine	*Escherichia coli*	Combinatorial overexpression of aromatic amino acid biosynthesis genes in L-tyrosine producing *E. coli* strains T1 and T2. Shikimate dehydrogenase (*ydiB*) and kinase (*aroK*), or their combination for overexpression resulted in the best L-tyrosine yields. Also *tyrB*, encoding aromatic amino acid transaminase, overexpression together with *ydiB* and *aroK* improved production	(Lutke-Eversloh and Stephano-poulos, 2008)

Production of chemicals and proteins new to the host organism

Chemical	Host Organism	Notes (pathway engineered or modification method applied)	Reference
Carotenoids	*Candida utilis*	Yeast cells transformed with the ctr operon from *Erwinia* coding for enzymes of carotenoid synthesis, redirect farnesyl pyrophosphate, an intermediate in the synthesis of ergosterol toward lycopene and β-carotene	(Shimada *et al.*, 1998)
Lycopene	*Escherichia coli*	Expression of the Pps (PEP synthase) and deletion of the PYK (pyruvate kinase) activities result in higher yields of lycopene production. This indicates that isoprenoid synthesis in *E. coli* is limited by the availability of glyceraldehyde3P	(Farmer and Liao, 2001)

(*Continued*)

Table 1.1 *(Continued)*

Chemical	Host Organism	Notes (pathway engineered or modification method applied)	Reference
Pregnenolone and progesterone	Yeast	Introduction of adrenodoxin (ADX), and adrenodoxin reductase (ADR) P450scc, β-OH steroid dehydrogenase-isomerase; Delta7 sterol reductase and disruption of delta22 sterol desaturase from the steroidogenic pathway from ergosterol	(Duport *et al.*, 1998)
Low molecular weight fructan	Sugar beet	Transformation with a sucrose fructosyl transferase from *Helianthus tuberosus*	(Sevenier *et al.*, 1998)
Globotriose and UDP galactose	Recombinant *Escherichia coli* and *Corynebacterium ammoniagenes*	*E. coli* cells overexpressing UDP-Gal biosynthetic genes with *C. ammoniagenes* able to produce globotriose from orotic acid and galactose	(Koizumi *et al.*, 1998)
Gamma linolenic acid	Tobacco plants	Transformation of tobacco plants with a cyanobacterial delta-6 desaturase that allows a change in the composition of fatty acids introducing polyunsaturated ones	(Reddy and Thomas, 1996)
Medium chain length polyhydroxyalkanoates	*Arabidopsis thaliana*	β-oxidation of plant fatty acids generate various R-3-hydroxyacyl-CoA that serve as precursors of polyhydroxyalkanoates in plants transformed with polyhydroxyalkanoate synthase (PhaC1) synthase from *Pseudomonas aeruginosa* and the gene products were directed to peroxisomes and glyoxysomes	(Mittendorf *et al.*, 1998)

(Continued)

Table 1.1   (*Continued*)

Chemical	Host Organism	Notes (Pathway engineered or modification method applied)	Reference
L-alanine	*Lactococcus lactis*	*Bacillus sphaericus* alanine dehydrogenase was introduced into a *L. lactis* strain deficient in lactate dehydrogenase shifting the carbon fate from lactate toward alanine	(Hols *et al.*, 1999)
1,2 Propanediol	*Escherichia coli*	NADH-linked glycerol dehydrogenase expressed together with methylglyoxal synthase improved up to 1.2 g/L	(Altaras and Cameron, 1999)
Isoprenoid	*Escherichia coli*	Isopentenyl diphosphate isomerase gene from yeast overexpressed in an *E. coli* strain that expressed *Erwinia* carotenoid biosynthetic genes resulted in increased accumulation of β-carotene	(Kajiwara *et al.*, 1997)
Vitreoscilla hemoglobin (VHb)	*Various microbial cells*	Improves growth, protein secretion, metabolite productivity, and stress resistance in various hosts	(Zhang *et al.*, 2007)
Heterologous proteins	*Escherichia coli*	Suppression of nonenzymatic gluconoylation/ phosphogluconoylation products of three different heterologous proteins by overexpression of phosphogluconolactonase	(Aon *et al.*, 2008)

Substrate	Host Organism	Notes (pathway engineered or modification method applied)	Reference
		Extension of substrate range for growth and product formation	
Pentoses such as xylose and arabinose	*Tetragenococcus halophila*	Mannose PTS (PEP:mannose phosphotransferase, PFK and glucokinase triple mutants are able to ferment pentoses in the presence of hexoses)	(Abe and Higuchi, 1998)

(*Continued*)

Table 1.1   (*Continued*)

Substrate	Host Organism	Notes (pathway engineered or modification method applied)	Reference
Xylose	*Saccharomyces cerevisiae*	Xylulokinase gene was introduced in a recombinant yeast strain that expresses the xylose reductase and xylitol dehydrogenase. The resulting strain is able to perform ethanolic fermentation from xylose as sole carbon source at high aeration levels	(Toivari *et al.*, 2001)
Starch	*Saccharomyces cerevisiae*	Yeast cells expressing an active glucoamylase from *Rhizopus oryzae* were able to use starch as C-source. The foreign protein was targeted to the cell wall through fusion with a yeast α-agglutinin	(Murai *et al.*, 1997)
Arabinose fermentation	*Zymomonas mobilis*	Arabinose isomerase, ribulokinase, ribulose 5 phosphate epimerase, transaldolase and transketolase from *E. coli* were introduced into *Z. mobilis* under the control of a constitutive promoter	(Deanda *et al.*, 1996)

Chemical	Host Organism	Notes (source and type of bioremediation genes)	Reference
Addition of new catabolic activities for detoxification, degradation, and mineralization of toxic compounds			
Mercury (II)	Yellow poplar engineered with the mercury reductase (MerA) gene	Converts the highly toxic Hg(II) to Hg(0) by transformation with a bacterial MerA gene	(Rugh *et al.*, 1998)

(*Continued*)

Table 1.1   (*Continued*)

Chemical	Host Organism	Notes (source and type of bioremediation genes)	Reference
Organopollutant degradation (toluene and trichloroethylene) and heavy metals arsenic, chromium, lead, cesium plutonium and uranium	*Deinococcus radiodurans*	TOD genes to degrade organopollutants (toluene deoxygenase, a flavoprotein, a ferredoxin, and a terminal oxygenase genes) expressed in radioactive environments	(Lange *et al.*, 1998)
Mercury (II)	*Deinococcus radiodurans*	The mercury resistance operon mer from *E. coli* were introduced in a highly radiation resistant bacterium with the aim of remediating radioactive waste contaminated with heavy metals. The mercury resistance levels correlated with the genic dose of the integrated operon	(Brim *et al.*, 2000)

Chemical	Organism	Notes (source and type of bioremediation genes)	Reference
Explosives	Tobacco plants	Transgenic plants expressing pentaerythritol tetranitrate reductase from *Enterobacter cloacae* are able to degrade glycerol trinitrate and potentially trinitrotoluene, pollutants commonly present in military sites	(French *et al.*, 1999)
Biphenyls	Pseudomonads	The biphenyl degradative pathway was introduced into pseudomonads living in the rhizosphere with the potential use to bioremediate polluted soil	(Brazil *et al.*, 1995)

(*Continued*)

Table 1.1 (*Continued*)

Chemical	Organism	Notes (source and type of bioremediation genes)	Reference
n-Butanol	*Saccharomyces cerevisiae*	An n-butanol biosynthetic pathway was introduced. Isozymes from different organisms (*S. cerevisiae*, *E. coli*, *C. beijerinckii*, and *Ralstonia eutropha*) were substituted for the Clostridial enzymes. Production of n-butanol was improved ten-fold (to 2.5 mg/L)	(Steen *et al.*, 2008)
Isobutyraldehyde	*Synechococcus elongatus*	A non-native metabolic pathway was engineered to produce the four carbonaldehyde isobutyraldehyde. Stopping at the aldehyde stage mitigated against toxicity of the alcohol and favored the recovery of the aldehyde through the gas phase	(Atsumi *et al.*, 2009)
Fatty acids (FAs)	*Escherichia coli*	Engineering to produce structurally tailored fatty esters, fatty alcohols, and waxes directly from simple sugars in strains made defective for FAs degradation. FA biosynthesis was deregulated with a plant thioesterase	(Lu *et al.*, 2008; Steen *et al.*, 2010)
1,3-propanediol (1,3-PD)	Several organisms	Multiple ME approaches and strategies for producing 1,3-PD are described as applied in both native producers and heterologous hosts with acquired ability to produce the diol	(Celinska, 2010)
L-threonine	*Escherichia coli, Corynebacterium glutamicum*	Multiple ME strategies directed to L-threonine biosynthetic pathway, intracellular consumption, and transmembrane export	(Dong *et al.*, 2011)

(*Continued*)

Table 1.1   (*Continued*)

Chemical	Organism	Notes (strategy and rationale of modification)	Reference
		Modification of cell properties	
Alteration of source–sink relations and carbon partitioning provoking direct effect on growth of plants	Tobacco plants	Plants transfected with the yeast invertase gene under the control of an ethanol inducible promoter avoid the deleterious effects on growth of a constitutive expression	(Caddick *et al.*, 1998)
Decrease of O_2 photosynthesis inhibition	Rice plants	Maize PEP carboxylase introduced into C3 plant enhances photosynthesis by acquisition of part of the metabolic machinery to concentrate CO_2 characteristic of C4 plants (maize)	(Ku *et al.*, 1999)
Herbicide resistance (glyphosate)	Tobacco cells	Introduction of the EPSPS from petunia into the chloroplast genome avoids escape and dissemination of the foreign genes because of their absence into pollen cells	(Daniell *et al.*, 1998)
Harvest index	*Nicotiana tabacum*	Phytochrome A (heterologous (oat) apoproteins) gene were introduced into tobacco under the control of 35S CaMV promoter	(Robson *et al.*, 1996)
Low temperature resistance of higher plants	Tobacco plants	Delta 9 desaturase from cyanobacteria introduced in tobacco exhibit reduced levels of saturated fatty acids in membrane lipids and increased chilling resistance	(Ishizaki-Nishizawa *et al.*, 1996)
Altered fermentation pattern with production of 2,3-butanediol and acetoin	*Serratia marcescens*	*Serratia marcescens* transformed with the bacterial (*Vitreoscilla*) hemoglobin gene (vgb) where growth is not necessarily improved, but fermentation pattern altered according to medium composition	(Wei *et al.*, 1998)

(*Continued*)

Table 1.1   (*Continued*)

Chemical	Organism	Notes (strategy and rationale of modification)	Reference
Enhanced growth and altered metabolite profiles	*Nicotiana tabacum*	Transgenic tobacco plants expressing Vitreoscilla hemoglobin gene exhibit better yield and faster growth and altered alkaloids contents (nicotine and anabasine)	(Holmberg *et al.*, 1997)
Enhanced potato tuber growth with altered sugar content	Potato plants	Yeast invertase expression either in the apoplast (extracellular space) or cytoplasm, allows increased or decreased, tuber size accompanied with lower or higher, tuber numbers per plant, respectively	(Sonnewald *et al.*, 1997)
Enhanced starch content	Potato (*S. tuberosum*), Maize (*Zea mays*), Wheat (*Triticum aestivum*), Rice (*Oryza sativa*)	Expression of a highly active, allosteric insensitive bacterial AGPase results in a higher flux of carbon into starch in potato tubers. Increased seed yield and starch content in maize, wheat, and rice	(Giroux *et al.*, 1996; Stark *et al.*, 1992) (Smidansky, 2002, p. 311; Smidansky, 2003, p. 312)
Starch synthesis and degradation	Potato (*S. tuberosum*)	Expression of a bacterial AGPase both increased starch synthesis and degradation	(Sweetlove *et al.*, 1999).
Increased fruit sugar content	Tomato (*Solanum lycopersicum*)	Introgression lines (chromosome segments) from wild species of tomato (*S. pennellii*) into the background of cultivated tomato. Increased invertase activity and of soluble solids	(Fridman *et al.*, 2004)

(*Continued*)

Table 1.1   (Continued)

Chemical	Organism	Notes (strategy and rationale of modification)	Reference
Altered fermentation pattern: shift from a homolactic to a mixed acid fermentation by perturbing the redox status of the cell	*Lactococcus lactis*	Transformation with the *Streptococcus* NADH oxidase (nox2) gene results in a mixed acid fermentation according to the expression level of NADH oxidase and to the associated redox status of the cell	(Lopez de Felipe *et al.*, 1998)
Hybridoma cells growth in glutamine-free media	Hybridoma cells	Glutamine synthetase from Chinese hamster was introduced into a hybridoma cell line achieving parental levels of antibody production in a glutamine-free medium	(Bell *et al.*, 1995)
Decreased yield and growth rate	*Escherichia coli*	Decreasing the number of lipoyl domains per lipoate acetyltransferase in pyruvate dehydrogenase resulted in adverse effects on growth and biomass yield	(Dave *et al.*, 1995)
Expression of *E. coli* glycine betaine synthetic pathway and yeast trehalose synthetic genes	Plants	Transgenic plants transformed with genes from choline-to-glycine betaine pathway enhances stress tolerance to cold and salt, whereas those carrying yeast TPS1 (UDP dependent trehalose synthesis pathway) exhibit higher draught tolerance but with negative side effects	(Strom, 1998)

(Continued)

Table 1.1   (*Continued*)

Chemical	Organism	Notes (strategy and rationale of modification)	Reference
Enhanced recombinant protein production	*Escherichia coli*	Acetolactate synthase from *Bacillus* was introduced in *E. coli* to drive the excess pyruvate from the glycolytic flux away from acetate to acetolactate which was then converted into acetoin, a less toxic metabolite from the point of view of heterologous protein production	(Aristidou *et al.*, 1995)
Suppressed acid formation	*Bacillus subtilis*	Mixed substrate consumption, glucose and citrate, the latter exerting likely regulatory roles on glycolytic enzymes, PFK and PK	(Goel *et al.*, 1995)
Anaerobic growth and improved ethanol yield	*Pichia stipitis*	The *S. cerevisiae* URA1 gene encodes a dihydroorotate dehydrogenase that uses fumarate as an alternative electron acceptor enabling anaerobic growth and fermentation in *P. stipitis* when transformed with this gene	(Shi and Jeffries, 1998)
Changes in sugar utilization pattern	*Escherichia coli*	Overexpression of pyrroloquinoline quinone glucose dehydrogenase resulted in increased sugar-dependent respiration according to the quality of the carbon source (either PTS or non-PTS sugar)	(Sode *et al.*, 1995)

Tools for Metabolic and Cellular Engineering

MCE requires the development and application of several skills from the various disciplines contributing to the field. The area of molecular biology needs:

(1) Transformation systems for microorganisms used in industrial production, or in bioprocesses (e.g. for *Corynebacterium*, commonly used for the production of aminoacids, or for *Pseudomonads*, used in the

degradation of xenobiotics) (Dong *et al.*, 2011; Keasling, 1999; Steen *et al.*, 2010).

(2) Promoters and special vectors used in such transformations: e.g. the yeast retrotransposon Ty3 employed for site-specific integration of heterologous genes with advantages of stability and high copy number (Wang and Da Silva, 1996), or the filamentous fungal vector *Agrobacterium* able to transform the genera *Neurospora*, *Trichoderma*, *Aspergillus*, and *Agaricus* (de Groot *et al.*, 1998).

(3) Multicistronic expression vectors to allow one-step multigene ME in mammalian cells (Fussenegger *et al.*, 1999).

(4) Methods for stabilizing cloned genes: e.g. by integration into the chromosomes of host organisms.

(5) Biochemical methods to search for and analyze metabolic pathways.

Microbiological techniques, analytical chemistry, and biochemistry allow the evaluation of the effectiveness of a modified metabolic pathway:

Culture of the modified microorganism. In this respect, the ideal culture system enabling the application of mathematical and computational tools (see below) is continuous culture. However, it may happen that the stability of the genetically modified microorganism precludes the possibility of continuous maintenance of the culture in the long-term. In this case either batch or fed-batch semi-continuous systems have to be used.

Optimization of growth medium suitable for the operation of the desired metabolic pathway. For instance, an organism (*Serratia* spp.) modified with the bacterial hemoglobin gene (*vgb*) that was designed to improve growth and to avoid by-product formation, was reported to display various desirable fermentation patterns and modified by the medium composition (Wei *et al.*, 1998).

Mass balance. This allows calculation of the yield of the desired product with respect to various substrates; by comparison with maximal theoretical yields, the shortfall from the thermodynamic limit may be evaluated.

Isotopic labeling and analysis of blocked mutants, as well as the determination of enzyme activities and metabolites, allow the determination of the effectiveness of operation of a given metabolic pathway that contributes to the consumption of a certain substrate, or the formation of a required product.

The employment of noninvasive methods (e.g. nuclear magnetic resonance, and flow cytometry) are preferred, since they allow a direct evaluation of the performance of the microorganism under conditions similar to or identical with those in the industrial bioprocess.

The following mathematical and computational tools are essential to current advances:

DNA data bases and operating software (Overbeek *et al.*, 2000; Covert *et al.*, 2001) (see also Table 5.1). Available online are: Kyoto Encyclopedia of Genes and Genomes (KEGG pathway database), EcoCyc and BioCyc (Biocyc: http://biocyc.org). EcoCyc is an extensively human-curated database specialized for *E. coli* K-12, whereas KEGG and BioCyc maintain databases for many organisms but with little human curation effort (Rosa da Silva *et al.*, 2008).

Metabolic pathways, databases, including kinetic and thermodynamic enzyme data. In this respect, several Internet sites are available (Covert *et al.*, 2001; Karp, 1998, 2009; Overbeek *et al.*, 2000). KEGG (Kanehisa *et al.*, 2006, 2010) and EcoCyc (Keseler *et al.*, 2005, 2009) provides information on metabolism, metabolic pathways, signal transduction, gene regulation, and cellular processes (see Chap. 5).

Tools designed for estimation of theoretical yields (e.g. from metabolic pathway stoichiometries). This point is developed in Chaps. 2 and 4.

Reconstruction of metabolic (constraint-based reconstruction and analysis, (Orth and Palsson, 2010)) and signaling (Bhalla, 2003; Bhalla and Iyengar, 1999; Weng *et al.*, 1999) networks from pathway stoichiometry. Analysis of elementary flux modes (Schuster *et al.*, 1999; Schuster *et al.*, 2000).

Tools for the design of engineered metabolic pathways. Several algorithms have been proposed for this purpose (Hatzimanikatis *et al.*, 1996).

Quantitative tools for the simulation, prediction, and performance analysis of modified microorganism (e.g. MCA, Biochemical System Theory (BST), and MFA). These methods encompass a series of stoichiometric and linear optimization procedures enabling the estimation of metabolic fluxes. Some of these tools and their applications are indicated in Chaps. 4 and 5.

Topological (Albert and Barabasi, 2002; Oltvai and Barabasi, 2002) and dynamic (Aon *et al.*, 2006a, 2008a) analysis of metabolic and subcellular networks (see Chap. 5).

Some of these developments are integrated into the Transdisciplinary Approach (TDA) which is developed in the next section.

Engineering Cells for Specific Biotransformations

Possibilities for redirection of substrate either to microbial products or to biomass may be achieved through modification of environmental parameters or by engineering the microorganism itself (Fig. 1.1) (Aon and Cortassa, 1997). Thus, we propose optimization of a specific biotransformation process by directed modification of the microorganism itself (rather than of process) with the dual aim of achieving higher yields of products of economic interest; improved environmental quality should also be considered.

The originality of the present approach is that it integrates several disciplines into a coordinated synthesis (i.e. microbial physiology and bioenergetics, thermodynamics and enzyme kinetics, biomathematics

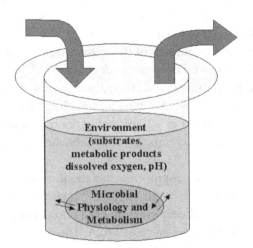

Figure 1.1. The microorganism as a target for bioengineering at metabolic, energetic, and physiological levels in chemostat cultures. The use of continuous cultures to study a microorganism in steady state provides a rigorous experimental approach for the quantitative evaluation of microbe's physiology and metabolism. Continuous cultures allow the definition of the phase of behavior that suits the aim of the engineering, e.g. output fluxes of metabolic by-products of interest. Thus, continuous culture is a fundamental requisite of the TDA approach for the rational design of cells (see Chaps. 4 and 5). An operational flow diagram is shown in Fig. 1.2.

and biochemistry, and genetics and molecular biology). Thus, it will be called a TDA. The TDA approach provides the basis for the rational design of microorganisms or cells in a way that has rarely been applied to its full capabilities. Progress in the area of a rational design of microorganisms has been hampered by the fact that few scientists can simultaneously master fermentation and recombinant DNA technologies along with mathematical modeling, i.e. MCA. In fact, in most cases researchers either apply sophisticated recombinant DNA technologies in a trial and error scheme, or they use mathematical techniques without the framework of molecular biology and physiology. The TDA approach can improve or optimize an existing process within an organism. The use of heterologous pathways for the production of new chemicals, or use various feedstock and cheap substrates or degrade xenobiotics, necessarily involves a previous modification of the organism using DNA recombinant techniques. In general, the TDA approach for MCE is iterative in nature and may be outlined as follows (Fig. 1.2):

Step (I). Physiological and bioenergetic studies are performed either in continuous batch or fed-batch cultures according to the nature of the process. The aim is to achieve a state known as "balanced growth" that allows

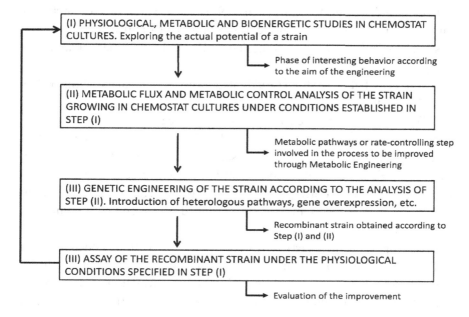

Figure 1.2. Engineering metabolic fluxes: The TDA approach.

the application of analytical tools such as MFA (see Chap. 2). Whenever a steady state is feasible, MCA can also be applied. Its expected outcome is determination of the most favorable behavior shown by the microorganism according to the aim of the engineering, e.g. ethanol production by *S. cerevisiae* at high growth rates in continuous culture (see Chap. 4).

Although stage (I) is a clear one, wherever we have a potentially useful microorganism, several considerations must be taken into account. Often, continuous cultures cannot be run with cheap substrates (e.g. molasses, whey) especially full-scale industrial processes. Under these conditions, batch or fed-batch cultures must be used and mathematical modeling techniques such as MFA applied (see step II).

The TDA approach can be initiated with a recombinant microorganism. Thereby it is possible to introduce heterologous metabolic pathways using DNA recombinant techniques to provide a microorganism with novel activities, e.g. xylitol or arabinose degradation in *Zymomonas mobilis* (i.e. to introduce into *Z. mobilis* the ability to degrade xylose or arabinose) or products such as biodiesel, alcohols, wax esters, and xylanase excretion for the utilization of hemicellulose in *E. coli* produced from non-native pathways (Steen *et al.*, 2010). In this case, if chemostat cultures cannot be used (e.g. due to plasmid instability) then batch or fed-batch cultures may be employed to explore the recombinant's ability to perform the desired task. This organism may then be subjected to subsequent quantitative analysis.

Step (II). Metabolic studies facilitated by mathematical modeling. MCA and MFA of the strain performed under the conditions described in step (I).

MFA, incorporating the elementary flux mode analysis, may help determine the theoretical as well as the actual yields of the metabolite or macromolecule the production of which we seek to optimize (see Chaps. 4 and 5). Moreover, elementary flux mode distribution and bioenergetic behavior may be investigated during the phase of interest (e.g. the growth rate at which the metabolite is maximally produced, excreted, or accumulated intracellularly).

When applying the matrix form of MCA, the intracellular concentrations of the intermediates of the target pathway must be measured before further ME. Enzyme kinetics must be investigated if information is not already available in the literature. The determination of the elasticity coefficients, their array in matrix form, and the matrix inversion may be another aim of this step. Such inversion provides a matrix of control coefficients both for flux and metabolite concentrations. Thus, this step allows the identification of the rate-controlling steps of the flux or metabolite level in a metabolic pathway (see Chaps. 4 and 5).

Step (III). Genetic engineering. Gene overexpression or up-modulation of all the enzymes which control the flux or all the activities participating in the elementary flux mode that gives the highest yield of the desired product. In the case of recombinant strains (e.g. those constructed by introduction of heterologous pathways) this step is still valid, since new rate-controlling steps of the specific biotransformation process may arise and become pivotal. The outcome of step (III) is a modified microorganism optimized for a specific biotransformation process.

Step (IV). This step iterates step (I): The engineered microorganism should be assayed under the physiological conditions defined in step (I). The assay should allow evaluation of the improvement achieved in the biotransformation process with the use of the engineered microorganism.

In Chap. 4 we describe in detail each of the steps as applied to examples of prokaryotes or eukaryotes.

Current Trends: Biofuels and Biomass Interconversion Processes

Although the production of fuels and chemicals have been central targets of ME, applications to biofuels and biochemical methods of biomass conversion have substantially increased only since 2005 (Stephanopoulos, 2008). As such, production of H_2 and other biofuels are at the forefront of research in the production of energy from renewable resources.

The importance of developing CO_2-neutral fuel sources has been highlighted by the detailed modeling of climate change effects, its global and national economic impacts, and the increasing competition for fossil fuel reserves (Stephens *et al.*, 2010a, 2010b). Of the clean energy technologies being developed, almost all target the electricity market (e.g. photovoltaic, solar thermal, geothermal, and wind and wave power), which currently only accounts for $\sim33\%$ of global energy demand. However, to secure future fuel

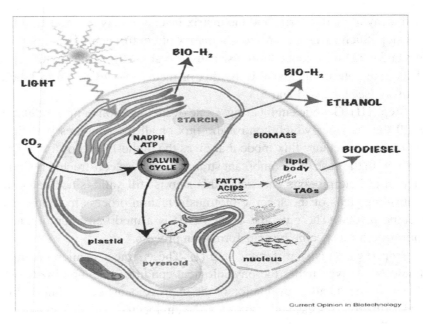

Figure 1.3. Photosynthesis is the fundamental driving force that supports all biofuel synthetic processes, converting solar energy into biomass, carbon storage products (e.g. carbohydrates and lipids), and/or H_2. The photosynthetic light reactions and the Calvin cycle produce carbohydrates that fuel mitochondrial respiration and cell growth.
Source: Beer *et al.* (2009). Reprinted from *Curr. Opin. Biotechnol.* 20, Beer LL, Boyd ES, Peters JW, Posewitz MC. (2009) Engineering algae for biohydrogen and biofuel production, 264–271. © (2009), with permission from Elsevier.

supplies (66% of global energy) biofuels represent almost the only viable option, at present.

Plants interconvert sunlight into chemical energy stored in, e.g. starch, sugars, and lipids. These feedstocks can be used for the production of biofuels (Fig. 1.3). A wide range of crop plants (sugarcane, oil palm, sugar beet, rapeseed, soya beans, wheat, and corn) have been used for a "first generation" of biofuels such as ethanol, diesel, and methane. These biofuels are most extensively used in Brazil, USA, South-East Asia, and Europe. At present, biofuel production from starch or triacylglycerides poses the competition dilemma between food and fuel. Solutions aimed at solving this ethical and practical issue are being explored. The big challenge is to generate biotools for processing lignocellulosic biomass, a highly abundant residual with limited nutritional value.

Algae and other marine organisms are responsible for the fixation of almost half of the inorganic carbon from the atmosphere (Field *et al.*, 1998) with a typical range of 1.0–1.8 of photosynthetic quotient (moles of oxygen released per mole carbon dioxide fixed) (Boyle and Morgan, 2009). In principle, solar-driven biohydrogen production is feasible. The combustion of the evolved H_2 yields only H_2O, thereby completing a clean energy cycle. A select group of photosynthetic organisms have evolved the ability to harness the huge solar energy resource to drive H_2 fuel production from H_2O. Photosystem II (PSII) drives the first stage of the process, by splitting H_2O into protons (H^+), electrons (e^-), and O_2.

Under anaerobic conditions, with mitochondrial oxidative phosphorylation largely inhibited, some organisms (e.g. *C. reinhardtii*) redirect the energy stored in carbohydrates to a chloroplast hydrogenase (H_2ase), likely using a NAD(P)H-PQ e^- transfer mechanism, to facilitate ATP production via photophosphorylation (see Chap. 4) (Beer *et al.*, 2009). The discovery of sustainable H_2 production by sulfur deprivation established a "two-stage photosynthesis and H_2 production" by temporally splitting O_2 production and H_2 production (Melis *et al.*, 2000). This intervention allows bypassing of the sensitivity of the reversible H_2ase to O_2 and represents an opportunity for ME (see Chap. 4). Hydrogenase essentially acts as H^+/e^- release valve by recombining H^+ (from the medium) and e^- (from reduced ferredoxin) to produce H_2 gas that is excreted from the cell.

Microalgae are efficient transducers of sunlight into chemical energy while able to grow in salt water year round under diverse conditions (Beer *et al.*, 2009). Microalgal biofuel systems produce fuels from single-celled microalgae (eukaryotes or cyanobacteria). Thus, they have the key potentiality of making a substantial contribution to global energy demand and CO_2 sequestration from the atmosphere, without increasing the pressure on arable land or important forest ecosystems (Stephens *et al.*, 2010b). Many microalgae can be grown in saline water and are able to produce a wide range of feedstocks for the production of biofuels (Fig. 1.4), including biodiesel, methane, ethanol, butanol, and hydrogen, based on their efficient production of starch, sugars, and oils (Beer *et al.*, 2009; Stephens *et al.*, 2010b). Because they absorb CO_2 during growth, from both atmospheric and (in some cases) industrial sources, microalgae can also contribute to carbon capture.

Solar energy conversion efficiency by microalgae can be improved at several levels through synergies between biology and engineering. Most

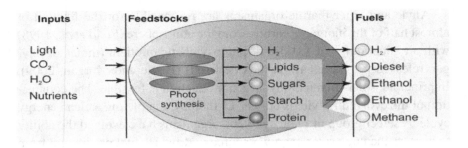

Figure 1.4.   Potential for photoautotrophic biofuel production from micro-algal systems. *Source*: Stephens *et al.* (2010b). Reprinted from *Trends Plant Science* 15, Stephens E, Ross IL, Mussgnug JH, Wagner LD, Borowitzka MA, Posten C, Kruse O, Hankamer B. Future prospects of microalgal biofuel production systems, 554–564. © (2010), with permission from Elsevier.

microalgal ponds have a solar energy conversion efficiency of 1–4% under normal operating conditions (Stephens *et al.*, 2010b), and higher efficiencies can be achieved with closed photobioreactor systems. The saturation of antenna systems under high-light conditions results in the photoprotective dissipation of excess energy as heat and fluorescence. Up to 95% of the incident solar energy can be dissipated (i.e. wasted) via these mechanisms. Biologically, strategies are aimed at reducing the size of the chlorophyll-binding photosynthetic light-harvesting antenna systems so that each photosystem obtains only the light that it needs, rather than an excess. Bioprocess engineering can help minimize light scattering at the bioreactor surface to maximize the light entering the bioreactor. Thus, a combined biological engineering strategy helps to optimize the utilization of the solar energy available at the illuminated culture surface of exposed microalgae, and specifically to their chlorophyll-binding light-harvesting antenna systems (Kruse and Hankamer, 2010). Rapid mixing cycles on a millisecond time scale, can be used to move cells between light and dark zones in the reactor (the "flashing light effect"), helping to overcome the light saturation of antenna systems, and improving the matching between energy captured and its utilization in carbon fixation.

Optimization of algal growth for temperature and dissolved O_2 and CO_2 levels must also be achieved for maximal and efficient transduction of incident sunlight into carbon. The high temperatures that typically accompany high incident-light levels often result in growth inhibition or death. Open ponds are effectively cooled by evaporation, limiting the upper temperature to about 40°C. Excess dissolved O_2 level must be carefully controlled because it can result in inhibition of carbon fixation due to the oxidase action of Ribulose Bisphosphate Carboxylase (RuBisCo). Dissolved CO_2 levels must also be carefully regulated (e.g. to maintain stable pH levels) (Stephens *et al.*, 2010b).

Metabolic Areas that have been Subjected to MCE

The main objective of this section is to update the work performed on MCE in the previous years (Table 1.1), to highlight the main areas of metabolism that have been improved (Fig. 1.5), and to point out those areas to which less effort has been devoted.

MCE has dealt with manipulation of existing pathways or reactions aimed at producing a certain metabolite or macromolecule, or the introduction of new pathways or reactions into host cells. The main examples reported in Table 1.1 have been classified into five groups (Cameron and Tong, 1993; Lee and Papoutsakis, 1999; Stephanopoulos *et al.*, 1998): (i) enhanced production of metabolites and other biologicals already produced by the host organism; (ii) production of modified or new metabolites and other biologicals that are new to the host organism; (iii) broadening the substrate utilization range for cell growth and product formation; (iv) designing improved or new metabolic pathways for degradation of various chemicals especially xenobiotics, and (v) modification of cell properties that facilitate bioprocessing (fermentation and/or product recovery).

The pathways shown in Fig. 1.5 have been subjected to extensive MCE: the pyruvate pathway for production of organic acids or flavor compounds in bacteria such as acetoin and diacetyl (Lopez de Felipe *et al.*, 1998; Platteeuw *et al.*, 1995) (see also Chap. 4); sugar transport (hexose transporters: see Ozcan and Johnston, 1999, for a review); phosphotransferase system (PTS): (Gosset *et al.*, 1996; Liao *et al.*, 1996), arabinose and xylose assimilating pathways (Deanda *et al.*, 1996; Zhang *et al.*, 1995) (see Chap. 4); pathways for production of propanediol (Cameron *et al.*, 1998), butanol (Papoutsakis and Bennett, 1999), lysine (Eggeling and Sahm, 1999; Vallino

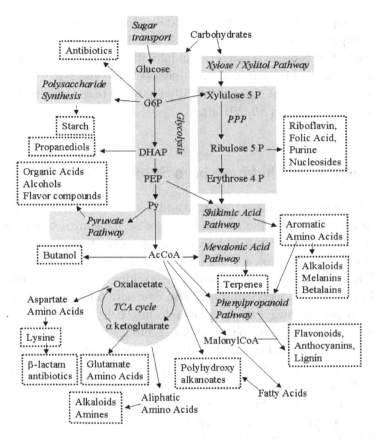

Figure 1.5.   Primary and secondary metabolic routes and their connections to central catabolic pathways through common metabolites. Schematic representation of some pieces of information detailed in Table 1.1. On gray boxes are represented multiple step pathways while end products are enclosed in dashed-rectangles. Several of the pathways shown have been subjected to intensive ME. The pathways depicted include those described for plants, bacteria, and fungi. The pathways for plants were taken from Morgan *et al.* (1999).

and Stephanopoulos, 1993), aromatic amino acids (Flores *et al.*, 1996; Gosset *et al.*, 1996; Liao *et al.*, 1996), β-lactam antibiotics (Khetan and Hu, 1999). The pathways from the aspartic family of amino acids for the production of L-threonine have been intensively engineered in *E. coli* and *Corynebacterium glutamicum* during the last decade (Dong *et al.*, 2011).

Sugar transport has been worked on either in relation to production of aromatic amino acids or modulation of catabolic pathways (Table 1.1).

The sugar transport PTS functions at expense of phosphoenolpyruvate (PEP) which is a precursor for the synthesis of aromatic amino acids. Thus, elimination of the PTS system enables an increase in aromatic amino acid yield (Flores *et al.*, 1996) (see Chap. 3).

Sode *et al.* (1995) achieved an increase of respiration rates in *E. coli* through engineering of the glucose transport mediated by the coupled enzyme system pyrroloquinoline quinone glucose dehydrogenase.

The NAD/NADH ratio does not merely reflect the cellular redox state but is centrally involved in its control. This fact was evidenced by a shift in the pattern of end-fermentation products from homolactic to mixed acid, including flavor compound production such as acetoin or diacetyl in *L. lactis* (Lopez de Felipe *et al.*, 1998).

Another important observation derived from Table 1.1 concerns the growth efficiencies of plants as suitable targets for MCE (see Chap. 7). Main objectives are to increase crop yields, stress resistance, produce new chemicals, e.g. degradable plastics, or act as factors that mitigate environmental pollution (Poirier, 1999). In this sense, two of the four examples covered in section "Addition of new catabolic activities for detoxification, degradation, and mineralization of toxic compounds," concern the engineering of plants to scavenge and eliminate by-products of industrial activity from natural environments. Another area of plant physiology in which progress has happened over the last decade is the ME of increased starch content. Among the four main steps targeted are: ADPglucose pyrophosphorylase, amyloplastidial adenylate transporter, NAD-malic enzyme, and the plastidial adenylate kinase (Lytovchenko *et al.*, 2007) (see also Table 1.1).

Despite considerable progress during the last decade, the following areas of metabolism and cellular functions still need further MCE:

(i) Production of novel carbohydrates (e.g. fructans, dextrans), and content modulation, quality modification (e.g. degree of branching) of storage and structural polysaccharides, e.g. glucans, xylans, and pectins.

(ii) Cellular transport functions (e.g. of protons, sodium, and calcium).

(iii) Cellular processes (e.g. cell division cycle, catabolite repression, and cell differentiation). In this respect, some progress has been reported in MCE of plant development showing altered flowering, early senescence, and tuber sprouting (see Lytovchenko *et al.*, 2007; and Refs. therein).

Genome Sequencing, Comparative Genomics, and Biological Complexity

When the first edition of this book was in press, the genome sequences of the human and major model organisms (yeast, worm, fly, bacteria) were completed and published (Rubin *et al.*, 2000; for a review). A revised and more accurate version of the human genome by the Human Genome Sequencing Consortium became available in 2004. A recent book by one of the major players in genome sequencing, Craig Venter, tells a personal and fascinating insider's story worth knowing (Venter, 2007). By January 2010, the genome sequencing of >350 species have been completed. These extraordinary achievements, thanks to the concerted effort of the scientific community throughout the planet, is a precious lasting legacy for mankind and future generations.

If there is one major change between the present edition of this book and the previous one in the Spring of 2002, it is the availability of the genetic makeup of many species. Now stored and systematized in databases, the knowledge about genes and proteins constitutes a fertile field for data mining, and the ground work for exploring genetic interrelationships within and between species, and their evolutionary meaning. The recent creation of the first synthetic bacterium (Gibson *et al.*, 2010) testifies for the enormous potential available. This work represents an outstanding step in synthetic biology as it demonstrates the feasibility of making a new living form by introducing a synthetic chromosome into a recipient bacterium, now controlled by the manmade genome. Complete genome sequence makes it possible to study networks of genes rather than individual genes or pathways. The new fields of Bioinformatics and Functional Genomics were born in the postgenomic era, as a result of all the new information available. The scientific starting point of the 21st century could not be stronger.

One of the first lessons learned from genome sequencing of species with diverse lineage and evolutionary paths was that the number of genes and core proteomes do not explain their apparent complexity (Rubin *et al.*, 2000). *Drosophila*, a complex multicellular metazoan with specialized cell types, sophisticated development, and a complicated nervous system has a core proteome only twice the size of that of yeast. Thus, from a genetic makeup perspective, an apparently more complicated organism like the fly, looks only more than twice as complicated as a single-celled yeast. Furthermore, despite the large differences between fly and worm in terms of development and morphology, they use a core proteome of similar size. *Arabidopsis, C. elegans,* and *Drosophila* have a similar range of 11,000–15,000 different types of proteins, suggesting this is the minimal complexity required by extremely diverse multicellular eukaryotes to execute development and respond to their environment (the Arabidopsis Genome Initiative, Adams *et al.,* 2000; Arabidopsis Genome Initiative, 2000; Rubin *et al.,* 2000).

Concerning the human genome, the best estimate is that the total number of protein-coding genes is in the range 20,000–25,000. The lower bound is based on the number of currently known genes (19,599), whereas the upper bound depends on estimates of the number of additional genes (International Human Genome Sequencing Consortium, Consortium, 2004). The core proteome of the human genome may not be much larger than that of the fly or worm. Presumably, the complex traits of a human being may be achieved using largely the same molecular components as the fly or the worm. These examples illustrate that there can be large differences in morphological and behavioral complexity among different organisms that have similar numbers of genes (Miklos and Maleszka, 2000; Miklos and Rubin, 1996). Thus, spatiotemporal organization of similar components interacting in different configuration patterns rather than gene number *per se* seems to be more essential for characterizing the organized complexity of living systems.

Alternative splicing, i.e. different ways in which a gene's protein-coding section (exons) can be joined together to create a functional messenger RNA molecule, can be a way by which more proteins are encoded per gene (Rubin, 2001). *Drosophila* has about 14,000 genes, as compared to about 18,000 in *Caenorhabditis elegans* and an estimated 20,000–25,000 in humans. At least a third of *Drosophila* transcripts are alternatively spliced, so the predicted number of proteins encoded by this genome is over 20,000. It is believed that human genetics complexity arises from alternative splicing. This could at least in part explain that the human genome is 30 times larger than that of the worm but the number of genes less than two times larger (Hartl and Jones, 2009) (see Table 1.2).

Sequence similarity comparisons consistently failed to give information about nearly one-third of the components that make every organism

Table 1.2   Number of genes and proteome cores in different organisms

Organism	Protein-Coding Genes	Proteome
Plants (*Arabidopsis thaliana*)	25,498	11,000
Yeast (*Saccharomyces cerevisiae*)	6,241	4,383
Worm (*Caenorhabditis elegans*)	18,424	9,453
Fly (*Drosophila melanogaster*)	13,601	8,065
Mouse (*Mus musculus*)	22,011	~10,000
Man (*Homo sapiens*)	20,000–25,000	~10,000

uniquely itself (Rubin *et al.*, 2000). Thus, approximately 30% of the predicted proteins in every organism bear no similarity to proteins in its own proteome or in the proteomes of other organisms. For instance, bigger and more complex proteins are present in the fly and worm as compared with yeast, including, not surprisingly, more proteins with extracellular domains involved in cell–cell and cell–substrate interactions (Rubin *et al.*, 2000). The population of multidomain proteins is somewhat larger and more diverse in the fly than in the worm. Basic intracellular processes, such as translation or vesicle trafficking, appear to be conserved across kingdoms, reflecting a common eukaryotic heritage. More elaborate intercellular processes, including physiology and development, use different sets of components. For example, membrane channels, transporters, and signaling components are very different in plants and animals, and the large number of transcription factors unique to plants contrasts with the conservation of many chromatin proteins across the three eukaryotic kingdoms.

From DNA Sequence to Biological Function

There is an interesting asymmetry between reductionists and modern integrationists in biological science. An integrationist, using rigorous systems-level analysis, does not need or wish to deny the power of successful reduction. Indeed he uses that power as part of his successful integration. Many reductionists, by contrast, seem for some reason to require intellectual hegemony.

Noble D. (2006) *The Music of Life: Biology Beyond the Genome.* Oxford University Press: New York.

In the past 50 years, we have witnessed outstanding scientific discoveries and technological achievements: (i) the elucidation of the mechanism

of gluconeogenic substrates (Aon and Cortassa, 1995; Monaco *et al.*, 1995. This effect was completely reversible on glucose addition (Aon and Cortassa, 1995, 1997).

Pleiotropy can arise if a protein is functionally required in different places, or at different times, or both (Miklos and Rubin, 1996). Pleiotropic effects may also arise through the strategic function of a gene product deeply nested in metabolic or regulatory networks which might affect several processes either simultaneously or in sequence. For example, in yeast, the products (proteins) of *SNF1* or *SNF4* genes have been postulated to be involved in a regulatory network which triggers the derepression of several gluconeogenic enzymes when yeast grows in the presence of nonfermentable carbon sources, e.g. ethanol, acetate, or glycerol (Schuller and Entian, 1987). The deletion of *SNF1* or *SNF4* genes within isogenic backgrounds, and the growth of the mutants in chemostat cultures, reveal in addition to their postulated effects, newly described pleiotropic consequences on cell cycle, fermentative behavior, and cellular energetics (Aon and Cortassa, 1998; Cortassa and Aon, 1998).

Systems Biology

Among the many revenues of the widespread genetic and high-throughput information now available, the re-emergence of physiology as *Systems Biology* stands out. Modern Biology takes for granted four "great ideas" (Nurse, 2003a): (i) the gene is the basis for heredity, (ii) the cell is the fundamental unit of organisms, (iii) biology is based on chemistry, and (iv) species evolve by natural selection. Systems Biology has been nominated as the "fifth great idea," tentatively summarized as "multi-scale dynamic complex systems formed by interacting macromolecules and metabolites, cells, organs, and organisms underlie most biological processes" (Vidal, 2009). As such, Systems Biology aims at a systems-level understanding of biological phenomena (Kitano, 2001).

As a result of the major undergoing transition between *analytical* and *integrative* approaches in Biology (Aon and Cortassa, 2005), the importance of complex systems and whole-system approaches has become paramount (Aon and Cortassa, 2009; Aon *et al.*, 2010; Lloyd and Rossi, 2008). Cells, organisms, social-, economic-, and eco-systems are complex because they

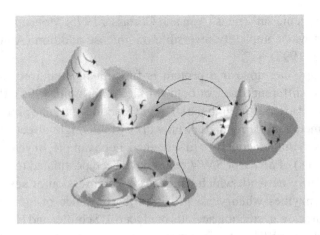

Figure 1.6.   A geometric interpretation of homeodynamics. Several types of attractors with their corresponding basins of attraction are represented. The putative trajectories followed by the indicated system's dynamics as well as the separatrices between basins are emphasized by arrows. The homeodynamic condition implies that the system's dynamics visualized as a fluid flowing around itself, may shift between attractors at bifurcation points where stability is lost. Thereby, the system's dynamics following a perturbation, flies away toward another attractor exhibiting either qualitative or quantitative changes in its behavior. The upper left 3-dimensional (3D) plot shows saddle and fixed points; the latter with different values, each one representing a different branch of steady states. Alternative occupancies of these states, following the change of a bifurcation parameter, gives a bistable switch with memory-like features. Also stable and unstable foci are depicted in the upper left 3D plot. The lower right 3D plot, shows a limit cycle with its basin of attraction that may be attained through an unstable focus, characteristic of oscillatory behavior (self-sustained or damped oscillations, respectively). The middle 3D plot depicts an attractor with three orbits embedded in it, with the potential for chaotic behavior.
Source: Lloyd *et al.* (2001).

consist of a large number of usually nonlinearly interacting parts; they also operate in multiple spatial and temporal scales. *Systems Biology* represents a systemic approach analyzing interconnections and their functional interrelationships rather than component parts (Aon and Cortassa, 2009; Ehrenberg *et al.*, 2009; Vidal, 2009). The systems approach has evolved over at least 50 years through theoretical biology, molecular physiology, ecology, and more recently functional genomics and bioinformatics. Its foundations can be traced back to the work on systems theory by von Bertalanffy (Skyttner, 2005; Von Bertalanffy, 1955). Common to the systems approach is that studies do not focus on the properties of individual units (molecules, cells, organs, organisms) but rather on those of modules and networks.

Systems Biology started to emerge as a distinct field with the advent of currently developing methods of high throughput, -omics, technologies, i.e. gen-, transcript-, prote-, and metabol-omics. Massive data gathering from -omics technologies, together with the growing capability of generating computational models of complicated systems, have made possible the massive integration and interpretation of information constituting the core of Systems Biology. As such, Systems Biology has the potential to allow us to gain insights into the fundamental nature of health and disease, along with their control and regulation (Aon and Cortassa, 2011; Lusis and Weiss, 2010).

Generally speaking, two approaches top-down and bottom-up can be identified within Systems Biology. The top-down approach involves the generation of different sort of networks, and their integration and simulation with computational models. Bottom-up approaches include the study of selected processes in cells, organs, or organisms, at high spatiotemporal resolution, which can also be simulated through computational modeling (Ehrenberg *et al.*, 2009). The data-driven top-down approach of Systems Biology represents a further development of Functional Genomics and Bioinformatics. The concept encompasses building of, sometimes genome- or organism-wide, networks of different types (e.g. protein, metabolite maps, phosphorylation, genetic interaction, and gene expression networks), and their integration into multiscale modeling. Then the properties of these networks are analyzed, employing suitable mathematical models and simulations. This area is developing driven by the rapidly increasing ability of generating at high-throughput "-omics" data (Ehrenberg *et al.*, 2009).

Although high throughput –omic technologies are important components of Systems Biology research, it is becoming increasingly clear that monitoring mRNA and protein levels through transcriptomics and proteomics, respectively, is insufficient to infer about the physiological status of cells (Mukhopadhyay *et al.*, 2008). These analyses need to be complemented with cell-wide studies involving metabolomics and flux analysis in order to understand and successfully engineer cells (Matthew *et al.*, 2009). Metabolomics, proteomics, and transcriptomics, each provide data on the concentrations of cellular biochemicals. Initial limitations of metabolome analyses given by the unequivocal identification of metabolites are being overcome by *metabolite profiling* of complex mixtures through improved separation techniques and high resolution detection systems (Fiehn *et al.*, 2000; Doebbe *et al.*, 2010; Roessner *et al.*, 2001).

Fluxomics measures the cellular rates of enzyme-catalyzed reactions (Cascante and Marin, 2008). These concentration and flux data, when overlaid on the metabolic network architecture, can be used to build and test quantitative, integrative models of microbes (Reaves and Rabinowitz, 2010). Current ongoing advances in all of these fields are happening.

Systems Biology and the Complex Systems Approach

Collective behavior of particles, agents, mitochondria, molecules, in networks present in, e.g. cells, markets, the atmosphere, is taking center stage in the analysis of complex systems.

Emergence in complex systems arises from self-organizing principles resulting from nonlinear mechanisms and the continuous exchange of energy, matter, and information with the environment (Aon and Cortassa, 1997, 2009; Haken, 1983; Lloyd *et al.*, 2001; Nicolis and Prigogine, 1977). Manifesting themselves as novel, and unexpected macroscopic spatiotemporal patterns, emergent behavior is the single most distinguishing feature of organized complexity in a system. Emergent organization cannot be anticipated in any way from the behavior of the isolated components of a system, and does not result from the existence of a "central controller." Self-organization is one of the main mechanisms that generate organized complexity, and this is expressed as disparate types of nonlinear outputs (e.g. oscillations, chaos) in a wide range of systems (Ball, 2004), including biological ones (Aon and Cortassa, 1997; Lloyd, 2009).

Integrated to the modern concepts of chaos, fractals, critical phenomena, and networks (Liebovitch and Todorov, 1996; Savageau, 1995; Smith and Lange, 1996; Sornette, 2000; Vicsek, 2001), Systems Biology takes part of what can be loosely called a Complex Systems Approach (Fig. 1.7). The *Complex Systems Approach* integrates Systems Biology to nonlinear dynamic systems analysis, involving a combination of experimental with mathematical techniques. One of the foundational concepts of the *Complex Systems Approach* is self-organization which is based on nonequilibrium

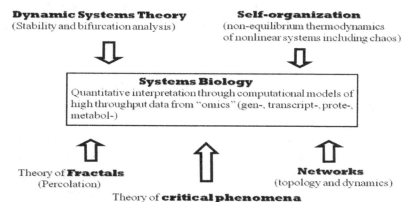

SYSTEMS BIOLOGY and the COMPLEX SYSTEMS APPROACH

Dynamic Systems Theory
(Stability and bifurcation analysis)

Self-organization
(non-equilibrium thermodynamics of nonlinear systems including chaos)

Systems Biology
Quantitative interpretation through computational models of high throughput data from "omics" (gen-, transcript-, prote-, metabol-)

Theory of **Fractals**
(Percolation)

Networks
(topology and dynamics)

Theory of **critical phenomena**
(criticality and renormalization)

Figure 1.7. The Complex Systems Approach. This approach, which emphasizes the analysis of interconnections and relationships rather than component parts, integrates Systems Biology to nonlinear dynamic systems analysis, comprising the modern concepts of chaos, fractals, critical phenomena, and networks.
Source: Aon and Cortassa (2009). Reproduced from Aon MA, Cortassa S. (2009) Chaotic dynamics, noise and fractal space in biochemistry. In: Meyers R (ed), *Encyclopedia of Complexity and Systems Science*. Springer, New York.

thermodynamics of nonlinear open systems, i.e. those that exchange energy and matter like cells, organisms, and ecosystems (Aon and Cortassa, 1997; Lloyd *et al.*, 1982; Nicolis and Prigogine, 1977; Schneider and Sagan, 2005; Yates, 1987). Due to the capability of complex biological systems to exhibit emergent behavior, detection and characterization of qualitative changes in spatiotemporal organization is crucial. For these, nonlinear dynamic stability and bifurcation analyses of computational models constitute essential tools for exploring the qualitative behavior of complex systems (see Chaps. 5 and 6). The Complex Systems Approach is nowadays pervasive, playing an increasing role in most of the scientific disciplines, including biology (cellular biochemistry, molecular biology, cellular, and developmental biology).

Systems Biology, integrated with the Complex Systems Approach, is needed since the focus in biology and medicine, and physiology in particular, is shifting toward studying the properties of complex networks of reactions and processes of different natures, and how these control the behavior of cells and organisms in health and disease (Almaas *et al.*, 2004; Aon, 2009, p. 34; Aon, 2010, p. 196; Aon and Cortassa, 2009; Aon *et al.*, 2007; Aon, 2007a, p. 33; Aon *et al.*, 2007b; Barabasi, 2004, p. 162; Barabasi and Oltvai, 2004; Saks *et al.*, 2007; Weiss *et al.*, 2006).

Networks in Systems Biology

Focus in biology and medicine are now shifting toward studying the *properties of networks* and how these control the behavior of cells and organisms in health and disease. A wide range of systems, including the Internet, power grids, ecological and economic and social webs of collaboration as well as metabolic maps and clustered cardiac mitochondria, can be described as networks. Despite obvious differences between these systems, they all share common features in terms of network properties. For instance, networks exhibit scale-free topologies, i.e. most of the nodes in a network will have only a few links and these will be held together by a small number of nodes exhibiting high connectivity (Barabasi, 2003; Barabasi and Oltvai, 2004) (Fig. 1.8). Also scale-free dynamics in evolutionarily widely divergent cellular types such as yeast and heart cells is possible. This scale-free dynamics, characterized as *dynamic fractals*, manifest as multioscillatory events with frequency outputs over a broad range of temporal scales spanning at least three orders of magnitude, and are described by inverse power laws (Aon *et al.*, 2006a, 2008b; Yates, 1992).

Four main directions have characterized the study of cellular networks (Aon *et al.*, 2007a; Xia *et al.*, 2004): (i) architectural (structural morphology); (ii) topological (connectivity properties); (iii) dynamical (time-dependent behavior) and (iv) molecular.

These distinctions attempt to widen the view of networks that at present is prevalently assimilated as a collection of nodes and edges as conceived in graph theory, a branch of mathematics (Albert and Barabasi, 2002) that emphasizes the topological aspects of network connectivity. In the molecular view of networks, the cell is conceptualized as a complex network of interacting proteins, nucleic acids, and other biomolecules. Biomolecular networks are topologically interpreted as a collection of nodes (representing

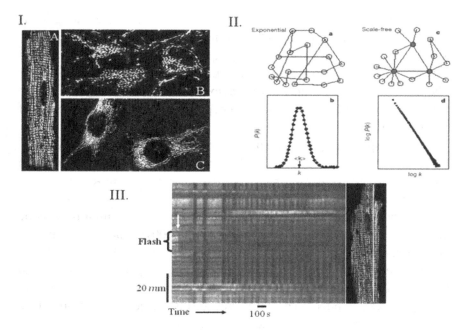

Figure 1.8.   The architectural, topological, and dynamic views of networks.

(I) The **architectural** approach is based on microscopic imaging as applied to intracellular structures, e.g. mitochondria, visualized through fluorescent probes in fixed specimens or living cells. In panels A–C, the different architecture of mitochondrial networks is shown after loading of cardiomyocytes (A), cortical neurons (B), and Hela (C) cells, respectively, with the cationic potentiometric dye tetramethylrhodamine ethyl ester (TMRE) used to monitor $\Delta\Psi$m.

(II) The **topological** approach conceives networks as a collection of nodes and edges. It describes network connectivity that can be characterized by the probability, $P(k)$, that a node has k links. For a *random network* $P(k)$ peaks strongly at $k = <k>$ and decays exponentially for a large k. For a *scale-free network* most nodes have only a few links, but a few nodes, called hubs (dark gray), have a very large number of links (Reproduced from Jeong *et al.*, 2000).

(III) The **dynamic** view emphasizes the autonomous self-organization of cellular (patho)physiological processes in time. In the example, a self-organized transition visualized by TMRE fluorescence as a cell-wide $\Delta\Psi_m$ depolarization (dark gray bands) after a laser flash, is shown. This synchronized $\Delta\Psi_m$ depolarization followed by sustained oscillations is triggered after the ongoing autonomous activity of ROS production in the cardiac mitochondrial lattice reaches a threshold level. Mitochondrial criticality, apparent under oxidative stress, is defined by the appeareance of a cluster of mitochondria with threshold levels of ROS that spans the whole cell (Aon *et al.*, 2004, 2006). For visualization of the spatiotemporal responses of $\Delta\Psi_m$ presented in panel (III), a 2 to 3 pixel wide line was drawn along the length of the myocyte (as shown in the cell to the right of panel (III)) and biomolecules), some of which are connected by links (representing interactions). Although useful, we interpret this view as too restrictive since, in a wider sense, networks also refer to temporal (different dynamics) aspects and not only to spatial (structural-molecular and topological) relationships. The dynamic aspects of network behavior involve functional relationships between processes, whose function occurs in widely different scales of space and time (see Chap. 5).

From a spatiotemporal perspective, networks have been analyzed as regular in structure and topology (e.g. lattice) and dynamics (e.g. each node in the network exhibiting stable fixed points or limit-cycle oscillations) or irregular in structure and topology (e.g. random or scale-free networks) and dynamics (e.g. each node in the network exhibiting chaotic attractors) (Strogatz, 2001) (Aon *et al.*, 2006a, 2007a, 2008b). Spatiotemporal organization of networks is at the basis of the collective and large-scale spreading consequences exhibited by complex phenomena (Aon and Cortassa, 2011; Aon *et al.*, 2007a, 2011; Weiss *et al.*, 2006). The collective dynamics exhibited by networks may result in emergent self-organized spatiotemporal behavior under certain conditions (Aon *et al.*, 2008b).

The concept of "molecular phenotypes" was introduced from the standpoint of cellular networks of interacting metabolites, proteins, and nucleic acids (Lusis and Weiss, 2010). Molecular phenotypes consist of levels of gene transcripts, proteins, and metabolites. These are the result of multiple genetic factors which depend on the type of disease (e.g. cardiovascular), and together with environmental factors contribute to disease risk. These "molecular phenotypes" are perturbed or changed by environmental and genetic factors which in turn determine (patho)physiological states that contribute to a disease. "Systems genetics" attempts to assess molecular phenotypes quantitatively and to identify underlying networks associated with clinical traits. Functional networks have been reconstructed

←───

Figure 1.8. (*Continued*)   the average fluorescence profile along the line was determined for the whole timeseries of images for a given experiment. A new image was then created, showing the line fluorescence as a function of time (timeline image).
Reproduced from Aon MA, Cortassa S, O'Rourke B. (2007a) On the network properties of mitochondria. In: Saks V (ed), *Molecular System Bioenergetics: Energy for Life*, pp. 111–135. Wiley-VCH, Weinheim, Germany.

using multiple, common genetic perturbations, for cardiovascular traits such as exercise endurance and body weight (Lusis and Weiss, 2010; Nadeau *et al.*, 2003; Shao *et al.*, 2008). The network constructed consisted of traits (nodes) and edges assigned based on significant correlations. This approach is based on the assumption that traits are correlated as a result of shared genetic determinants. As a validation criterion of the network constructed, interactions as well as functional relationships were confirmed through physiological studies and single-gene mutants or treatment with pharmacological agents. A similar strategy has been utilized to model an inflammatory network associated with atherosclerosis (Gargalovic *et al.*, 2006). The approach of System Genetics helps to characterize functional relationships in complex biological systems.

Network organization of subcellular organelles (e.g. mitochondria) is becoming a relevant topic. The view of mitochondria as networks includes emergence, self-organized behavior, and scale-free organization (Aon, 2010). The importance of the network organization of mitochondria is reinforced by the emerging role as signaling organelles (Aon *et al.*, 2004a; Zhou *et al.*, 2010). Mitochondria function as a source of signaling molecules (e.g. reactive oxygen species, ROS) or as a target of second messengers such as Ca^{2+} or intermediary metabolites such as ADP (Balaban, 2002; Cortassa *et al.*, 2006; Saks *et al.*, 2006). In this context, interaction of mitochondria with other organelles, e.g. SR, ER, becomes crucial (Maack *et al.*, 2006). In the cell, mitochondria behave as a network and as such they exhibit collective behavior. The intense "flickering" exhibited by mitochondria loaded with membrane potential sensors, initially attributed to random type of (white) noise, in fact corresponds to low amplitude, high frequency oscillations (Aon *et al.*, 2006a). Recent experimental evidence suggests that mitochondria behaves as a coupled network of coupled oscillators. Each individual mitochondrion senses neighboring mitochondria in the network. This interdependency, which can be quantified statistically, becomes crucial because under stressful conditions these networks can collapse (Aon *et al.*, 2003, 2004a, 2006a). In an ischemia/reperfusion (I/R) scenario (especially on reperfusion after ischemic injury), the mitochondrial network of cardiac cells can collapse. This collapse is followed by an escalation of failures from the (sub)cellular to the organ level. Ultimately, the electrical activity of the cardiomyocyte is altered, and in the tissue level gives rise to arrhythmias, or block of electrical conduction (Akar *et al.*, 2005; O'Rourke *et al.*, 2005). In the acute heart failure, self-organized emergent behavior reveals another view of Systems Biology. This dynamic view of networks goes beyond the initial definition given to Systems Biology: "the quantitative interpretation of high throughput data through mathematical models." The main addition is that the collective dynamics of networks can give rise to complex, emergent spatiotemporal patterns of behavior. The qualitative behavior of those patterns (e.g. type of dynamics: oscillations, chaos, spreading waves) and the circumstances in which they arise become paramount.

Temporal and Spatial Scaling in Cellular Processes

Biological processes at subcellular, cellular, and supracellular levels scale in space and time (Aon and Cortassa, 1993, 1997, 2009; Aon *et al.*, 2008c; Lloyd *et al.*, 1992; West, 1999). Scaling appeals to the interaction between the multiple levels of organization exhibited by cells and organisms, thus linking the spatial and temporal aspects of their organization.

Biologists have the notion of scaling from long ago, especially with respect to the relationship between the size of organisms and their metabolic rates, or the so-called allometric relationship (Aon and Cortassa, 1997; Enquist *et al.*, 2000). The concept of the design of organisms in terms of fractal geometry as an underlying explanation of the link between size and metabolism has reinvigorated this research field (Brown *et al.*, 2000).

Scaling refers to a quantity that depends on its argument through a power law. Thus, if a variable value changes according to a *power law* when the parameter it depends on is growing linearly, we say it *scales*, and the corresponding exponent is called a *critical exponent* (Aon and Cortassa, 2009). The concept of scaling has been extended to the fractal description and characterization of the dynamic performance of systems in general (e.g. biological, economical, geophysical) exhibiting self-organized (chaotic) or (colored) noise behavior. A great insight there has been to show that in time series from, e.g. market prices or mitochondrial membrane potential, the short-term fluctuations are intrinsically related to the long-term trends, appealing to the notion of networks (see Chap. 5).

Why is the concept of scaling so pervasive and fundamental for biology and biochemistry? Because it shows the intrinsic interdependence among the different levels of organization exhibited by living systems, which is expressed by the correlation among the different scales of space and time involved (Aon and Cortassa, 1997; Bassingthwaighte *et al.*, 1994; Brown *et al.*, 2000; Lloyd and Murray, 2005). Geometric and dynamic fractals capture the essence of this crucial feature through scaling and self-similarity (Bassingthwaighte *et al.*, 1994).

The concept of fractals arose to explain the predominantly irregular geometry of objects in the real world (Mandelbrot, 1983; Schroeder, 1991). The discovery of chaos in dynamics (Lorenz, 1963; see Gleick, 1988, for a historical review) and criticality in phase transitions (Sornette, 2000; Wilson, 1979) emphasized that *scaling* is a common feature of chaos and criticality as it is with fractals. Thus, scaling appears in the description of fractals, in the theory of critical phenomena and in chaotic dynamics (West, 1990).

Networks exhibit scale-free topologies, i.e. most of the nodes in a network will have only a few links and these will be held together by a small number of nodes exhibiting high connectivity (Barabasi, 2003; Wagner and Fell, 2001). This is a manifestation of the scaling exhibited by network organization in the topological and dynamical sense. The statistical distribution of the connectivity within scale-free networks (given by the number of links exhibited by the nodes of a network) exhibits a continuous hierarchy of nodes, spanning from rare hubs to numerous tiny nodes, rather than having a single, characteristic scale (Fig. 1.8) (Barabasi, 2003). The scale-free topology exhibited by networks can be explained by *growth* and *preferential attachment* among those nodes having higher probability of expanding their links, concepts largely derived from studies on network topology of the Internet. These two concepts introduced a more dynamic view of the emergence of topology of network connectivity, explaining the origin of hubs and power laws (Barabasi, 2003).

Emergent properties of cell function arise from transitions between levels of organization at bifurcation points in the dynamics of biological processes (Aon and Cortassa, 1997). Thereby, a dynamic system loses its global stability and behavioral changes occur. These modifications may be quantitative, qualitative, or both (Fig. 1.6; see also Chap. 5). Quantitatively, it may happen that the system dynamics moves at a limit point to a different branch of steady-state behavior, (lower or higher), e.g. as in bistability.

Under these conditions, the system does not change its qualitative behavior, i.e. it continues to be at a point attractor, either a stable node or a focus. However, at some bifurcation points, drastic qualitative changes occur; the system evolves from a monotonic operation mode toward periodic behavior, or a more complex (e.g. chaotic) trajectory (Abraham, 1987; Aon and Cortassa, 1991; Nicolis and Prigogine, 1977; 1989). Thus, dynamically organized phenomena are *homeodynamic* (Fig. 1.6), and are visualized as demonstrating spatiotemporal coherence.

Under *homeodynamic* conditions a system (e.g. network of reactions or cells), may exhibit emergent spatiotemporal coherence, i.e. dynamic organization (Lloyd *et al.*, 2001). A graphic analogy of the concept of *dynamic organization* under *homeodynamic* conditions is shown in Fig. 1.9.

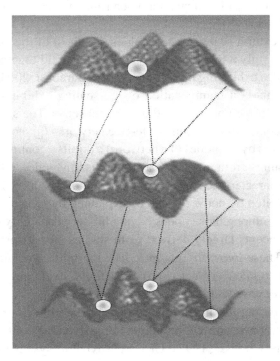

Figure 1.9.   A graphic analogy of the concept of *dynamic organization* under *homeodynamic* conditions. *Dynamic organization* in, e.g. cells or tissues, is an emergent property arising from transitions between levels of organization at bifurcation points in the dynamics of biological processes.
Source: Lloyd *et al.* (2001).

Dynamic organization in cells or tissues is thus an emergent property arising from transitions between levels of organization at bifurcation points in the dynamics of biological processes. In Fig. 1.9, the different landscapes represent the dynamic trajectories of subcellular processes (e.g. enzyme activity, synthesis of macromolecules, cell division; all indicated as spheres in the plot), resulting from the functioning of those processes on different spatiotemporal scales, (levels of organization). The dotted lines that link the spheres (different subcellular processes), indicate the coupling between them. The coupling between processes that function simultaneously on different spatiotemporal scales, *homeodynamically* modifies the system trajectories (the landscapes' shape), as represented by the movement of the spheres across peaks, down slopes and in valleys. The sphere of the landscape on top symbolizes a process occurring at a higher level of organization (higher spatial dimensions and longer relaxation times), i.e. a macroscopic spatial structure (waves, macromolecular networks, subcellular organelles, etc.). Indeed, the functioning of the system is coordinated and coupling occurs top-bottom as well as bottom-up. The interdependent and coupled crosstalk between the two flows of information crosses levels of organization through and beyond each level (Lloyd *et al.*, 2001; Yates, 1993).

Scaling in microbial and biochemical systems

The spatiotemporal scaling shown by sub- and supracellular processes allows interpretation of the balanced growth exhibited by microorganisms. Any two state variables of an exponentially growing microbial system are related by an allometric law (Aon *et al.*, 2008b; Rosen, 1967, 1970). In the case of balanced growth these two state variables may represent N, a population of microbes (biomass), and M, a population of macromolecules (protein, carbohydrate) synthesized in a constant proportion with the microbial biomass. Growth is balanced when the specific rate of change of all metabolic variables (concentration or total mass) is invariant (see Eq. (3.1) in Chap. 3) (Barford *et al.*, 1982; Cooper, 1991; Cortassa *et al.*, 1995; Roels, 1983).

Hence the fact that several sub- and supracellular systems scale their functioning in space and time exponentially, suggests that: (i) a defined relationship exists between the whole (cell) and each of its constituents (macromolecules); (ii) processes happen over broad time scales, given by their relaxation times following perturbation (see below).

Temporal scaling in microbial and biochemical systems implies that these react with different relaxation times toward a perturbation. We prefer the term scaling to hierarchy because in biological systems in general, and cellular ones in particular, there are vertical flows of information which coexist with horizontal ones. These mutual and reciprocal interactions are effected through coupling of sequential and parallel processes, respectively. This is clearly the case for a spatially highly interconnected and dynamically coupled system (i.e. a cell or an organism) that exhibits sequential and parallel processing. The consequences of the temporal scaling for cell function are diverse and important. In coupled processes, a variable in one dynamic subsystem because of its fast relaxation toward perturbations may act as a parameter of another dynamic subsystem that relaxes slowly (Von Bertalanffy, 1950).

Significant simplifications through reduction in the number of variables may be achieved by analysis for relaxation times. Essentially, the system's description is reduced to the slow variables (i.e. long relaxation times) (Heinrich *et al.*, 1977; Reich and Sel'kov, 1981; Roels, 1983). Otherwise stated, the overall dynamics of a system is governed by the relaxation times of the slow processes even though the system contains rapid motions.

By comparing the relaxation times of the intracellular processes with respect to those characterizing relevant changes in environmental conditions, it may help to decide whether the changes in the environment occur much faster than the mechanism by which the organism is able to adjust its activities or vice versa (Esener *et al.*, 1983; Roels, 1983; Vaseghi *et al.*, 1999). As an example, the concentrations of ATP and NADH exhibit rapid relaxation times because of their high turnover rates. In other words, the rate of adaptation of the mechanisms involved in keeping the intracellular concentrations of ATP and NADH is rapid as compared with the characteristic times of the changes in substrate concentration in the environment. Hence, the behavior of the system can be directly expressed in terms of the substrate concentration in the environment; this is a consequence of a pseudo-steady state assumption (Roels, 1983).

Temporal scaling allows us to understand that some cellular process (e.g. a metabolic pathway) may achieve a "balanced growth condition" before others. Thus, it has been suggested that a microorganism may apportion its total energy-producing capacity rapidly between fermentation and respiration modes long before adjustment of the specific rates of oxygen uptake or carbon dioxide production (Barford *et al.*, 1982). Thus, the respiratory quotient (CO_2/O_2) attains a "balanced condition" before either of its components.

Views of the Cell

Black and gray boxes: Levels of description of metabolic behavior in microorganisms

The chemical reaction equation for the growth of a microorganism is a complex one. The following chemical equation represents the amount of carbon, NADPH, NADH, NH_4^+, and CO_2 required or produced during the synthesis of 1 gm of yeast cell biomass from glucose as carbon source:

$$7.4\,C_6H_{12}O_6 + 7.2\,NH_3 + 7.9\,NADPH + 14.5\,NAD^+ \longrightarrow$$
$$9.9\,C_4H_{7.5}O_{1.7}N_{0.73} + 4.7\,CO_2 + 7.9\,NADP + 14.5\,NADH + 18\,H_2O.$$

The arrow which indicates the direction of the reaction "hides" a complex network of around 1,000 chemical reactions occurring inside an organism or cell, e.g. in bacteria (Bailey and Ollis, 1977; Stephanopoulos and Vallino, 1991; Stouthamer and Van Verseveld, 1987; Varma and Palsson, 1994), yeast (Cortassa *et al.*, 1995; Vanrolleghem *et al.*, 1996), or mammalian cells (Vriezen and van Dijken, 1998; Zupke *et al.*, 1995).

The description of a complex network of chemical reactions may be achieved at different levels of explanation or detail. Schematically, there are two levels of description: black and gray boxes. In the black box approach only the input(s) and output(s) of the system (e.g. microorganisms) are specified (see Chaps. 2 and 3). If we progressively incorporate details of what is occurring inside the box (e.g. mechanisms, reactions), it then becomes a gray box (see Chap. 4). The gray level of the box will be darker or lighter, depending on how deep is our knowledge of the physiological and dynamic conditions of the system under study (see Chaps. 3, 4, and 6).

The complex multilayered cellular circuitry shown in Fig. 1.10 intends to depict the complexity that we face in MCE. On the one hand, each line connecting two nodes (metabolites) of the metabolic network (bottom layer) is

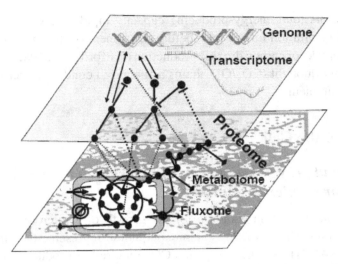

Figure 1.10.   A view of cells as multilayered mass–energy-information networks of reactions. Metabolic reactions (metabolome, fluxome) embedded in the cytoplasmic scaffolds are shown diagrammatically with each chemical species represented by a filled circle (bottom layer). Central catabolic pathways (glycolysis and the TCA cycle) are sketched. The chemical reaction network for the synthesis of a microorganism, (e.g. bacteria, unicellular fungi or algae), is a complex one comprising around 1,000 chemical reactions (Alberts *et al.*, 1989) (see text for further explanation). A typical mammalian cell synthesizes more than 10,000 different proteins (proteome), a major proportion of which are enzymes that carry out the mass–energy transformations (bottom layer). The information-carrying networks (genome, transcriptome, proteome) of cells are shown at the top layer. The signaling networks (represented with filled circles linked through dashed, regulatory lines) intertwine the multiple layers through activation or repression. The information-carrying networks are shown in a different layer just for the sake of clarity and its presence on top does not imply hierarchy. On the contrary, crosstalk connections existing between layers are emphasized by arrows (activatory, with arrowheads, or inhibitory, with a dash). Each network has its own set of components and mechanisms of interactions, e.g. the nodes in signaling networks may be proteins, second messengers, or transcriptional factors. The proteins are continuously synthesized through transcription, translation mechanisms of gene expression, and exert feedback regulation (e.g. DNA-binding proteins) on its own or the expression of other proteins, e.g. enzymes (depicted as double arrows on top layer). Moreover, the proteins taking part of cytoskeleton either exert feedback regulation on its own expression (Cleveland, 1988), or influence metabolic fluxes through epigenetic mechanisms (e.g. (de)polymerization, dynamic instability) (see Aon *et al.*, 2000a, 2000b, for reviews; Aon and Cortassa, 2002; Lloyd *et al.*, 2001) (see text for further explanation).

catalyzed by an enzyme, the amount of which is defined by gene expression (top layer) (transcription, translation, post-translational modifications). The enzyme activity (rate at which the conversion of one intermediate into another proceeds in the bottom layer) will be determined either by the intrinsic reaction dynamics, e.g. substrate inhibition, product activation (bottom layer), or intracellular signaling pathways, e.g. allosteric regulation, or covalent modification (top layer).

A typical mammalian cell synthesizes more than 10,000 different proteins, a major proportion of which are enzymes that carry out the mass–energy transformations of the network shown in the bottom layer. The information-processing function of cells is supervised by other networks (top layer), i.e. gene expression and intracellular signaling pathways, the latter intertwined with the bottom layer (Fig. 1.10). Many proteins in living cells function primarily as transfer and information-processing units (Bray, 1995). At least one-third of cellular proteins are dynamic structural elements of cells (Penman *et al.*, 1982). Compelling experimental evidence suggests that metabolism is strongly associated with cellular scaffolds, and that these insoluble matrices and their interactions in turn deeply influence the dynamics of chemical reactions, i.e. they also belong to the top layer (Clegg, 1984, 1991; Cortassa and Aon, 1996; Ovadi and Srere, 2000) and (Aon *et al.*, 2000a, 2001; Saks *et al.*, 2001). The latter emphasizes the multidimensional character of physiological responses *in vivo*. Elucidation of the "Reactome" indicates many unsuspected protein–protein interactions (e.g. by two protein hybrid assessment). The discovery of signaling molecules that interact with microtubules as well as the multiple effects on signaling pathways of drugs that destabilize or hyperstabilize microtubules, indicate that cytoskeletal polymers are likely to be critical to the spatial organization of signal transduction (Gundersen and Cook, 1999).

Together with the metabolic and signaling networks shown in Fig. 1.10, we must consider the regulatory circuitry required for gene expression, involving transcriptional activators, suppresors, *cis-* or *trans*-acting factors, most of them consisting of proteins, or DNA-binding proteins (Fig. 1.10, top layer). This regulatory net mainly determines the transcriptional level of gene expression. Often these transcriptional regulatory schemes are deduced from qualitative studies of the molecular biology of recombinant microorganisms (e.g. bacteria, yeast) grown on agar plates. These recombinant microorganisms are frequently poorly characterized (either metabolically or physiologically) under incompletely specified environmental conditions (e.g. when grown with rich undefined media).

Changes observed in the transcriptome, the proteome, or the metabolome do not always correspond to phenotypic alterations. The total set of fluxes in the metabolic network of a cell is not static but a dynamic function emerging from the interaction between the multilayered transducing networks of mass–energy-information (Fig. 1.10) and a certain environment. Whole pathway fluxes are sensitive to macroscopic network cues such as topological, morphological, and dynamic organization (Aon *et al.*, 2007a, 2008b).

Fluxes are determined at particular nodes of the biochemical network by local factors (e.g. enzymatic activity levels involving substrate affinities and concentrations of reactants, products and activator or inhibitor). The information network further influences the activity level of enzymes through gene expression, translation, transcription, and post-translational protein modifications. The fluxome, or the total set of fluxes in the metabolic network of a cell, represents integrative information on several cellular processes, and hence it is a unique phenotypic characteristic of cells. Flux analysis provides a true dynamic picture of the phenotype because it captures the metabolome in its functional interactions with the environment and the genome (Cascante and Marin, 2008).

Several reported data show that large variation of fluxes in metabolism can be achieved through a change in growth conditions, even at a constant growth rate. These *in vivo* flux changes are, however, only partially reflected in changes of enzyme levels. Only large changes lead to differences in enzyme levels, but these differences are much smaller than the variations in the fluxes (Vriezen and van Dijken, 1998). Even though fluxes sometimes differed by a factor of 45, the maximum difference found in the enzyme levels was only a factor of 3. Both the largest flux differences and the largest variations in enzyme levels were observed for the glycolytic enzymes. On the bases of the following observations: (i) a higher level of all glycolytic enzymes detected in low oxygen chemostat cultures, and (ii) the levels of hexokinase and pyruvate kinase were diminished only in low glucose chemostat cultures, Vriezen and van Dijken (1998) concluded that the down-regulation of glycolysis is only partially accomplished at the level of enzyme synthesis. Furthermore, flux modulation is primarily effected via concentrations of substrates, activators, and inhibitors (bottom layer according to Fig. 1.10). Similar conclusions were reached by Sierkstra *et al.* (1992), who measured mRNA levels and activities of glycolytic enzymes (HK, PFK, PGI, PGM), glucose-6-phosphate dehydrogenase, and glucose-regulated enzymes (pyruvate decarboxylase, pyruvate dehydrogenase, invertase, alcohol dehydrogenase), in glucose-limited continuous cultures of an industrial strain of *S. cerevisiae* at different dilution rates. The analysis showed that there is no clear correlation between enzyme activity and mRNA levels, despite the fact that PGI1 mRNA fluctuated and a slight decrease in PGI was registered at increasing growth rates. Although increased dilution rate led to an increase in glycolytic flux, the activity of most enzymes remained constant. According to these results, Sierkstra *et al.* (1992) suggested that glycolytic flux is not regulated at either transcriptional or translational levels under the conditions studied, but rather invokes effectors of enzymes (i.e. allosteric or protein phosphorylation mechanisms, Fig. 1.10, bottom layer).

During mixed-substrate cultivation of *S. cerevisiae*, the observed differential regulation of enzyme activities indicated that glucose is preferentially used as the starting material for biosynthesis. On the other hand ethanol serves as a dissimilatory substrate for energy production (and a source of acetyl-CoA for biosynthesis). Thus, only when the ATP requirement for glucose assimilation was completely met by oxidation of ethanol, were the glyoxylate cycle enzymes required for assimilation of ethanol into compounds with more than two carbon atoms (de Jong-Gubbels *et al.*, 1995). In contrast to the coordinated expression pattern that was observed for the key enzymes of ethanol assimilation, activities in cell-free extracts of the glycolytic enzymes, phosphofructokinase and pyruvate kinase, exhibited little variation with the ethanol to glucose ratio. This is consistent with the view that regulation of these enzyme activities is largely controlled by the concentrations of substrates and products and/or by allosteric enzyme modification (de Jong-Gubbels *et al.*, 1995).

Thus, a major area of research in future must determine whether the transcriptional level of regulation operates effectively under defined environmental conditions. As well as this, its articulation with the regulatory network of metabolism (e.g. allosteric or covalent modification) remains a major area of study. Unsolved problems include the following: (i) Under which conditions do the transcriptional regulatory networks exert control and (ii) for which metabolic blocks or specific reactions? (iii) How do main biological processes (division and differentiation) influence these regulatory circuits? (iv) Are they always the same or (v) do they change following cell division or differentiation? All these are open questions which remain to be elucidated.

Expression levels of more than 6,200 yeast genes using high-density oligonucleotide arrays for monitoring the expression of total mRNA populations, has been performed (Wodicka *et al.*, 1997). More than 87% of all yeast mRNAs were detected in *S. cerevisiae* cells grown in rich medium. The expression comparison between cells grown in rich and minimal media have identified a relatively small number of genes with dramatically different expression levels. Many of the most highly expressed genes are common to cells grown under both conditions, including genes encoding well-known "house-keeping" enzymes (e.g. *PGK1, TDH3, ENO2, FBA1*, and *PDC1*), structural proteins such as actin, and many ribosomal proteins (Wodicka *et al.*, 1997). However, many of the most highly expressed genes and those with the largest differences, under the conditions explored by Wodicka *et al.* (1997), are of unknown function. Another large-scale screen of genes that express differentially during the life cycle of *S. cerevisiae* at different subcellular locations, has been performed in diploid strains containing random *lacZ* insertions throughout the genome (Burns *et al.*, 1994). Powerful as they are for knowing which genes are expressed as mRNA species under defined environmental conditions, these approaches do not give information about the effectiveness of the functioning of the products of those genes in cells, e.g. actual reaction rates, enzyme kinetics, activation, inhibition (Fig. 1.10).

Another timely topic for MCE, concerns the relationship between the different cellular processes, although these are frequently treated separately. Although fragmentary, some experimental evidence exists that shows a link, e.g. between cell division, cell differentiation, and catabolite repression in yeast (Aon and Cortassa, 1995, 1998; Cortassa *et al.*, 2000; Cortassa and Aon, 1998; Monaco *et al.*, 1995). We will deal with this subject more thoroughly later (see Chaps. 4 and 7).

Transduction and Intracellular Signaling

Cells may be viewed as multilayered mass–energy-information networks of reactions (Aon and Cortassa, 2005) (Fig. 1.10). Metabolic networks and the energy machinery encompassing phosphorylation and redox potentials, and electrochemical ionic gradients constitute the mass–energy transduction machinery (Aon and Cortassa, 1997; Cortassa *et al.*, 2002). Signaling cascades carry out the information flow in those reaction networks intermediary between the genome and the transcriptome, proteome and metabolome (or phenotype, in general) (Fig. 1.10). These signaling networks pervade all levels between the genome–transcriptome–proteome–metabolome to ensure their connection.

Cells exhibit many distinct signalling pathways that allow them to react to environmental stimuli. Proteins with or without enzymatic activity, metabolites, and ions are main constituents of signaling networks. Their information output is comprised of concentration levels of intracellular metabolites (second messengers such as cAMP, AMP, phosphoinositides, reactive oxygen or nitrogen species), ions, proteins or small peptides, growth and transcriptional factors (Alliance, 2002; Kim *et al.*, 2009). According to their levels intracellular messengers act as allosteric effectors (positive or negative) on enzymes whose action reverberate on whole metabolic pathways that take part in crucial cellular mechanisms in response to environmental challenges (e.g. oxygen or substrate shortage) or cues (e.g. light, temperature).

Cells respond to sudden environmental changes such as a glucose pulse to steady-state cultures of *S. cerevisiae* (Vaseghi *et al.*, 2001) or ischemia in perfused rat heart (Marsin *et al.*, 2002), storage polysaccharides during light–dark transitions (Gomez-Casati *et al.*, 2003), or modulation of gene expression under oxidative stress mediated by reactive oxygen species (Morel and Barouki, 1999, for a review).

Through signaling networks cells modulate, suppress or activate, gene expression (transcription, translation), whole metabolic pathways, or certain enzymatic reactions within them (see Chap. 10). At the cellular level, the operation of signaling networks results in the regulation of, e.g. proliferation, hormone secretion, migration, and differentiation (Wiley *et al.*, 2003).

Increasing experimental evidence favors the idea that much of metabolism occurs strongly associated to and influenced by the dynamic cellular scaffolds, and that these insoluble matrices and their dynamics are in turn influenced strongly by the dynamics of biochemical reactions (Aon *et al.*, 2000b, 2004b; Welch and Clegg, 2010). At this point, transduction and coherence merge since the cytoskeleton fulfills all the requirements for systems to self-organize, and as a prevailing and ubiquitous (macro)molecular cytoplasmic network, may function as a link between, e.g. the stress-sensing and the stress-transduction mechanisms operating as the organism interacts with its environment.

Intracellular signaling is orchestrated through a large number of components by way of their interactions and their spatial relationships (Araujo and Liotta, 2006; Kholodenko, 2006; Weng *et al.*, 1999). Networking and nonlinearity of the input–output transfer characteristics results in several emergent properties that the individual pathways by themselves do not have. In this way the supply of an additional store of information within the intracellular biochemical reactions of signaling pathways becomes possible. Although mutations or altered gene expression can result in persistent activation of protein kinases, connections between pre-existing pathways may also result in persistently activated protein kinases capable of eliciting biological effects. Based on the considerations mentioned above the following properties arise: (i) extended signal duration; (ii) activation of feedback loops that confer on the system the ability to regulate output for considerable periods; this is achieved by allowing coupling between fast and slow responses; (iii) definition of threshold stimulation for biological effects, as signals of defined amplitude and duration are required to evoke a physiological response; (iv) simultaneous operation of multiple signal outputs that provide a filter mechanism to ensure that only appropriate signals are translated into alterations in biological behavior (Bhalla and Iyengar, 1999).

In systems where two signaling pathways interact through a feedback loop, the frequency and amplitude characteristics as well as the duration of the extracellular signal may determine the sustained activation of the system (Bhalla and Iyengar, 1999; Weng *et al.*, 1999). The system may behave as a self-organized bistable that defines a threshold level of stimulation, provided an autocatalytic loop (the nonlinearity) exists in the network (Aon and Cortassa, 1997). These bistable systems may be also deactivated, thus the emergent properties of this feedback system define not only the amplitude and duration of the extracellular signal required to be activated but the magnitude and duration of deactivation as well (Bhalla and Iyengar, 1999). Robustness is therefore one of the emergent features of these networks because of their ability to deliver, once activated, a constant output in a manner unaffected by small fluctuations caused by activating or deactivating events.

Systems pharmacology reveals that, rather than via specific action at a target site, many drugs have multiple perturbative actions across the entire network, so as to result not only in the desired therapy, but also in adverse side-effects. According to this view, the effects of the drug both therapeutic actions and adverse events, are the result of perturbation of the whole network (Berger and Iyengar, 2009; Wist *et al.*, 2009; Xie *et al.*, 2009). A complementary perspective comes from recent studies utilizing MCA of complex biochemical and transport networks in cells. In these studies it was shown that apparently unrelated processes control each other through *diffuse loops* (Cortassa *et al.*, 2009a, b). The existence of *control by diffuse loops* allows to understand the secondary effects of pharmacological agents as an action on a complex, interconnected network of reactions that brings about changes in processes without direct mechanistic links between them (Aon and Cortassa, 2011).

Self-Organized Emergent Phenomena

> I gave the title "The music of life" to this book because music also is a process, not a thing, and it has to be appreciated as a whole.
> Noble D. (2006) *The Music of Life: Biology Beyond the Genome.* Oxford University Press: New York.

Even in integrated form, the genomic databases will in themselves never allow us to go from DNA sequence to function directly; this arises as a consequence of the extreme complexity and multifaceted dynamic and self-organized nature of cells and organisms. As already stressed, major emergent behavior of organisms relegates the importance of genome-based information per se (see Aon and Cortassa (1997), Lloyd (1992, 1998), for reviews). Emergent properties, those single most distinguishing features of complex behavior in a system, manifest themselves as novel, and unexpected, macroscopic spatiotemporal patterns. Most importantly, they cannot be anticipated in any way from the behavior of the isolated components of that system. There is nothing inexplicable about emergence in complex systems: it arises from self-organizing principles resulting from nonlinear mechanisms and the continuous exchange of energy, matter, and information with the environment (Aon and Cortassa, 2009; Haken, 1983; Lloyd *et al.*, 2001; Nicolis and Prigogine, 1977; Yang *et al.*, 2008, 2010). Self-organization is one of the main mechanisms that generate organized complexity, and this is expressed as disparate types of nonlinear outputs (e.g. oscillations, chaos) in a wide range of systems (Ball, 2004), including biological ones (Lloyd, 2009, 2001).

In the second half of the twentieth century, many of the fundamental details of the spatiotemporal organization of living systems became established. For the first time, we have at hand biophysical theories with which to approach a qualitative analysis of the organized complexity that characterize living systems. Two main foundations of this biophysical theory are self-organization (Haken, 1983; Kauffman, 1989, 1995; Nicolis and Prigogine, 1977), and Dynamic Systems Theory (Abraham, 1987; Rosen, 1970). Self-organization is deeply rooted in nonequilibrium thermodynamics (Nicolis and Prigogine, 1977), and the kinetics of nonlinear systems, whereas Dynamic Systems Theory derives from the geometric theory of dynamical systems created by Poincaré (Abraham and Shaw, 1987).

By applying this biophysical theory of biological organization formulated to successively more complicated systems (i.e. artificial, artificial–biological-oriented, or biological) (Aon and Cortassa, 1997; Lloyd, 1992), it became clear that self-organization is a fundamental and necessary property of living systems. For the appearance of self-organization, systems must (Nicolis and Prigogine, 1977; Von Bertalanffy, 1955):

(1) be open to fluxes of energy and matter;
(2) operate via some coupled processes through some common intermediate;
(3) have at least one process that exhibits a kinetic nonlinearity.

Kinetically, biological systems in general are nonlinear because of multiple interactions between their components, e.g. protein–protein, feedback (e.g. substrate) inhibition, feedforward (product) activation, cross-activation or cross-inhibition. In the case of chemical reactions, these may be arranged in linear, branched, cyclic pathways, or as combinations of these basic configurations. Within this framework, metabolic pathways with different topologies (linear, circular) and sources of nonlinear kinetics (allostery, stoichiometric autocatalysis) were compared as energy converters for thermodynamic efficiency (Fig. 1.11). The converters were investigated for

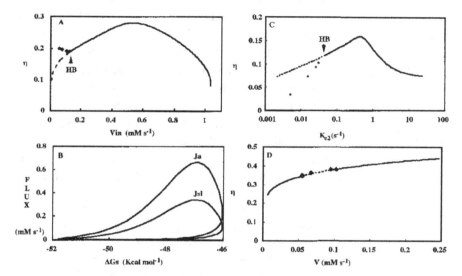

Figure 1.11. Thermodynamic performance of metabolic pathways with different topologies and feedback mechanisms as a function of the rate of substrate uptake by the cell. The putative advantage of oscillatory dynamics on thermodynamic performance was investigated on three different metabolic pathways: (A–B) A stoichiometric model of glycolysis in which the nonlinear mechanism is the autocatalytic feedback exerted on glucose phosphorylation by the stoichiometry of the ATP production of the anaerobic functioning of the glycolytic pathway. The glycolytic pathway is also an example of a *linear-branched topology*. (A) The stability analysis as a function of the substrate input V_{in} provided the steady-state concentrations of the metabolites from which the thermodynamic functions could be calculated. The kinetic parameters used were: $k_P = 0.7516 \, \text{mM s}^{-1}$, $k_1 = 0.0949 \, \text{mM}^{-1} \text{s}^{-1}$, $k_3 = 0.1 \, \text{mM}^{-2} \text{s}^{-1}$, $k_5 = 0.1 \, \text{s}^{-1}$; $k_9 = 0.05 \, \text{s}^{-1}$ and $k_7 = 0.1 \, \text{s}^{-1}$. The values of the constants C_A and P_t were 10 mM and K_M for the ATPase activity was 2 mM. ΔG_a and ΔG_s were 7.3 and 47 Kcal mol^{-1} and RT 0.596 Kcal mol^{-1}. (B) Phase space relations of the fluxes (J_a, ATP synthesis; J_{sl}, glucose degradation) with the input force (ΔG_s, the chemical potential of glucose conversion) in a limit cycle, which is included to explain the reversion of the declining tendency of thermodynamic efficiency in (a), as glycolysis approaches a Hopf bifurcation (HB). The parameters correspond to those described in (a) for the lowest V_{in} values represented by diamonds. (C) A model of the PTS in bacteria that represents a

possible energetic advantage under oscillatory dynamics as compared with asymptotic (steady state) behavior, as the rate of substrate uptake was systematically varied. Essentially, oscillatory dynamics arising from stoichiometric autocatalysis allowed an advantage in thermodynamic efficiency, whereas there was not such an advantage for oscillations where the underlying nonlinear mechanism was allostery under conditions of identical pathway topology. Furthermore, efficiency decreased in the case of a cyclic topology with a similar nonlinear mechanism as the source of the oscillations. These experiments demonstrate that the topology of the metabolic pathway also has a pivotal influence on the putative energetic advantage exerted by periodic dynamics (Aon and Cortassa, 1991; Cortassa *et al.*, 1990). The higher thermodynamic efficiency recorded in the linear pathway with a stoichiometric autocatalytic mechanism is due to a phase shift between the simultaneous maxima in catabolic and anabolic fluxes and a minimum in the chemical potential of substrate conversion (Fig. 1.11(b)).

Thus due to the intrinsic built-in nonlinearities in the network of chemical reactions, and also separation in distinct subcellular compartments,

←——

Figure 1.11. (*Continued*)　global *circular topology* including a covalent cycle and branches. The parameter k_{e2} represents the rate at which the phosphate group of the covalent cycle A–A~P is transferred to glucose when transported into the cell (Aon and Cortassa, 1997). Thermodynamic efficiency was analyzed as a function of k_{e2} and the following parameters were used: $k_{e1} = 0.9\,(s^{-1})$; $k_1\,(mM^{-1}\,s^{-1}) = 0.3$; $k_3\,(mM^{-2}\,s^{-1}) = 0.1$; $k_5\,(s^{-1}) = k_7\,(s^{-1}) = 0.1$; $k_9\,(s^{-1}) = 0.05$; τ. dp. $V_{max} = 0.4141$; $K_g = 1.0$ (mM); K_{pe} (mM) 0.1; K_{pk} (mM) 0.5; $K_p = 1.0$. The kinetic parameters of the proton pump were: $K_M = 2$ mM; $V_M = 0.5\,Mm\,s^{-1}$. The total nucleotide (C_A) and phosphate pool (P_T) were both 10 mM, whereas the total sum of substrate A involved in the covalent modification cycle (*TA*) was 1 mM. (D) An allosteric model described in Goldbeter (1996). The simulations were performed with the following parameter values: $k_1 = 2.9093$ mM s $^{-1}$; $k_2 = 0.15\,s^{-1}$; $k_3 = 0.1$ mM s $^{-1}$; $k_s = 3.0$ mM; $k_{adp} = 5.0$ mM; $k_p = 2.7088\,10^{-1}$ mM; $C_n = 10.0$ mM; $n = 2$; $L = 5000$. Thermodynamic efficiency was studied as a function of V, the rate of substrate input.

For selected points (diamonds in (a), (c), (d)) in the oscillatory domain (dashed lines) the simulation of the temporal evolution of the system enabled the computation of the thermodynamic efficiency, η, in order to examine the energetic advantage of oscillatory dynamics in all models represented in (a), (c), (d) (see Eqs. in Aon and Cortassa, 1997). In all figures, the plain and dashed lines represent stable and unstable steady states, respectively.

Source: Reproduced from Dynamic Biological Organization, 1997, pp. 310–311, Ch. 8, Fig. 8.11, Aon and Cortassa, Chapman & Hall, London, with kind permission from Kluwer Academic Publishers.

biological systems are able to change their dynamic behavior, i.e. to bifurcate toward new steady states, or attractors. The example shown in Fig. 1.11 illustrates the fact that biological systems are able to display diverse mechanisms of energetic adaptation when challenged under different environmental conditions.

These common features exhibited by different cellular systems, as well as their basic dynamic nature, emphasize that a relatively small proportion of the information on the spatiotemporal organization in cells and organisms is encoded in the DNA. Furthermore, the catalytic role of cytoskeletal polymers beyond their well-known structural one (Aon and Cortassa, 1995; Aon *et al.*, 2000b; Clegg, 1984, 1991), and the crowded nature of the intracellular environment (Aon *et al.*, 2000b; Gomez-Casati *et al.*, 1999, 2001), add new feasible behavioral possibilities. These are the appearance of self-organized behavior in concerted dynamic subsystems, e.g. microtubule polymerization dynamics and coupled-enzymatic reactions (Aon *et al.*, 1996, 2004b; Cortassa and Aon, 1994b), or the triggering of ultrasensitive catalytic behavior in a crowded milieu (Aon *et al.*, 2001; Gomez-Casati *et al.*, 1999, 2003).

In summary, (sub)cellular systems are able to show coherent as well as emergent properties (for a review, Aon *et al.* (2008b, 2004b), Lloyd and Rossi (2008)). On this basis, for example, enzymatic activities can show threshold effects (e.g. bistability or multiple stationary states) during entrainment by another autocatalytic process (e.g. cytoskeleton polymerization). Indeed, the observation that the dynamics of enzymatic reactions in turn may be entrained by the intrinsic dynamic instability of cytoskeleton polymers (Mitchison and Kirschner, 1984) may throw further light on the physiological role of so-called structural elements (Cortassa and Aon, 1996). Subcellular networks of organelles, e.g. mitochondria, connected through chemical messengers, e.g. reactive oxygen species, can give rise to stunning spatiotemporal organization at "critical" states (Aon *et al.*, 2003, 2004a). When close to the "critical" state, these systems exhibit threshold behavior, i.e. a small additional perturbation translates into an emergent self-organized behavior that escalates from the subcellular to the whole organ level (Akar *et al.*, 2005; Aon *et al.*, 2006b; O'Rourke *et al.*, 2005). The combinatorial binding of transcription factors to other proteins as well as to high and low affinity DNA sites, or that the spacing between DNA binding sites is altered (Miklos and Rubin, 1996), points out other example of nonlinear outputs. Thus, large responses may result from small changes in the concentrations of transcriptional components, or phosphorylation of transcriptional factors (Aon *et al.*, 2001; Gomez-Casati *et al.*, 1999).

Knowledge of all the putative regulatory mechanisms and the intrinsic nonlinear nature of biological processes should be taken into account when designing a modified organism for a specific biotransformation. In this respect, mathematical simulation of the desired process by attempting to incorporate into the model formulation as many variables as necessary to account for the known dynamical behavior will enhance the predictive power of the analysis and enlighten the decisions (see Chap. 5). While synthesizing biological information into a coherent scheme, there is a trade-off to be accomplished between *completeness* and *understandability*. We have greatly developed our ability to be more *complete* at integrating biological data. However, by doing so, we may reach levels of trade-off between completeness and understandability that are beyond reach of our present knowledge. We believe that there is much more room to accomplish the mentioned trade-off. However, at this point it is useful to remember that the *complete* simulation of a cell could involve a program whose description is *as large and complex as the cell itself.* As Barrow (1999) put it: "It is like having a full-scale map, as large as the territory it describes: extraordinarily accurate; but not so useful, and awfully tricky to fold up."

Homeodynamics and Coherence

Biological systems operate under *homeodynamic* conditions (see Chaps. 5 and 6). Thus continuous motion of the dynamics of the system, and its propensity for shifting between attractors at bifurcation points, based on its intrinsic dynamic properties underlies its complex behavior (Fig. 1.6) (Lloyd and Murray, 2005, 2006; Lloyd and Rossi, 2008; Lloyd *et al.*, 2001). This goes some way toward explaining the characteristics of its self-organization.

Models based on the stoichiometry of metabolic pathways for flux calculations apply for stable (asymptotic) steady states, and currently do not account for regulatory mechanisms (e.g. product inhibition, effector activation). Thus, the dynamic behavior of metabolic networks cannot be assessed solely by these types of models. In Chaps. 3–6, we introduce basic concepts and tools for dealing with the dynamics of biological processes.

Relevant to the functional behavior of cells or tissues, are the mechanisms by which dynamic organization under homeodynamic conditions is achieved. Unlike static controls, dynamic ones are able to sense cellular conditions and modulate the pathway flux based on the sensed information (Holtz and Keasling, 2010). Increasing enzyme expression levels is one of the main approaches used in achieving high fluxes with static controls. However, regulatory feedbacks in response to, e.g. limiting accumulation of a metabolic intermediate through a pathway in the face of changing conditions, represent an appealing situation for engineering dynamic control of fluxes. As an example, excess glucose flux and the diversion of carbon to acetate formation reduce the productivity of lycopene-generating *E. coli* cultures. An acetyl phosphate-activated transcription factor and promoter senses excess glucose flux. The accumulation of acetyl phosphate inside the cell is sensed by the engineered system that up-regulates the transcription of two genes (*pps* and *idi*) diverting the flux from acetate production to lycopene (Farmer and Liao (2000) quoted in Holtz and Keasling (2010)). More recent examples using similar principles have been reported including rewiring of signaling pathways (Kiel *et al.*, 2010).

Coherence may arise from the synchronization in space and time of molecules, or the architecture of supramolecular or supracellular structures, through self-organization, in an apparent, "purposeful," functional way. Under coherent behavioral conditions, spatially distant, say cytoplasmic regions, function coordinately, spanning spatial coordinates larger than the molecular or supramolecular realms, and show temporal relaxation times slower than those usually considered typical of molecular or supramolecular components (Aon *et al.*, 2000a, 2003, 2004a, 2004b; Lloyd *et al.*, 2001). Thus, a major functional consequence is brought about: the qualitative behavior of the system changes through the scaling of its spatiotemporal coordinates (Aon *et al.*, 2008a,b).

It has been proposed that the ultradian clock has timing functions that provide a time base for intracellular coordination (Aon *et al.*, 2000b; Klevecz *et al.*, 2004; Lloyd, 1992; Lloyd and Murray, 2005, 2006). In the latter sense, ultradian oscillations are a potentially coherence-inducer of the top-bottom type.

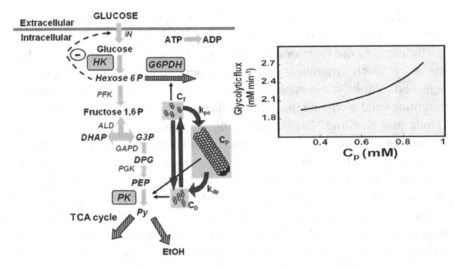

Figure 1.12 Depicted is a scheme of a metabolic network comprising sugar transport, glycolysis, the branch to the Pentose Phosphate (PP) pathway, the cycle of assembly–disassembly of microtubular protein (MTP), and their interactions. A mathematical model of this scheme has been formulated based upon experimental data obtained with yeast *in vitro*, *in situ*, and *in vivo* (Aon and Cortassa, 2002). The concentrations of polymerized or nonpolymerized (GTP-bound) MTP, and the oligomeric status of pyruvate kinase (PK) (Aon and Cortassa, 2002; Lloyd *et al.*, 2001) were taken into account. The computational model, that couples the cycle of assembly–disassembly of MTP and the metabolic network, was subjected to stability and bifurcation analyses as a function of the depolymerization and polymerization constants of MTP. The inset shows the phase plane analysis of the glycolytic flux as a function of the level of polymerized MTP, C_p, highlighting the fact that the degree of cytoskeleton polymerization may exert systemic effects on the glycolytic flux.
Source: Redrawn from Aon and Cortassa (2002).

We previously stated that dynamically organized phenomena are visualized as being spatiotemporally coherent. Several mechanisms are at the origin of this bottom-up coherence that we briefly describe. Waves of, e.g. second messengers or ions, may arise through a combination of amplification in biochemical reaction networks and spatial spreading through diffusion or percolation. Amplification may arise at instabilities in the dynamics of biochemical reactions given by autocatalysis through allosteric or ultrasensitive mechanisms (Aon *et al.*, 2000a, 2001, 2004b; Gomez-Casati *et al.*, 1999, 2003). Under these conditions, transduction (sensitivity amplification) and coherence (spatiotemporal waves) mutually cooperate.

Biochemically and thermodynamically, cellular metabolism may be represented as a set of catabolic and anabolic fluxes coupled to each other through energy-transducing events. In this framework, metabolic flux coordination is an expression of the dynamics of cellular organization. Flux coordination was shown in the regulation of cell growth and division or sporulation in *S. cerevisiae*, involving subcellular structural remodeling (Aon and Cortassa, 1995, 1997, 2001; Cortassa *et al.*, 2000). Dynamic organization of cytoskeleton components is among the cellular mechanisms of flux coordination. Supramolecular organization of cytoskeleton components gives rise to sophisticated spatial organization and intrincate fractal geometries in cells. The cellular cytoskeleton fulfills all the requirements for systems to self-organize, i.e. it is open to fluxes of matter (proteins) and energy (GTP). Furthermore, nonlinearity is provided by autocatalysis during polymerization or by the so-called dynamic instability that implies the catastrophic depolymerization of microtubules (Kirschner and Mitchison, 1986). An interpretation of microtubular dynamic instability as an example of a bistable irreversible transition has been proposed. The dynamic coupling between changes in the geometry of cytoskeleton organization and of enzymatic reactions taking place concomitantly produces entrainment of one system by the other in a global bistable switch (Aon *et al.*, 1996; Cortassa and Aon, 1996). Figure 1.8 shows that the degree of cytoskeleton polymerization may exert systemic effects on the glycolytic flux. Stability and bifurcation analyses were performed on a mathematical model that couples the dynamics of assembly–disassembly of microtubular protein (MTP) to the glycolytic pathway and to the branch of the tricarboxylic acid cycle, ethanolic fermentation, and the pentose phosphate pathways. The level of polymerized MTP was changed through variation of parameters related with MTP dynamics, i.e. the rate constant of microtubules depolymerization, k_{dp}, or polymerization, k_{pol} (Fig. 1.12). On the one hand, the enzymatic rates and metabolites concentration are entrained by the MTP polymeric status (Aon and Cortassa, 2002), whereas phase plane analysis shows that the flux through glycolysis coherently changes with the level of polymerized MTP (Fig. 1.12, inset) as reflected by augmenting or diminishing the glycolytic flux concomitantly with an increase or decrease of polymer levels. Further studies showed that high levels of the polymer decrease the negative control exerted by the branch to the PP pathway over glycolysis, thus allowing higher fluxes through the network (Aon and Cortassa, 2002). The glycolytic flux coordinately increased or decreased with increased or decreased polymeric MTP (C_P), respectively. High levels of polymerized MTP bestow (i) *coherence*, as reflected by the coordinated increase or decrease of the glycolytic flux induced by an increase or decrease of polymeric MTP (C_P), respectively (Fig. 1.12), and (ii) *robustness*, conferred by a significant increase in the range of stable steady-state behavior exhibited by the metabolic network (Aon and Cortassa, 2002; Aon *et al.*, 2004b). Overall, robustness in both prokaryotes and eukaryotes is achieved by negative and positive feedback regulation. However, the ability to predict consequences of altering feedbacks is highly dependent on computer simulation, since even simple networks can show counter-intuitive responses (Kiel *et al.*, 2010).

Functions of Rhythms and Clocks

Unicellular and multicellular organisms match the time dependencies of their internal environments with the periodicities of the external world (i.e. the elaboration of annual, seasonal, daily, and tidal rhythms), and optimize for tolerance to changes in the external environment (Hildebrandt, 1982; Kitano, 2004; Lloyd *et al.*, 1982; Yates, 1993; Zhou *et al.*, 2005). However, more fundamental is the timekeeping and synchronization necessary for the maintenance of coherence of all intracellular processes (the "internal harmony" of the organism) (Aon *et al.*, 2000b; Brodsky, 1998; Hildebrandt, 1982; Yates, 1992, 1993). For instance, the provision of energy, biosynthetic pathways, assembly of multimeric proteins, membranes and organelles, stress responses, cell differentiation, migration, and cell division require temporal organization on many time scales simultaneously (Chandrashekaran, 2005; Lloyd and Murray, 2005, 2006, 2007). This complex biological timing requires more than circadian organization; coordination on the ultradian domain (i.e. faster time scales where clocks cycle many times in a day) is essential. Examined more closely, it is evident that additional clocks are required, for instance, a circahoralian clock provides a time base on a scale of hours (Brodsky, 2006) while faster rhythms or oscillations measured in minutes (Berridge and Galione, 1988; Chance *et al.*,

1973; Gerasimenko *et al.*, 1996; Petersen and Findlay, 1987; Petersen and Maruyama, 1984), seconds (Roussel *et al.*, 2006) or milliseconds (Aon *et al.*, 2006a) abound in biological systems.

The ability of biological systems, either unicellular or multicellular, to exhibit rhythmic behavior in the ultradian (less than 24 h) domain is a fundamental property because of its potential role as the source of coherence, e.g. entrainment and coordination of intracellular functions (Lloyd, 1992; Lloyd *et al.*, 2008; Lloyd and Rossi, 2008). In systems exhibiting chaos, many possible motions are simultaneously present. In fact, since a chaotic system traces a strange attractor in the phase space, in principle a great number of unstable limit cycles are embedded in this attractor, each characterized by a distinct number of oscillations per period (see Chap. 6) (Aon *et al.*, 2000b; Peng *et al.*, 1991; Shinbrot *et al.*, 1992). Thus, a controlled chaotic attractor can, potentially, be a multioscillator that can be tuned into different time domains. Synchronization of multioscillatory states, implying controlled chaotic behavior of selected orbits (Lloyd and Lloyd, 1993, 1994, 1995) appears to be an essential property governing cell division (Lloyd, 1992), the circadian clock (Lloyd and Lloyd, 1993), and the coordination of metabolism and transcription across a population of single-celled organisms, as shown for continuous cultures of *S. cerevisiae* (Lloyd, 2009; Murray and Lloyd, 2007; Roussel and Lloyd, 2007). The ultradian (circahoralian) clock with ~40 min period described in the yeast *S. cerevisiae* growing aerobically in continuous culture, has been proposed to function as an intracellular timekeeper coordinating biochemical activity (Keulers *et al.*, 1996a; Lloyd *et al.*, 2002; Satroutdinov *et al.*, 1992). Respiratory oscillations are (among many) a main physiological manifestation of the ~40 min temperature-compensated periodicity (Lloyd, 2008). Microarray analysis showed a genome-wide oscillation in transcription, with expression maxima at three temporally separated clusters on the 40 min time-base (Klevecz *et al.*, 2004). During the first temporal cluster (reductive phase, low respiration) ~4,700 transcripts (of ~5300) are expressed whereas the remaining ~600 were detected in the second temporal cluster (oxidative phase, high respiration). The detection of synchronous bursts of DNA replication, coincident with the phase of decreasing respiration rates, suggests temporally separated windows for biochemical reactions otherwise incompatible because of the sensitivity to ROS and oxidative stress exhibited by DNA synthesis (Lloyd *et al.*, 2008). This evident timekeeping function of the ultradian clock uncovers its powerful biological role, which appears to have been evolutionarily conserved across a wide range of phyla (Lloyd and Murray, 2007).

Chapter 2

Matter and Energy Balances

Mass Balance

Bioprocess optimization through manipulation of metabolism requires a quantitative analysis of metabolic flux performance under defined conditions. As an integrated network of chemical reactions, metabolism obeys the general laws of chemistry, e.g. mass and energy conservation.

Material balance relies on the first law of thermodynamics. All the mass entering a system must be recovered, transformed or unchanged, at the output of a system, or at the end of a process. There are essentially two approaches to mass and energy balances. The first approach is to perform a global balance that takes into account the overall input and output of the system while disregarding the nature of the transformations taking place inside the living organism participating in the bioprocess. This approach is called "Black Box," and is the most commonly used by process engineers to evaluate the performance of a biotransformation. A second approach, let us call it "Light Gray," considers most reactions taking part in the metabolic machinery of an organism that participates in a particular bioprocess, and calculate energy and material balances thereof. This chapter will apply the concepts of mass balance from the level of the whole process taking place in the bioreactor ("black box") to the "light gray" description of microbial growth associated with the metabolic changes occurring within growing cells (see Chap. 1 Section: "Black and gray boxes: levels of description of metabolic behavior in microorganisms").

From the point of view of the exchange of matter and energy, most bioprocesses operating either at laboratory or industrial scales, may be classified as closed or open. A bioprocess that operates under *batch* conditions corresponds to a closed system that does not exchange matter except gases (usually O_2 and CO_2). Another operational mode is the *fed-batch* fermenter that is again open to gas exchange, but additionally allows the input of matter (e.g. substrates, vitamins). Completely open systems are those continuous processes consisting of input of matter and energy and exit of biomass, products, and energy. Substrates in excess with respect to the limiting one also flow out the system. In this sense, cells are nonequilibrium systems that depend on the rates at which energy or matter, or both, are fed into the system (Aon and Cortassa, 1997) (see Chaps. 3 and 6). To support steady-state operation, cells must dissipate energy and be continuously fed with matter.

The operation costs of continuous processes are much higher than batch or fed-batch ones. In each case, an evaluation of the yield is required to make a decision about the type of operating mode that should be applied. As an example, streptomycin production by *Streptomyces griseus* or antibodies from hybridoma cells are currently run in batch cultures. Ethanol fermentations from sugarcane or beet molasses are carried out in fed-batch cultures, whereas continuous cultures are run for acetic acid production from ethanol by *Acetobacter aceti* (Doran, 1995).

A cell participating in a bioprocess behaves as an open system because of the continuous uptake of substrates and excretion of metabolic products. Even under the condition of "endogenous" metabolism (when internal stores are consumed because of exhaustion of nutrients available in the environment), the cellular system excretes products. The substrates and products in the environment of cellular systems either may be freely changing or "clamped" depending on the experimental conditions. For instance, in batch culture the medium is continuously changing, whereas in a chemostat culture operating at steady state, both external and internal metabolite concentrations are fixed. This latter condition facilitates the rigorous application of some quantitative techniques germane to the purpose of MCE, i.e. MFA and MCA (see Chaps. 4 and 5).

General formulation of mass balance

The first law of thermodynamics states that during any chemical or physical change (subatomic reactions are excluded), there is no modification of the total amount of matter participating. Expressed in formal terms:

$$M_{input} + M_{prod} - M_{output} - M_{consum} = M_{accumul}, \qquad (2.1)$$

where M_{input} stands for the input of matter to the system; M_{prod} is the matter produced during the process; M_{output} stands for the output of matter from the system; M_{consum} is the matter consumed; $M_{accumul}$ is the matter accumulated in the system during the process.

Equation 2.1 is of general validity, irrespective of the type of system or process under study. It can be applied to chemical species in a fermentor, the mass of a molecular species, or to elemental balances such as those for carbon. It may also be applied to processes occurring exclusively inside cells, e.g. synthesis of cytoplasmic proteins or those undergoing processes that comprise intra- and extra-cellular stages, as is the case for sucrose transformation into ethanol by yeast alcoholic fermentation. The latter involves the cleavage of sucrose by secreted invertase, followed by its catabolism through cytoplasmic glycolysis and end-product excretion.

There are particular conditions under which simplifications of Eq. (2.1) can be considered.

(i) When there are no reactions involved, but only physical changes, i.e. there is no matter being produced or consumed, then M_{prod} and M_{consum} cancel out, and Eq. (2.1) can be reduced to:

$$M_{input} - M_{output} = M_{accumul}. \qquad (2.2)$$

This condition is without interest for the processes under consideration in this book, and will not be further analyzed.

(ii) Under steady-state conditions, in which the concentration of chemicals and the other system variables are constant, the amount of accumulated matter, $M_{accumul}$, is zero. The mass balance expression can then be written as follows:

$$M_{input} + M_{prod} - M_{output} - M_{consum} = 0. \qquad (2.3)$$

Equation (2.3) is also valid for the total mass in a closed system, as well as for the elemental balance of nitrogen or sulphur, in batch processes with chemoorganotroph organisms; it is valid for carbon, but more difficult to evaluate, since CO_2 is a gaseous compound and leaves the system.

Integral and differential mass balances

Different methods are used to perform a mass balance, depending on the operational mode of the process being considered: (i) in batch and fed-batch processes; the total and/or initial input of matter feeding the system is readily available, and a so-called integral balance can be performed; (ii) in continuous processes, where the fluxes of matter being exchanged by the system are calculated, a "differential mass balance" is then performed. This can be formalized as follows:

$$
\begin{bmatrix} \text{matter} \\ \text{input to} \\ \text{the system} \end{bmatrix} + \begin{bmatrix} \text{matter} \\ \text{generated in} \\ \text{the system} \end{bmatrix} - \begin{bmatrix} \text{matter} \\ \text{consumed in} \\ \text{the system} \end{bmatrix} - \begin{bmatrix} \text{matter} \\ \text{output from} \\ \text{the system} \end{bmatrix}
$$

$$
= \begin{bmatrix} \text{matter} \\ \text{accumulated} \\ \text{in the system} \end{bmatrix} \tag{2.4}
$$

that in the steady state takes the form:

$$
\begin{bmatrix} \text{matter} \\ \text{input to} \\ \text{the system} \end{bmatrix} + \begin{bmatrix} \text{matter} \\ \text{generated in} \\ \text{the system} \end{bmatrix} = \begin{bmatrix} \text{matter} \\ \text{consumed in} \\ \text{the system} \end{bmatrix} + \begin{bmatrix} \text{matter} \\ \text{output from} \\ \text{the system} \end{bmatrix}
$$

$$
\tag{2.5}
$$

A note of caution should be raised at this point. While performing a mass balance (as well as other type of calculations), care should be taken to keep the units consistent. Analysis of units is very helpful for the detection of sources of error or missing information.

As our focus is on MCE, we will leave here the general formulation of material balance to focus on processes involving metabolic transformations in cells. For a deeper explanation on the general matter of bioprocess balancing, the reader is referred to the book of Doran (1995) for useful examples.

Growth stoichiometry and product formation

As any process operating in closed or open systems, the growth of a cell or its division to generate another cell may be considered from the point of view of mass balances. From this viewpoint, the two approaches defined under Section: *Material and energy balances*, will be developed. On the one hand, the "black box" approach, which just looks at the overall inputs and outputs of cells without taking into account the biochemical transformations occurring inside them. The overall mass balance equation for growth of a chemoorganotroph organism without product formation (Fig. 2.1), may be expressed as follows:

$$C_w H_x O_y N_z + aO_2 + bH_g O_h N_i \Rightarrow cCH_\alpha O_\beta N_\delta + dCO_2 + eH_2O, \quad (2.6)$$

where a, b, c, d, and e are the stoichiometric coefficients of the overall biomass synthesis reaction, and $\alpha, \beta, \delta, g, h, i, w, x, y$, and z, the elemental formula coefficients. Figure 2.1 illustrates the "black box" approach for biosynthesis of heterotrophic, phototrophic, or chemolithotrophic microbial mass (Fig. 2.1, top, middle, and bottom panels, respectively). The carbon and nitrogen substrates, the energy source, and the electron acceptor, used by the microorganism (O_2 or an organic compound) are taken into account. It also shows the by-products associated with biomass production, e.g. CO_2 and H_2O, as in the case of chemoorganotrophs. Such by-products differ for the microbial processes in phototrophic organisms, performing photosynthesis, and thereby O_2 instead of CO_2 is evolved. On the other hand, chemolitotrophic organisms use reduced inorganic compounds, e.g. SH_2 as electron donor and energy source (Fig. 2.1).

Other elements are used to produce microbial biomass, e.g. P, S, Na, K, Ca, and trace elements (Fe, Cu, Mn), but the uptake of these elements may be considered negligible in mass balance calculations. When performing elemental analysis, a small fraction of ash is usually formed. This fraction corresponds mainly to P and S residues, although other elements may also account for ash.

Roels (1983) has collated the compositions of a number of microorganisms, and derived values for a mean "chemical formula" (Table 2.1). It can be appreciated that a "chemical formula" is not a distinctive feature of an organism as different organisms may have the same "formula." Moreover, different elemental formula coefficients may occur within a species when it grows under different environmental conditions, e.g. with substrates of different degree of reduction, under aerobiosis or anaerobiosis, or with different nitrogen sources.

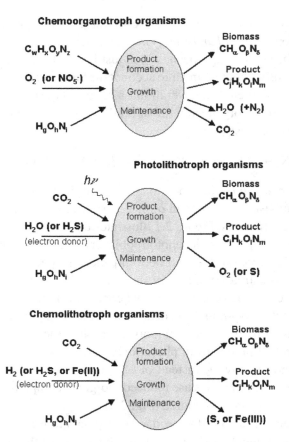

Figure 2.1.   Black box scheme of growth and product formation exhibited by different kind of organisms. In the scheme are emphasized the nature of the carbon source (i.e. organic compound in top panel, or inorganic, CO_2, in middle and bottom panels), the nitrogen source ($H_gO_hN_i$), the electron donor (in top panel the C-source, H_2O or H_2, SH_2 or Fe(II) in middle and bottom panels) and the products of metabolism. Chemoorganotrophs use O_2 as electron acceptor; alternatively they may use NO_3^-, whereas chemolithotrophs and photolithotrophs use endogenous organic compounds.

As an example, the chemical formula of *Pseudomonas mendocina* growing under microaerophilic conditions was calculated from its elemental composition (C, 43.5%; H, 6.5%; N, 11.5%; O, 30.3%) as follows (Verdoni *et al.*, 1992):

$$C: \frac{43.5}{12} = 3.6; \quad H: \frac{6.5}{1} = 6.5; \quad N: \frac{11.5}{14} = 0.82; \quad O: \frac{30.3}{16} = 1.89$$

Table 2.1 Elemental composition and generalized degree of reduction of biomass with respect to various nitrogen sources

Organism	Elemental formula	Degree of reduction (γ)		
		NH_3	HNO_3	N_2
Candida utilis	$CH_{1.83}O_{0.54}N_{0.10}$	4.45	5.25	4.75
Candida utilis	$CH_{1.87}O_{0.56}N_{0.20}$	4.15	5.75	4.75
Candida utilis	$CH_{1.83}O_{0.46}N_{0.19}$	4.34	5.86	4.91
Klebsiella aerogenes	$CH_{1.75}O_{0.43}N_{0.22}$	4.23	5.99	4.89
Klebsiella aerogenes	$CH_{1.73}O_{0.43}N_{0.24}$	4.15	6.07	4.87
Klebsiella aerogenes	$CH_{1.75}O_{0.47}N_{0.17}$	4.30	5.66	4.81
Saccharomyces cerevisiae	$CH_{1.64}O_{0.52}N_{0.16}$	4.12	5.40	4.60
Saccharomyces cerevisiae	$CH_{1.83}O_{0.56}N_{0.17}$	4.20	5.56	4.71
Saccharomyces cerevisiae	$CH_{1.81}O_{0.51}N_{0.17}$	4.28	5.64	4.79
Paracoccus denitrificans	$CH_{1.81}O_{0.51}N_{0.20}$	4.19	5.79	4.79
Paracoccus denitrificans	$CH_{1.51}O_{0.46}N_{0.19}$	4.02	5.54	4.59
Escherichia coli	$CH_{1.77}O_{0.49}N_{0.24}$	4.07	5.99	4.79
Pseudomonas C12B	$CH_{2.00}O_{0.52}N_{0.23}$	4.27	6.11	4.96
Aerobacter aerogenes	$CH_{1.83}O_{0.55}N_{0.25}$	3.98	5.98	4.73
Average	$CH_{1.79}O_{0.50}N_{0.20}$	4.19	5.78	4.79
		(3%)	(4.5%)	(2.1%)

The degree of reduction was calculated as indicated by Eq. (2.16). The differences between ammonia, nitrate, and N_2 arise because of their γ values of -3, $+5$, and 0, respectively are taken as null. In brackets are indicated the standard deviation of the degree of reduction expressed as percentage of the average.
Source: Reprinted from Roels (1983, pp. 30–41) © copyright 1983 Elsevier Science, with permission of the author.

The formula $C_{3.6}H_{6.5}O_{1.89}N_{0.82}$ can be recalculated as that of a molecule containing a single atom carbon: $CH_{1.81}O_{0.53}N_{0.23}$, thereby corresponds to a molecular weight of 25.5.

To find out the stoichiometric coefficients of biomass synthesis, once the chemical formula is known, an "elemental balance" must be performed. In such an elemental balance only the main biomass components are considered, i.e. C, N, O, and H:

C balance	$w = c + d$	(2.7a)
N balance	$z + bi = c\delta$	(2.7b)
O balance	$y + 2a + bh = c\beta + 2d + e$	(2.7c)
H balance	$x + bg = c\alpha + 2e$	(2.7d)

In order to calculate the stoichiometric coefficients, the system of Eq. (2.7) cannot be determined because five coefficients are unknown, and only four equations are available. A fifth equation is provided by the respiratory quotient, RQ (Eq. (2.8)), that in physiological terms tells us about the type of metabolism displayed by the organism, i.e. respiratory or respirofermentative.

$$RQ = \frac{\text{moles of } CO_2 \text{ produced}}{\text{moles of } O_2 \text{ consumed}} = \frac{d}{a} \tag{2.8}$$

Respiratory quotients range from 1.0 for oxidative breakdown of carbohydrates to larger values for fermentative metabolism, and in ethanolic fermentation can attain values of 10. However, the minimal value depends on the degree of reduction of the substrate being oxidized. For instance, in methane or ethanol it falls to values of 0.5 and 0.67, respectively.

In practice, the stoichiometric coefficient e (Eqs. (2.7c) and (2.7d)) is usually very difficult to measure, due to the large excess of water in the aqueous systems where most bioprocesses are studied. A change in water concentration is subjected to a large experimental error, because it is negligible in comparison with the water present in the medium. A way to circumvent this problem is to perform a balance of the reducing power of available electrons (see below).

If growth is associated with other organic products, as happens under anaerobic conditions (when the electron acceptor is an oxidized metabolic intermediate), then:

$$C_w H_x O_y N_z + b H_g O_h N_i$$
$$\Rightarrow c CH_\alpha O_\beta N_\delta + d CO_2 + e H_2O + f C_j H_k O_l N_m, \tag{2.9}$$

where $C_j H_k O_l N_m$ stands for a by-product (e.g. ethanol, lactic acid or a secondary metabolite).

The formation of product adds a right-hand term to the system of Eq. (2.7); this additional term contains an unknown coefficient, f.

The balance equation for biomass and product formation for photosynthetic organisms (e.g. higher plants, green algae, and photosynthetic bacteria, e.g. cyanobacteria) reads as follows:

$$a CO_2 + b H_g O_h N_i + c H_2O \Rightarrow d CH_\alpha O_\beta N_\delta + e O_2 + f C_j H_k O_l N_m. \tag{2.10}$$

While for the chemolitotrophic organisms, such as sulphur iron or hydrogen bacteria (Fig. 2.1, bottom panel) is:

$$aCO_2 + bH_gO_hN_i + cSH_2 \Rightarrow d\,CH_\alpha O_\beta N_\delta + eS + fC_jH_kO_lN_m.$$

$$(2.11)$$

Biomass and product yields

There are some coefficients that, in spite of being dependent on the physiological status of the growing organism, may be assessed from a complete "black box" approach. We are referring here to biomass and product yields with respect to the consumed carbon substrate.

The biomass yield is defined as the amount of biomass synthesized per mole or gram of carbon substrate. In formal terms and expressed on a mass basis:

$$Y_{XC} = \frac{\text{grams of produced cells}}{\text{grams of consumed substrate}} = \frac{c(MW\ \text{cells})}{(MW\ \text{substrate})}. \qquad (2.12)$$

If product is also formed:

$$Y_{PC} = \frac{\text{grams of synthesized product}}{\text{grams of consumed substrate}} = \frac{f(MW\ \text{product})}{(MW\ \text{substrate})}. \qquad (2.13)$$

Y_{XC} and Y_{PC} are, in fact, phenomenological coefficients. The mechanistic stoichiometric coefficients (deduced in, e.g. a "light gray" approach) may be higher, depending upon culture conditions and the degree of coupling between anabolism and catabolism (see Chap. 3). In fact, part of the consumed substrate is directed toward biomass synthesis, and part is devoted to energy generation to drive biosynthetic reactions. Thus, growth yields are meaningful parameters that can also be derived from the biochemistry of microbial growth (see Chap. 4).

The maximal yield of product obtained from a given substrate, may be predicted in the black box approach from an electron balance (see below). In physiological experiments, the yield may be obtained from the slope of the specific rate of substrate consumption as a function of the growth rate, as indicated by the following expression:

$$\frac{dS}{dt} = m_s + \frac{\mu}{Y_{XS}}, \qquad (2.14)$$

the left-hand term being the instantaneous rate of substrate consumption, and in the right-hand term, m_S, the maintenance coefficient, μ, the growth rate, and Y_{XS}, the yield of biomass on substrate S. We shall not go into details of Eq. (2.14) here; a more detailed treatment of the relationship between substrate consumption, growth, and yield is addressed in Chap. 3.

Electron balance

In metabolic networks a large number of reactions imply changes in the redox state of the compounds involved. In fact, catabolism includes either complete or partial oxidation of reduced compounds, coupled to the synthesis of high-energy-transfer potential intermediates, e.g. ATP. Thus, the electron balance is at the interphase between mass and energy balances.

Reducing power is another quantity subjected to conservation laws. In fact, what is actually conserved is the number of "available" electrons; i.e. those able to be transferred to oxygen during the combustion of a substance to CO_2, H_2O, and a nitrogenous compound (Doran, 1995). For many organic compounds, this is an exergonic process (i.e. ΔG is negative, see below section: *Energy balance*) as exemplified in Table 2.2 (Roels, 1983). It is convenient to calculate the available electrons (valence) from the elemental balance: $C(+4)$, $O(-2)$, $H(+1)$. In the case of nitrogen, its valence depends on the reference compound, whether it is NH_3 (valence -3 for N), NO_3^- ($+5$ for N), or molecular nitrogen N_2 (0 for N).

The degree of reduction of an organic compound is defined as the mole numbers of available electrons contained in the amount corresponding to one gram-mole C of the compound (Table 2.2). This is a very useful measurement, since it tells us whether a substrate can act as electron donor (and thereby supply energy for biomass growth) with a negative thermodynamic efficiency, i.e. without requiring catabolism to operate in order to fulfill the energy requirement of anabolism. If a compound is more reduced than biomass (degree of reduction is *ca.* 4 with NH_3 as the reference nitrogen compound), a negative thermodynamic efficiency of biomass growth will result (Table 2.2, see also section: *An energetic view of microbial metabolism*). The degree of reduction is calculated from the addition of the product of the coefficients of each element in the chemical formula times the valence of that element, as follows:

$$\gamma_S = 4w + x - 2y - 3z \quad \text{for a substrate,} \tag{2.15}$$

Table 2.2 Heat of combustion and free energy of combustion at standard conditions (unit molality, 298°K, 1 atmosphere) and at a pH 7

Compound	Formula	Degree of reduction, γ	ΔG_c°	ΔH_c° (kJ mole^{-1})	$\Delta G_c^\circ/\gamma$	$\Delta H_c^\circ/\gamma$
Formic acid	CH_2O_2	2	281	255	140.5	127.5
Acetic acid	$C_2H_4O_2$	4	894	876	111.8	109.5
Propionic acid	$C_3H_6O_2$	4.67	1533	1529	109.4	109.2
Butyric acid	$C_4H_8O_2$	5	2173	2194	108.7	109.7
Valeric acid	$C_5H_{10}O_2$	5.2	2813	2841	108.2	109.3
Palmitic acid	$C_{16}H_{32}O_2$	5.75	9800	9989	106.5	108.6
Lactic acid	$C_3H_6O_3$	4	1377	1369	114.8	114.1
Gluconic acid	$C_6H_{12}O_7$	3.67	2661		121	
Pyruvic acid	$C_3H_4O_3$	3.33	1140		114	
Oxalic acid	$C_2H_2O_4$	1	327	246	163.5	123
Succinic acid	$C_4H_6O_4$	3.5	1599	1493	114.2	106.7
Fumaric acid	$C_4H_4O_4$	3	1448	1337	120.7	111.3
Malic acid	$C_4H_6O_5$	3	1444	1329	120.3	110.8
Citric acid	$C_6H_8O_7$	3	2147	1963	119.3	109.1
Glucose	$C_6H_{12}O_6$	4	2872	2807	119.7	117
Methane	CH_4	8	818	892	102.3	111.5
Ethane	C_2H_6	7	1467	1562	104.8	111.6
Propane	C_3H_8	6.67	2108	2223	105.4	111.1
Pentane	C_5H_{12}	6.4	3385	3533	105.8	110.4
Ethene	C_2H_4	6	1331	1413	110.9	117.8
Ethyne	C_2H_2	5	1235	1301	123.5	130.1
Methanol	CH_4O	6	693	728	115.5	121.3
Ethanol	C_2H_6O	6	1319	1369	109.9	114.1
iso-Propanol	C_3H_8O	6	1946	1989	108.1	110.5
n-Butanol	$C_4H_{10}O$	6	2592	2680	108	111.7
Ethylene glycol	$C_2H_6O_2$	5	1170	1181	117	118.1
Glycerol	$C_3H_8O_3$	4.67	1643	1663	117.4	118.7
Glucitol	$C_6H_{14}O_6$	4.33	3084	3049	118.6	117.3
Acetone	C_3H_6O	5.33	1734	1793	108.4	112.1
Formaldehyde	CH_2O	4	501	572	125.3	142.9
Acetaldehyde	C_2H_4O	5	1123	1168	112.3	116.8
Butyraldehyde	C_4H_8O	5.5	2407		109.4	
Alanine	$C_3H_7NO_2$	5	1642	1707	109.5	113.8
Arginine	$C_6H_{14}N_4O_2$	5.67	3786	3744	111.3	110
Asparagine	$C_4H_8N_2O_3$	4.5	1999	1936	111.1	107.6
Glutamic acid	$C_5H_9NO_4$	4.2	2315	2250	110.2	107.1
Aspartic acid	$C_4H_7NO_4$	3.75	1686	1608	112.4	107.2
Glutamine	$C_5H_{10}N_2O_3$	4.8	2628	2570	109.5	107.1
Glycine	$C_2H_5NO_2$	4.5	1011	974	112.3	108.2

(Continued)

Table 2.2 (*Continued*)

Compound	Formula	Degree of reduction, γ	ΔG_c°	ΔH_c° (kJ mole^{-1})	$\Delta G_c^\circ/\gamma$	$\Delta H_c^\circ l\gamma$
Leucine	$C_6H_{13}NO_2$	5.5	3565	3588	108	108.7
Isoleucine	$C_6H_{13}NO_2$	5.5	3564	3588	108	108.7
Lysine	$C_6H_{14}N_2O_2$	5.67		3684		108.3
Histidine	$C_6H_9N_3O_2$	4.83		3426		118.2
Phenylalanine	$C_9H_{11}NO_2$	4.78	4647	4653	108	108.2
Proline	$C_5H_9NO_2$	5		2735		109.4
Serine	$C_3H_7NO_3$	4.33	1502	1455	115.6	112.0
Threonine	$C_4H_9NO_3$	4.75	2130	2104	112.1	110.7
Tryptophane	$C_{11}H_{12}N_2O_2$	4.73	5649	5632	108.6	108.3
Tyrosine	$C_9H_{11}NO_3$	4.56	4483	4437	109.2	108.1
Valine	$C_5H_{11}NO_2$	5.4	2920	2920	108.2	108.2
Thymine	$C_5H_6N_2O_2$	4.4		2360		107.3
Adenine	$C_5H_5N_5$	5		2780		111.2
Guanine	$C_5H_5N_5O$	4.6	2612	2500	113.6	108.7
Cytosine	$C_4H_5N_3O$	4.75		1828		96.2
Uracil	$C_4H_4N_2O_2$	4		1688		105.5
Hydrogen	H_2	2	238	286	119	143
Graphite	C	4	394	394	98.5	98.5
Carbon monoxide	CO	2	257	283	128.5	141.5
Ammonia	NH_3	3	329	383	109.7	127.7
Ammonium ion	NH_4^+		356	383		
	NO	-2	86.6	22	-43.3	-11
Nitrous acid	HNO_2	-3	81.4		-27.1	
Nitric acid	HNO_3	-5	7.3	-30	-1.5	6
Hydrazine	N_2H_4	4	602.4	622	150.6	155.5
Hydrogen sulfide	H_2S	2	323	247	161.5	123.5
Sulfurous acid	H_2SO_3	-4	-249.4	-329	62.4	82.3
Sulfuric acid	H_2SO_4	-6	-507.4	-602	84.6	100.3
Biomass	$CH_{1.8}O_{0.5}N_{0.2}$	4.8	541.2	560	112.8	116.7

Free enthalpy corresponds to combustion to gaseous CO_2, liquid water, and N_2 at unit molality of all reactants. The values in the sixth and seventh columns were obtained from the fourth and fifth columns, respectively, divided by the degree of reduction and the number of C atoms in the molecular formula.

Source: Reprinted from Roels (1983, pp. 30–41) ©copyright 1983, Elsevier Science, with permission of the author.

and

$$\gamma_B = 4 + \alpha - 2\beta - 3\delta \quad \text{for biomass.} \tag{2.16}$$

Ammonia is considered here as the N source. The degree of reduction of CO_2, or H_2O, is zero. The electron balance is a useful criterion for the identification of the kind of by-products that can be excreted to the extra-cellular medium when the carbon balance of a given culture is investigated.

Theoretical oxygen demand

The oxygenation of a bioprocess is an important condition for cell growth. Under oxygen limitation, bioprocesses operate under anaerobic or microaerophilic conditions and thereby the type of metabolism displayed by the organism.

Additionally, the stoichiometry of the energy coupling of the respiratory chain may vary according to the oxygen tension level maintained under particular growth conditions. The P:O ratio, namely the mole numbers of ATP synthesized per mole of electron pairs transported across the respiratory electron transport chain, is difficult to assess experimentally since several different possibilities of electron transport stoichiometries exist (see below). For instance, in chemostat cultures of *P. mendocina* at very low oxygen tensions, a P:O ratio of 3 had to be considered for closing energy balance calculations (Verdoni *et al.*, 1992).

An oxygen growth yield has also been defined as oxygen is also a substrate, mainly a catabolic one, acting as the final electron acceptor of the respiratory chain; Y_{O2} is a physiological parameter containing implicit information about the efficiency of oxidative phosphorylation (Stouthamer and van Verseveld, 1985).

If the stoichiometric coefficients of Eqs. 2.6 and 2.9 are known, the conservation law of available electrons can be written as

$$w\gamma_S - 4a = c\gamma_B + j\gamma_P f, \tag{2.17}$$

where a depicts the stoichiometric coefficient for oxygen (the oxygen demand for the growth reaction) whether associated or not with product formation. In Eq. 2.17, apart from CO_2 and O_2; the degree of reduction of

NH_3 is also taken as zero. In that case, the degree of reduction of biomass and product should be computed with respect to NH_3 as reference compound.

The electrons provided by the substrate acting as electron donor may be transferred to oxygen, biomass, and the product. The distribution of the electrons transferred may be calculated from the following expression (Doran, 1995):

$$1 = \frac{4a}{w\gamma_S} + \frac{c\gamma_B}{w\gamma_S} + \frac{fj\gamma_P}{w\gamma_S}. \tag{2.18}$$

Assuming that all electrons were transferred to biomass, the maximal amount that could be formed from a substrate may be calculated as follows:

$$c_{max} = \frac{w\gamma_S}{\gamma_B}. \tag{2.19}$$

The c_{max} can be never attained; as in all chemoorganotrophic organisms a part of the carbon substrate has to be oxidized to fuel the biological system with ATP to drive biosynthesis; thus, not all their electrons can be transfered to biomass.

Likewise, if no biomass is synthesized, the maximal amount of product that could be formed is given by

$$f_{max} = \frac{w\gamma_S}{j\gamma_P}. \tag{2.20}$$

Equation (2.20) should not be used alone, but together with the stoichiometric elemental balance given by Eq. (2.7). Aberrant results may otherwise be obtained, such as a value of c_{max} of almost 2 for methane (limited in fact to 1 because of its C content, (Erickson et al., 2000).

Opening the "Black Box": Mass balance as the basis of metabolic flux analysis

Anabolic fluxes

Metabolic Flux Analysis (MFA) is an approach fundamental to MCE. MFA is based on mass balance and stoichiometry of metabolic pathways (Bonarius et al., 1996; Christensen and Nielsen, 2000; Cortassa et al., 1995;

Otero and Nielsen, 2010; Palsson, 2009; Vallino and Stephanopoulos, 1993; Varma and Palsson, 1994).

In this section, we would like to introduce the basis of a method for estimating metabolic fluxes, in order to illustrate another application of material balances. This introduction will allow us to develop later the rationale followed for anabolic and catabolic flux calculations.

First, the overall metabolic network is divided into anabolic and catabolic pathways. The former takes into account all reactions leading to biomass synthesis. Otherwise stated, anabolism is the ensemble of reactions providing carbon, nitrogen, phosphorous, and sulphur from substrates to the macromolecules that constitute every cellular structure (Fig. 2.2). The strategy for computing the amount of substrate directed to macromolecules is to further divide the anabolism into two steps; in the first one, the substrates are converted into key intermediary metabolites precursors of macromolecules (Table 2.3). In the second step, the key intermediary metabolites are converted into monomer precursors that are polymerized to form the macromolecules (Fig. 2.2). For instance, to account for the synthesis of alanine from glucose as C source, two lumped steps are considered. In the first one, pyruvate is synthesized from glucose through the glycolytic pathway; second, pyruvate is converted into alanine through a specific pathway. The reader is invited to consult a biochemistry textbook for the details of reactions and metabolic pathways involved.

Although central metabolic pathways may show peculiar differences in some organisms, their major features are shared by nearly all of them. Table 2.3 summarizes the reactions leading from carbon substrates commonly used by *S. cerevisiae* to each of the key intermediates: Hexose-6-P, Triose-3-P, Ribose-5-P, Erythrose-4-P, 3-P-glycerate, P-enolpyruvate, Pyruvate, Oxalacetate, α-ketoglutarate, and Acetyl Co-A. Together with the stoichiometry for the conversion of a C source into each key intermediate, the relative amounts of ATP, NAD(H) or NADP(H), CO_2 and Pi are balanced. The formation of Ribose-5-P and Erythrose-4-P is concomitant with the reduction of NADP, since the oxidative branch of the pentose phosphate pathway is involved when glucose or glycerol act as C source. With ethanol, a large amount of NADPH is also synthesized together with all key intermediates because of the cofactor requirement of acetaldehyde dehydrogenase. The demand of phosphorylation energy required to fuel the anabolic step I

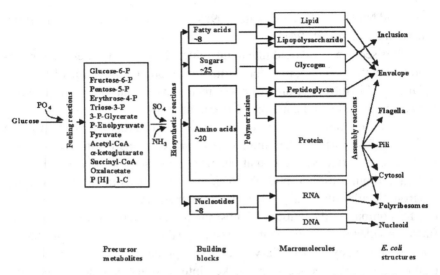

Figure 2.2. Overall scheme of metabolism in leading to the synthesis of *E. coli* from glucose as C source. The figure emphasized four levels of "reactions" according to their function in the organization of a bacterial cell: (i) Fueling reactions are those producing the key precursors metabolites (Step 1 of anabolism) as well as those supplying ATP and reducing power for biosynthesis. The latter involves catabolic and amphibolic pathways. (ii) Biosynthesis reactions refer to those pathways leading from the key intermediary metabolites to the monomers precursors of macromolecules, e.g. amino acids, nucleotides, sugars, and fatty acids. These pathways constitute what is called the Step 2 of anabolism. (iii) Polymerization reactions refer to the synthesis of the polymeric chains of proteins, nucleic acids, polysaccharides, and lipids. (iv) Assembly reactions that involve the modification, transport of macromolecules to specific locations in the cell, and their association to form cellular structures. The size of the boxes is proportional to their participation in the biomass.
Source: Reproduced from Ingraham *et al.* (1983) with © copyright permission from Sinauer Associates, Inc.

is rather less with glucose or glycerol than on the other C sources, a comparison made on the bases of a similar amount of key intermediates obtained. It should be noted that from the elemental composition of biomass, only carbon is balanced. Table 2.4 shows the second step of anabolism that transforms key intermediates into the monomer precursors of macromolecules (aminoacids, nucleotides, monosaccharides, and fatty acids).

Nitrogen, sulphur, and phosphorous are involved in the reactions of the second step (Step II of anabolism, Table 2.4). Oxygen and hydrogen are not balanced, as H_2O or H^+ were not included in the equations of Table 2.4 due

Table 2.3 Anabolic reactions in step I from fermentable and gluconeogenic carbon sources

Glucose + ATP	$\Rightarrow$	Hexose-6-P + ADP
Glucose + ATP + 2 NADP$^+$	$\Rightarrow$	Ribose-5-P + ADP + CO$_2$ + 2 NADPH
Glucose + ATP + 4 NADP$^+$	$\Rightarrow$	Erythrose-4-P + ADP + 2 CO$_2$ + 4 NADPH
Glucose + 2 ATP	$\Rightarrow$	2 Triose-P + 2 ADP
Glucose + 2 NAD$^+$	$\Rightarrow$	2 3-P-glycerate + 2 NADH
Glucose + 2 NAD$^+$ + 2 Pi	$\Rightarrow$	2 P-Enolpyruvate + 2 NADH
Glucose + 2 ADP + 2 NAD$^+$ + 2 Pi	$\Rightarrow$	2 Pyruvate + 2 ATP + 2 NADH
Glucose + 2 ADP + 4 NAD$^+$ + 2 CoA	$\Rightarrow$	2 Acetyl CoA + 2 ATP + 4 NADH + 2 CO$_2$
Glucose + 2 CO$_2$ + 2 NAD$^+$	$\Rightarrow$	2 Oxalacetate + 2 NADH
Glucose + ADP + 3 NAD$^+$ + NADP$^+$	$\Rightarrow$	α-ketoglutarate +3 NADH + NADPH + ATP + CO$_2$
2 Glycerol + 2 ATP + 2 NAD$^+$	$\Rightarrow$	Hexose-6-P + 2 ADP + 2 NADH
2 Glycerol + 2 ATP + 2 NAD$^+$ + 2 NADP$^+$	$\Rightarrow$	Ribose-5-P + 2 ADP + CO$_2$ + 2 NADPH + 2 NADH
2 Glycerol + 2 ATP + 2 NAD$^+$ + 4 NADP$^+$	$\Rightarrow$	Erythrose-4-P + 2 ADP + 2 CO$_2$ + 4 NADPH + 2 NADH
Glycerol + ATP + NAD$^+$	$\Rightarrow$	Triose-P + ADP + NADH
Glycerol + 2 NAD$^+$ + Pi	$\Rightarrow$	3-P glycerate + 2 NADH
Glycerol + 2 NAD$^+$ + Pi	$\Rightarrow$	2 P-Enolpyruvate + 2 NADH
Glycerol + ADP + 2 NAD$^+$ + Pi	$\Rightarrow$	Pyruvate + ATP + 2 NADH
Glycerol + ADP + 3 NAD$^+$ + CoA + Pi	$\Rightarrow$	Acetyl CoA + ATP + 3 NADH + CO$_2$
Glycerol + CO$_2$ + 2 NAD$^+$	$\Rightarrow$	Oxalacetate + 2 NADH
2 Glycerol + ADP + 5 NAD$^+$ + NADP$^+$	$\Rightarrow$	α-ketoglutarate + 5 NADH + NADPH + ATP + CO$_2$
2 Pyruvate + 6 ATP + 2 NADH	$\Rightarrow$	Hexose-6-P + 6 ADP + 2 NAD$^+$ + 5 Pi
5 Pyruvate + 15 ATP + 5 NADH	$\Rightarrow$	3 Ribose-5-P + 15 ADP + 12 Pi + 5 NAD$^+$
4 Pyruvate + 12 ATP + 4 NADH	$\Rightarrow$	3 Erythrose-4-P + 12 ADP + 9 Pi + 4 NAD$^+$
Pyruvate + 3 ATP + NADH	$\Rightarrow$	Triose-P + 3 ADP + NAD$^+$ + 2 Pi
Pyruvate + 2 ATP	$\Rightarrow$	3-P glycerate + 2 ADP + Pi
Pyruvate + 2 ATP	$\Rightarrow$	2 P-Enolpyruvate + 2 ADP + Pi
Pyruvate + NAD$^+$ + CoA	$\Rightarrow$	Acetyl CoA + NADH + CO$_2$
Pyruvate + CO$_2$ + ATP	$\Rightarrow$	Oxalacetate + ADP + Pi
2 Pyruvate + ATP + NAD$^+$ + NADP$^+$	$\Rightarrow$	α-ketoglutarate + NADH + NADPH + ADP + Pi + CO$_2$
2 Lactate + 6 ATP + 2 NADH + 4 Fe(III) cyt b$_2$	$\Rightarrow$	Hexose-6-P + 6 ADP + 2 NAD$^+$ + 5 Pi + 4 Fe(II) cyt b$_2$
5 Lactate + 15 ATP + 5 NADH + 10 Fe(III) cyt b$_2$	$\Rightarrow$	3 Ribose-5-P + 15 ADP + 12 Pi + 5 NAD$^+$ + 10 Fe(II) cyt b$_2$

(*Continued*)

Table 2.3   (*Continued*)

4 Lactate + 12 ATP + 4 NADH + 8 Fe(III) cyt b_2	$\Rightarrow$	3 Erythrose-4-P + 12 ADP + 9 Pi + 4 NAD^+ + 8 Fe(II) cyt b_2
Lactate + 3 ATP + NADH + 2 Fe(III) cyt b_2	$\Rightarrow$	Triose-P + 3 ADP + NAD^+ + 2 Pi + 2 Fe(II) cyt b_2
Lactate + 2 ATP + 2 Fe(III) cyt b_2	$\Rightarrow$	3-P glycerate + 2 ADP + Pi + 2 Fe(II) cyt b_2
Lactate + 2 ATP + 2 Fe(III) cyt b_2	$\Rightarrow$	2 P-Enolpyruvate + 2 ADP + Pi + 2 Fe(II) cyt b_2
Lactate + 2 Fe(III) cyt b_2	$\Rightarrow$	Pyruvate + 2 Fe(II) cyt b_2
Lactate + NAD^+ + CoA + 2 Fe(III) cyt b_2	$\Rightarrow$	Acetyl CoA + NADH + CO_2 + 2 Fe(II) cyt b_2
Lactate + CO_2 + ATP + 2 Fe(III) cyt b_2	$\Rightarrow$	Oxalacetate + ADP + Pi + 2 Fe(II) cyt b_2
2 Lactate + ATP + NAD^+ + $NADP^+$ + 4 Fe(III) cyt b_2	$\Rightarrow$	α-ketoglutarate + NADH + NADPH + ADP + Pi + CO_2 + 4 Fe(II) cyt b_2
4 Acetate + 8 ATP + 4 NAD^+ + 2 FAD^+	$\Rightarrow$	Hexose-6-P + 4 ADP + 4 AMP + 4 NADH + 2 FADH + 11 Pi + CO_2
10 Acetate + 20 ATP + 10 NAD^+ + 5 FAD^+	$\Rightarrow$	3 Ribose-5-P + 10 ADP + 10 AMP + 27 Pi + 10 NAD^+ + 5 FADH + 5 CO_2
8 Acetate + 16 ATP + 8 NAD^+ + 4 FAD^+	$\Rightarrow$	3 Erythrose-4-P + 8 ADP + 8 AMP + 21 Pi + 8 NAD^+ + 4 FADH + 4 CO_2
2 Acetate + 4 ATP + 2 NAD^+ + FAD^+	$\Rightarrow$	Triose-P + 2 ADP + 2 AMP + 5 Pi + 2 NADH + FADH + CO_2
2 Acetate + 3 ATP + 2 NAD^+ + FAD^+	$\Rightarrow$	3-P glycerate + 2 AMP + 4 Pi + ADP + 2 NADH + FADH + CO_2
2 Acetate + 3 ATP + 2 NAD^+ + FAD^+	$\Rightarrow$	P-enolpyruvate + 2 AMP + 4 Pi + ADP + 2 NADH + FADH + CO_2
2 Acetate + 2 ATP + $NADP^+$ + 2 NAD^+ + FAD^+	$\Rightarrow$	Pyruvate + 2 AMP + 4 Pi + NADPH + 2 NADH + FADH + CO_2
Acetate + ATP + CoA	$\Rightarrow$	Acetyl CoA + AMP + 2 Pi
2 Acetate + 2 ATP + 2 NAD^+ + FAD^+	$\Rightarrow$	Oxalacetate + 2 AMP + 4 Pi + 2 NADH
3 Acetate + 3 ATP + 2 NAD^+ + $NADP^+$ + FAD^+	$\Rightarrow$	α-ketoglutarate + 2 NADH + NADPH + 3 AMP + 6 Pi + CO_2
4 Ethanol + 8 ATP + 8 NAD^+ + 4 $NADP^+$ + 2 FAD^+	$\Rightarrow$	Hexose-6-P + 4 ADP + 4 AMP + 8 NADH + 4 NADPH + 2 FADH + 11 Pi + CO_2
10 Ethanol + 20 ATP + 20 NAD^+ + 10 $NADP^+$ + 5 FAD^+	$\Rightarrow$	3 Ribose-5-P + 10 ADP + 10 AMP + 27 Pi + 20 NADH + 10 NADPH + 5 FADH + 5 CO_2
8 Ethanol + 16 ATP + 16 NAD^+ + 8 $NADP^+$ + 4 FAD^+	$\Rightarrow$	3 Erythrose-4-P + 8 ADP + 8 AMP + 21 Pi + 16 NADH + 8 NADPH + 4 FADH + 4 CO_2
2 Ethanol + 4 ATP + 4 NAD^+ + 2 $NADP^+$ + FAD^+	$\Rightarrow$	Triose-P + 2 ADP + 2 AMP + 5 Pi + 4 $NADH^+$ + 2 NADPH + FADH + CO_2

(*Continued*)

Table 2.3 *(Continued)*

2 Ethanol + 3 ATP + 4 NAD$^+$ + 2 NADP$^+$ + FAD$^+$	$\Rightarrow$	3-P glycerate + 2 AMP + 4 Pi + ADP + 4 NADH + 2 NADPH + FADH + CO$_2$
2 Ethanol + 3 ATP + 4 NAD$^+$ + 2 NADP$^+$ + FAD$^+$	$\Rightarrow$	P-enolpyruvate + 2 AMP + 4 Pi + ADP + 4 NADH + 2 NADPH + FADH + CO$_2$
2 Ethanol + 2 ATP + 3 NADP$^+$ + 4 NAD$^+$ + FAD$^+$	$\Rightarrow$	Pyruvate + 2 AMP + 4 Pi + 3 NADPH + 4 NADH + FADH + CO$_2$
Ethanol + ATP + NADP$^+$ + NAD$^+$ + CoA	$\Rightarrow$	Acetyl CoA + AMP + 2 Pi + NADPH + NADH
2 Ethanol + 2 ATP + 4 NAD$^+$ + 2 NADP$^+$ + FAD$^+$	$\Rightarrow$	Oxalacetate + 2 AMP + 4 Pi + 4 NADH + 2 NADPH + FADH
3 Ethanol + 3 ATP + 6 NAD$^+$ + 4 NADP$^+$ + FAD$^+$	$\Rightarrow$	α-ketoglutarate + 6 NADH + 4 NADPH + 3 AMP + 6 Pi + CO$_2$

The overall reactions leading to the synthesis of key intermediary metabolites from glucose, glycerol, lactate, pyruvate, acetate. and ethanol are depicted with emphasis in ATP, reducing power, inorganic phosphate. and CO$_2$ participation.

to the fact that they are very difficult to measure (H$_2$O) and highly dependent on pH.

In *S. cerevisiae*, all pathways in step II are the same, irrespective of the C-source, with the exception of the synthesis of serine and glycine from acetate (Cortassa *et al.*, 1995). Different phosphorylation and redox potentials are necessary for the synthesis of monomer precursors. For instance, amino acids synthesis requires both ATP and reducing equivalents as NADPH, whereas in nucleotide synthesis ATP demand is much larger, and in lipid (fatty acid) synthesis NADPH dependence is greater.

The pathways in step II are exclusively anabolic ones, whereas the reactions of step I are shared between anabolism and catabolism; the pathways to which they belong are thereby called amphibolic pathways.

To compute the amount of substrate or of each of the intermediates required to build up a gram of biomass, both the relative amounts of macromolecules and their monomeric composition, should be known. The required amount of a monomer precursor of macromolecules is related, through the corresponding stoichiometric coefficient, to the amount of the key precursor intermediate (Table 2.5). Likewise, the amount of a key intermediary metabolite is related to the C source through the corresponding stoichiometric coefficient (Table 2.6). Tables 2.5 and 2.6 constitute the basis for the construction of stoichiometric matrices in MFA (see Chap. 4).

Table 2.4   Anabolic reactions in step II from each key intermediary metabolite

α-ketoglutarate + NADPH + NH_4^+	$\Rightarrow$ Glutamate + $NADP^+$
α-ketoglutarate + NADPH + ATP + 2 NH_4^+	$\Rightarrow$ Glutamine + ADP + Pi + $NADP^+$
α-ketoglutarate + 3 NADPH + ATP + NH_4^+	$\Rightarrow$ Proline + ADP + Pi + 3 $NADP^+$
α-ketoglutarate + CO_2 + 4 NADPH + 5 ATP + 4 NH_4^+	$\Rightarrow$ Arginine + 4 ADP + AMP + 4 Pi + PPi + 4 $NADP^+$
α-ketoglutarate + AcCoA + NAD^+ + 3 NADPH + ATP + 2 NH_4^+	$\Rightarrow$ Lysine + CO_2 + AMP + PPi + 3 $NADP^+$ + NADH
Oxalacetate + NADPH + NH_4^+	$\Rightarrow$ Aspartate + $NADP^+$
Oxalacetate + 2 ATP + NADPH + 2 NH_4^+	$\Rightarrow$ Asparagine + 2 ADP + 2 Pi + $NADP^+$
Oxalacetate + 2 ATP + 3 NADPH + NH_4^+	$\Rightarrow$ Threonine + 2 ADP + 2 Pi + 3 $NADP^+$
Oxalacetate + Formate + 4 ATP + NADH + 7 NADPH + $SO_4^=$ + NH_4^+	$\Rightarrow$ Methionine + AMP + 3 Pi + PPi + 3 ADP + 7 $NADP^+$ + NAD^+
Pyruvate + NADPH + NH_4^+	$\Rightarrow$ Alanine + $NADP^+$
2 Pyruvate + 2 NADPH + NH_4^+	$\Rightarrow$ Valine + 2 $NADP^+$ + CO_2
Pyruvate + Oxalacetate + 2 ATP + 5 NADPH + NH_4^+	$\Rightarrow$ Isoleucine + 5 $NADP^+$ + CO_2 + 2 ADP + 2 Pi
2 Pyruvate + AcCoA + 2 NADPH + NAD^+ + NH_4^+	$\Rightarrow$ Leucine + 2 $NADP^+$ + NADH + 2 CO_2
3-P-glycerate + NAD^+ + NADPH + NH_4^+	$\Rightarrow$ Serine + $NADP^+$ + NADH + Pi
3-P-glycerate + NAD^+ + NH_4^+ + ADP	$\Rightarrow$ Glycine + NADH + ATP + Formate
3-P-glycerate + NAD^+ + 5 NADPH + $SO_4^=$ + NH_4^+	$\Rightarrow$ Cysteine + 5 $NADP^+$ + NADH + Pi
2 PEP + Ery4P + ATP + 2 NADPH + NH_4^+	$\Rightarrow$ Phenylalanine + 4 Pi + 2 $NADP^+$ + CO_2 + ADP
2 PEP + Ery4P + ATP + NAD^+ + 2 NADPH + NH_4^+	$\Rightarrow$ Tyrosine + 4 Pi^+ + NADH + 2 $NADP^+$ + CO_2 + ADP
2 PEP + Ery4P + R5P + 3 ATP + NAD^+ + 2 NADPH + 2 NH_4^+	$\Rightarrow$ Tryptophan + 4 Pi + NADH + 2 $NADP^+$ + CO_2 + 2 ADP + AMP + PPi + Pyruvate
R5P + 5 ATP + 3 NH_4^+ + 2 NADPH + 2 NAD^+ + Formate	$\Rightarrow$ Histidine + 2 AMP + 3 ADP + 2 PPi + 2 NADH + 2 $NADP^+$ + 3 Pi
Ribose-5-P + 3-P-glycerate + 9 ATP + NAD^+ + 5 NH_4^+ + 2 NADPH + CO_2 + Formate	$\Rightarrow$ AMP + AMP + 8 ADP + PPi + 9 Pi + 2 $NADP^+$ + NADH
Ribose-5-P + 3-P-glycerate + 10 ATP + 2 NAD^+ + 5 NH_4^+ + NADPH + CO_2 + Formate	$\Rightarrow$ GMP + 2 AMP + 9 ADP + 2 PPi + 10 Pi + $NADP^+$ + 2 NADH
Ribose-5-P + Oxalacetate + NAD^+ + 4 ATP + 2 NH_4^+ + NADPH	$\Rightarrow$ UMP + AMP + 3 ADP + NADH + 3 Pi + Ppi

(Continued)

Table 2.4   (*Continued*)

Ribose-5-P + Oxalacetate + NAD^+ + 8 ATP + 3 NH_4^+ + NADPH	$\Rightarrow$	CTP + AMP + 7 ADP + NADH + 7 Pi + Ppi
Glucose 6 P + ATP + $Glucan_{n-1}$	$\Rightarrow$	$Glucan_n$ + ADP + Ppi
Glucose 6 P + ATP + $Manan_{n-1}$	$\Rightarrow$	$Manan_n$ + ADP + Ppi
Glucose 6 P + ATP + $Glycogen_{n-1}$	$\Rightarrow$	$Glycogen_n$ + ADP + Ppi
6 AcCoA + 5 ATP + 5 NADPH	$\Rightarrow$	Lauric acid(12) + 5 ADP + 5 $NADP^+$ + 5 Pi
8 AcCoA + 7 ATP + 8 NADPH	$\Rightarrow$	Palmitoleic acid(16:1) + 7 ADP + 8 $NADP^+$ + 7 Pi
9 AcCoA + 8 ATP + 9 NADPH	$\Rightarrow$	Oleic acid(18:1) + 8 ADP + 9 $NADP^+$ + 8 Pi
Triose-P + NADH	$\Rightarrow$	Glycerol-3-P
Glycerol-3-P + Fatty acid + ATP	$\Rightarrow$	Acyl-glycerol + Pi + AMP + Ppi

Anabolic reactions leading from the key intermediary metabolites to the synthesis of monomers precursor of macromolecules, e.g. amino acids, nucleotides, fatty acids, and sugars are depicted with emphasis in ATP, reducing power, ammonia, sulphate, inorganic phosphate, and CO_2 participation.

Table 2.7 presents some selected examples of the amount of intermediates, and/or C substrate, needed to synthesize an amino acid (arginine), a nucleotide (UTP), and a monosaccharide (glucose); these are a direct function of the growth rate. In chemostat cultures, the latter is equal to the dilution rate, D, when the culture is at steady state (see Chaps. 3–5).

Catabolic fluxes

Catabolic reactions account for the provision of energy and reduced intermediates fueling anabolism. A large diversity of catabolic pathways implying different stoichiometries are known to occur in nature. Among chemoorganotrophs (Fig. 2.1, top panel), a large range of substrates may act as energy sources from the very reduced, e.g. as methane, to oxidized ones, e.g. oxalic acid (Fig. 2.3; see also Table 2.2) (Linton and Stephenson, 1978). The search for microorganisms that degrade xenobiotics has shed light on pathways that use aromatic compounds, e.g. toluene, xylene, or other aromatics, as carbon and energy sources (Wilson and Bouwer, 1997).

Depending on the presence of exogenous or endogenous electron acceptors, aerobic or anaerobic metabolism occurs. Table 2.8 shows the most

Table 2.5 Stoichiometries of production of key intermediary metabolites from different carbon sources

Carbon substrate or intermediate	Cost of making 1 μmol of each key intermediary metabolite[a] (μmol/μmol)									
	G6P	R5P	E4P	TP	3PG	PEP	PY	AcCoA	Oaa	α kg
Glucose	1[b]	1	1	0.5	0.5	0.5	0.5	0.5	0.5	1
NADH	0	0	0	0	−1[b]	−1	−1	−2	−1	−3
NADPH	0	−2	−4	0	0	0	0	0	0	−1
ATP	1	1	1	1	0	0	−1	−1	0	−1
CO$_2$	0	−1	−2	0	0	0	0	−1	1	−1
Glycerol	2	2	2	1	1	1	1	1	1	2
NADH	−2	−2	−2	−1	−2	−2	−2	−3	−2	−5
NADPH	0	−2	−4	0	0	0	0	0	0	−1
ATP	2	2	2	1	0	0	−1	−1	0	−1
CO$_2$	0	−1	−2	0	0	0	0	−1	1	−1
Pyruvate	2	1.67	1.33	1	1	1	1	1	1	2
NADH	2	1.67	1.33	1	0	0	0	−1	0	−1
NADPH	0	0	0	0	0	0	0	0	0	−1
ATP	6	5	4	3	2	2	0	0	1	1
CO$_2$	0	0	0	0	0	0	0	−1	1	−1
Lactate	2	1.67	1.33	1	1	1	1	1	1	2
NADH	2	1.67	1.33	1	0	0	0	−1	0	−1
NADPH	0	0	0	0	0	0	0	0	0	−1
ATP	6	5	4	3	2	2	0	0	1	1
CO$_2$	0	0	0	0	0	0	0	−1	1	−1
Ethanol	4	3.33	2.67	2	2	2	2	1	2	3
NADH	−4	−3.33	−2.67	−2	−3	−3	−2	−1	−3	−4
NADPH	−4	−3.33	−2.67	−2	−2	−2	−3	−1	−2	−4
FADH	−2	−1.67	−1.33	−1	−1	−1	−1	0	−1	−1
ATP	12	10	8	6	5	5	4	2	4	6
CO$_2$	−2	−1.67	−1.33	−1	−1	−1	−1	0	0	−1
Acetate	4	3.33	2.67	2	2	2	2	1	2	3
NADH	0	−1.67	−1.33	0	−1	−1	0	0	−1	−1
NADPH	0	0	0	0	0	0	−1	0	0	−1
FADH	−2	−1.67	−1.33	−1	−1	−1	−1	0	−1	−1
ATP	12	10	8	6	5	5	4	2	4	6
CO$_2$	−2	−1.67	−1.33	−1	−1	−1	−1	0	0	−1

[a]Carbon, ATP, NAD(P)H, FADH, and CO$_2$ demand for production of 1 mol of each key intermediary metabolite is computed according to the balance of intermediates through the pathways indicated in Table 2.3.

[b]positive (negative) numbers indicate consumption (production) of the intermediate together with the key intermediary metabolite.

Fluxes of carbon, phosphorylation, and redox intermediates during growth of S. cerevisiae on different carbon sources.

Source: Cortassa et al. (1995) ©copyright 1995 John Wiley & Sons, Inc. Reprinted by permission of Wiley-Liss, Inc., a subsidiary of John Wiley & Sons, Inc.

Table 2.6 Monomer composition of each macromolecular fraction of yeast cells

Monomer	Relative amounts present in *S. cerevisiae*[a]		Cost of making 1 μmol of each of the macromolecular monomers (μmol/μmol)					
	(mol)	Metabolite[b]	NADPH	NADH	$\sim$P[c]	NH_3	CO_2	SO_4
Amino acid								
alanine	1.000	1 PY	1	0	0	1	0	0
arginine	0.351	1 αKG	4	−1	6	3	1	0
asparagine	0.222	1 oaa	1	0	2	2	0	0
aspartate	0.647	1 oaa	1	0	0	1	0	0
cysteine	0.015	1 3PG	4	−1	3	1	0	1
glutamate	0.658	1 αKG	1	0	0	1	0	0
glutamine	0.229	1 αKG	1	0	1	2	0	0
glycine	0.632	1 3PG	1	−1	0	1	0	0
glycine[d]	0.632	1 AcCoA	1	−1	0	1	0	0
histidine	0.144	2 R5P,1 3PG	4	−5	16	7	1	0
isoleucine	0.421	1 OAA, 1 PY	5	0	2	1	−1	0
leucine	0.645	2 PY, 1 AcCoA	2	−1	0	1	−2	0
lysine	0.623	1 αKG, 1 AcCoA	3	−1	2	2	−1	0
methionine	0.111	1 OAA, 1 AcCoA	5	0	4	1	0	1
phenylalanine	0.292	1 E4P, 2 PEP	2	0	1	1	−1	0
proline	0.360	1 αKG	2	1	1	1	0	0
serine	0.403	1 3PG	1	−1	0	1	0	0
serine[d]	0.403	1 AcCoA	1	−1	0	1	0	0
threonine	0.416	1 oaa	3	0	2	1	0	0
tryptophane	0.061	1 E4P, 1 PEP, 1 R5P	3	−2	3	2	−1	0
tyrosine	0.222	1 E4P, 2PEP	2	−1	1	1	−1	0
valine	0.577	2 PY	2	0	0	1	−1	0
Nucleotides			NADPH	NADH	$\sim$P	NH_3	CO_2	PO_4
ATP	0.754	1 R5P, 1 3PG	3	−3	13	5	1	1
GTP	0.754	1 R5P, 1 3PG	2	−3	15	5	1	1
UTP	1.000	1 R5P, 1 OAA	1	0	7	2	0	1
CTP	0.738	1 R5P, 1 OAA	1	0	8	3	0	1
Lipids			NADPH	NADH	$\sim$P			
Lauric	0.674	5 AcCoA	10	0	5			
Palmitoleic	1.000	7 AcCoA	14	1	7			
Oleic	0.403	8 AcCoA	16	1	8			
Glycerol P	0.667	1 TP	0	1	0			

(Continued)

Table 2.6 (*Continued*)

Monomer	Relative amounts present in *S. cerevisiae*[a]		Cost of making 1 μmol of each of the macromolecular monomers (μmol/μmol)					
	(mol)	Metabolite[b]	NADPH	NADH	$\sim$P[c]	NH$_3$	CO$_2$	SO$_4$
Polysaccharides			NADPH	NADH	$\sim$P			
Glycogen	0.741	G6P	0	0	2			
Glucan	0.857	G6P	0	0	2			
Mannan	1.000	G6P	0	0	2			

[a]Based on the composition reported (Bruinenberg *et al.*, 1983);
[b]key intermediary metabolites;
[c]phosphorylation energy required;
[d]in acetate, glycine, and serine are synthesized through a different pathway.
Fluxes of carbon, phosphorylation, and redox intermediates during growth of *S. cerevisiae* on different carbon sources.
Source: Cortassa *et al.* (1995) ©copyright 1995 John Wiley & Sons, Inc. Reprinted by permission of Wiley-Liss, Inc., a subsidiary of John Wiley & Sons, Inc.

common catabolic reactions either with oxygen or endogenous C compounds as electron acceptors, and their mass and energy balances. In aerobic metabolism depending on the nicotinamide adenine nucleotide (NADH or FADH) that acts as electron donor, the resulting amount of ATP synthesized will differ (Table 2.8). The resulting P:O ratio, i.e. the mole numbers of ATP synthesized in oxidative phosphorylation per mole atom of oxygen consumed, will be two for FADH, or three for NADH. Thus, the yield of ATP will vary depending on the substrate being oxidized and on the type of redox intermediate synthesized from its oxidation. As an example when succinate is acting as energy source, electrons are transferred via FADH so that the maximun P:O ratio is 2.

In the 1990s, alternative respiration mechanisms have been described according to which a NADH oxidase transfers electrons to molecular oxygen without translocating protons across the mitochondrial inner membrane; thereby, no ATP is synthesized. Thus during microbial growth the resulting ATP yield, as shown by the P:O ratio will result from the various stoichiometries of the electron transport pathways.

In the next section we further develop the main concepts about the quantification of energy balance.

Table 2.7 Amounts of precursors required for the synthesis of macromolecular monomers

	Monomer precursor of macromolecules		
	Arginine	Uracil 5' PPP	Glucose (glycogen)
Amount in biomass (mmol g^{-1} dw)	0.187	0.123	0.548
Key intermediary precursor	α-ketoglutarate	Ribose 5P	Glucose 6 P
		Oxalacetate	
Amount of key intermediary precursor	0.187	0.123 (Ribose 5P)	0.548
		0.123 (Oxalacetate)	
Amount of glucose as carbon source required to obtain the key intermediate	0.187	0.185 (0.123.1.5)[b]	0.548
NADPH required	0.748	0.123	—
NADH produced	0.187	—	—
Energy as ATP	1.122	0.861	0.548
CO_2 released	−0.187[a]	—	—

[a]The CO_2 is in fact required to synthesize oxalacetate from pyruvate;
[b]One or 0.5 moles of glucose are required for the synthesis of 1 mole R5P or oxalacetate, respectively.

Energy Balance

In a bioprocess, the energy balance may be regarded from two different viewpoints: from that of the chemical engineer, who will consider the process as a whole, or from that of the metabolic engineer. The chemical engineer pays special attention to the amount of energy released by the microorganisms. In fact, microorganisms evolve heat during growth and this is relevant because of the importance of temperature control in bioprocesses. Energy is therefore usually provided to cool the bioreactor and various heat exchanger models have been designed. Biochemical engineering textbooks provide further detail.

The metabolic engineer's viewpoint emphasizes the microorganism and its metabolic and physiological behavior. This is the perspective adopted in the present book. Metabolism involves a set of chemical reactions each one

Table 2.8 Main catabolic routes rendering ATP and reduction power operative in chemoorganotrophic organisms

$C_6H_{12}O_6 + 2\,ADP + 2\,NAD^+ + 2\,Pi$	$\Rightarrow$	$2\,Py + 2\,ATP + 2\,NADH(^a)$
$C_6H_{12}O_6 + ATP + 12\,NADP^+$	$\Rightarrow$	$6\,CO_2 + ADP + Pi + 12\,NADPH(^b)$
$Py + 1\,ADP + Pi + FAD^+ + 4\,NAD^+$	$\Rightarrow$	$3\,CO_2 + ATP + 4\,NADH + FADH(^c)$
$Py + NADH$	$\Rightarrow$	$EtOH + CO_2 + NAD^+ (^d)$
$Py + NADH$	$\Rightarrow$	$Lactate + NAD^+ (^e)$
$NADH + 3\,ADP + 3\,Pi + 1/2\,O_2$	$\Rightarrow$	$NAD^+ + 3\,ATP + H_2O(^f)$
$FADH + 2\,ADP + 2\,Pi + 1/2\,O_2$	$\Rightarrow$	$FAD^+ + 2\,ATP + H_2O(^g)$

[a]Glycolysis via the Embden–Meyerhof pathway. The ATP stoichiometry is 1 mole ATP mole-1 glucose if the Entner–Doudoroff pathway is used as in some prokaryotes.
[b]Pentose phosphate pathway operating in a cyclic mode to generate reduction equivalents in the form of NADPH to fuel biosynthetic redox reactions.
[c]Tricarboxylic acid overall stoichiometry including the pyruvate dehydrogenase catalyzed step.
[d]Ethanolic fermentation.
[e]Lactic fermentation.
[f]Oxidative phosphorylation indicating the mechanistic stoichiometry of 3 from NADH (P:O ratio = 3) The actual stoichiometry is usually lower than the mechanistic one.
[g]Oxidative phosphorylation indicating the mechanistic stoichiometry of 2 from FADH (P:O ratio = 2).

with an associated energetic change. These cannot be treated separately from mass balance, even in the black box approach. Since catabolism implies the oxidation of a reduced organic compound, energy is released and ATP synthesis coupled to it. In the first part of this section, we will deal with some general thermodynamic concepts applicable to bioprocesses in general, either via a modular approach to metabolism or to each of the individual reactions.

Forms of energy and enthalpy

Cellular systems are dissipative in the sense that they sustain fluxes of matter and energy, interconverting the nature of the forces (e.g. chemical, radiant, electrostatic) driving such fluxes. In cells, free energy stores are exchanged essentially between the chemical potentials, electrochemical gradients, and conformational energy of macromolecules.

There are several forms of energy, most of them being interconvertible. These forms of energy are: kinetic, potential, internal, mixing work, flow work, and finally heat.

Kinetic energy is associated with movement due to the translational velocities of the different components of a system. Potential energy is linked to the position of a system in a force field (e.g. gravitational, electromagnetic). In bioprocess engineering, mixing (usually referred to as shaft work) and flow (energy required to pump matter into and out of the system) work, are important terms. Internal energy is associated with molecular and atomic interactions and is measured as changes in the enthalpy of the system. Thermodynamics deals with these energy forms, and a series of state functions that represent them.

The most commonly used state functions are enthalpy and free energy, the former being the most useful to evaluate heat exchange in biotechnological processes. Enthalpy is defined as the sum of the internal energy of a system, U, plus the compression or expansion work, pV:

$$H = U + pV. \tag{2.21}$$

Thermodynamics textbooks provide definitions and examples of enthalpy in detail. The book by Roels *Energetics and Kinetics in Biotechnology* (1983) should be considered as a most useful work for the thermodynamic approach to biotechnological problems.

The general equation of energy balance is equivalent to the one of material balance. It is based on the first law of thermodynamics, and states that in any physicochemical process (except those involving subatomic reactions), the total energy of the system is conserved:

$$E_{input} - E_{output} = E_{accumul}, \tag{2.22}$$

where E_{input} stands for the flow of energy into the system; E_{output}, the output flow of energy from the system; $E_{accumul}$, the energy accumulated in the system during the process.

More specifically, Eq. (2.22) can be written as follows:

$$\sum_{input} M(u + e_k + e_P + pv) - \sum_{output} M(u + e_k + e_P + pv) - Q + W_s = \Delta E.$$

$$\tag{2.23}$$

In most bioprocesses, the contribution of kinetic and potential energy may be considered negligible. Then, Eq. (2.23) reads:

$$\sum_{input} Mh - \sum_{output} Mh - Q + W_s = \Delta E. \tag{2.24}$$

Some special cases of interest are the steady state, or when the system is thermically isolated in an adiabatic device (Eq. 2.25). For instance at the steady state, $\Delta E = 0$ or under adiabatic conditions, $Q = 0$, then Eq. (2.23) becomes:

$$\sum_{\text{input}} Mh - \sum_{\text{output}} Mh + W_s = \Delta E. \qquad (2.25)$$

These thermodynamic functions are called *state properties*. This means that they are only dependent on the state of the system, and that their values are independent of the pathway followed by the system to attain the corresponding state. The latter permits us to compute the value of a state function (e.g. enthalpy) by following a path that allows the use of tabulated state function changes. To exemplify the latter, let us suppose that we are interested in calculating the enthalpy change associated with a particular reaction occurring at high temperature, T_i. A strategy will be to find the tabulated change in enthalpy at standard pressure and temperature, and add to it the enthalpy change to cool the reactants from T_i to the standard conditions plus the enthalpy change associated with heating the reaction products to T_i. From a thermodynamic point of view, this procedure is equivalent to the reaction occurring at T_i because enthalpy is a state property.

State functions do not have absolute values, but are relative to a reference state chosen by the experimenter. Conventionally, in biological systems, the reference state corresponds to 25°C temperature, 1 atm of pressure, pure water (i.e. 55.5 M instead of 1 M), and pH 7.0.

Enthalpy changes may occur associated with: (i) temperature; (ii) physical aspects such as phase changes, mixing or dissolution; and (iii) reaction.

To compute the change of enthalpy associated with a change in the temperature of the system, a property called "sensible heat" is defined. *Sensible heat* is the enthalpy change associated with a temperature shift. The heat capacity of the system should be known to determine the sensible heat. The specific heat, C_p, which is the heat capacity expressed per unit mass, can be used to calculate the sensible heat as follows:

$$\Delta H = MC_P(\Delta T) = MC_P(T_2 - T_1) \qquad (2.26)$$

The concepts of *sensible heat* and *heat capacity* are of fundamental relevance for the work with calorimetry, and in their application to unravel microbial energetics (see below).

Enthalpy changes associated with metabolic reactions can be expressed as the heat of reaction ΔH_{rxn}, as follows:

$$\Delta H_{rxn} = \sum_{products} Mh - \sum_{reactants} Mh \quad \text{(expressed in terms of mass)}$$

$$(2.27)$$

$$\Delta H_{rxn} = \sum_{products} nh - \sum_{reactants} nh \quad \text{(expressed in mole terms)}$$

$$(2.28)$$

Calorimetric studies of energy metabolism

Calorimetry was initially developed to quantitate heat production during chemical reactions and by microbial fermentations. Actually bacterial thermogenesis was examined in the first report on cellular catabolism (see Lamprecht (1980) and references therein).

Calorimeters evaluate the amount of heat evolved (or used) by a culture; this equals the ΔH associated with a particular process under observation. The amount of heat evolved, as expected, depends on the nature of the energy source and is proportional to its degree of reduction (Table 2.2). The output of a calorimeter is the rate of heat production (dQ/dt). The latter was found to be proportional to the acceleration of growth rather than to the growth rate itself (Lamprecht, 1980). Thus, thermograms indicate the rate at which metabolism is operating, and the type of substrate breakdown, e.g. anaerobic catabolism evolves much less heat than aerobic catabolism for the same substrate. According to Table 2.9, *S. cerevisiae* releases 139 kJ mol^{-1} substrate growing on glucose on complex media under anaerobic conditions, whereas under aerobiosis the ΔH increases to 2,663 kJ mole^{-1} substrate. Large differences were also observed in synthetic media with yeast cells growing on either C-, or N- or C- and N-limited chemostat cultures (Larsson *et al.*, 1993). In the respiro-fermentative mode of glucose breakdown the heat yield, i.e. the heat production per gram of biomass formed, decreased to 7.3 kJ g^{-1} under glucose limitation. The largest value of heat yield, 23.8 kJ g^{-1}, was verified in N-limited cultures at low dilution rates (Larsson *et al.*, 1993).

The respiratory quotient (see section: *Growth stoichiometry and product formation*) has served as a useful basis for estimation of enthalpy

Table 2.9　Calorimetrically determined parameters of growing yeast cultures under varying conditions. Enthalpy changes ΔH (kJ mol^{-1}) substrate specific rate of heat production $d\Delta h/dt$ (W g^{-1}), enthalpy change per formed biomass $\Delta H'$ (kJ g^{-1}), substrate growth yield Y (g dry weight g^{-1} substrate). Some values are calculated from the data in the literature. AN, anaerobic; (AN), nearly anaerobic; AE, aerobic; (AE), nearly aerobic

Yeast	Condition	Medium	Substrate	ΔH	$d\Delta h/dt$	ΔH	Y
a. Anaerobic batch cultures							
Brewer's yeast	AN	Complex	Sugar	98.4	—		
Yeast (undefined)	(AN)	Complex	Cane sugar	2 × 110	—	2 × 27.9	0.0218
Yeast (undefined)	(AN)	Complex	Maltose	2 × 112	—	2 × 32.1	0.0193
S. cerevisiae	AN	Synthetic	Glucose	96.3	—	6.28	0.085
Baker's yeast	AN	Complex	Glucose	82.5	—	1.93	0.237
S. cerevisiae Y Fa	AN	Complex	Glucose	119	—	5.25	0.126
S. cerevisiae 211	(AN)	Complex	Glucose	99.2	—	4.23	0.131
S. cerevisiae 211	(AN)	Complex	Glucose	89.6	0.17	4.48	0.111
S. cerevisiae	(AN)	Complex	Glucose	63.9	—	—	—
Baker's yeast	AN	Complex	Glucose	129	0.34	—	—
S. cerevisiae	AN	Complex	Fructose	124	—	4.78	0.144
S. cerevisiae	AN	Complex	Galactose	125	—	4.82	0.144
S. cerevisiae	AN		Maltose	2 × 126	—	2 × 6.31	0.111
b. Aerobic batch cultures							
Rhodotorula sp.	AE	Complex	Glucose	536		1.05	7.61
C. intermedia	(AE)		Glucose	1695		11.4	0.826
S. cerevisiae 211	AE	Solid medium	Glucose	1674		—	—
Baker's yeast	AE		Glucose	—	1.97	—	—
Kl. aerogenes	AE	Synthetic	Glucose	—	3.0 ~ 4.2	—	0.19
D. hansenii	AE	Synthetic	Glucose	1130–169	0.4	12.6–18.8	0.5
S. cerevisiae	AE	Complex	Ethanol	854	—	33.7	0.55
Baker's yeast	AE	Complex	Ethanol	—	1.23	—	—

(*Continued*)

Table 2.9   (*Continued*)

Yeast	Condition	Medium	Substrate	ΔH	$d\Delta h/dt$	ΔH	Y
Hansenula sp.	AE		Ethanol	544	2.09	16.7	0.61
S. cerevisiae	AE	Complex	Acetic acid	678	—	45.2	0.25
c. Continuous cultures							
S. cerevisiae	AN	Complex	Glucose	139	0.2	10.4	0.074
S. cerevisiae	AE	Complex	Glucose	2663	2.15	14.5	1.017
Kl. aerogenes	AE	Synthetic	Glucose	—	0.13	—	—
S. cerevisiae	AE	Synthetic	Glucose	1070	1.54	10.4	0.572
S. cerevisiae	AE	Synthetic	Ethanol	616	—	28.3	0.474
Hansenula sp.	AE		Ethanol	607	—	18	0.73

Source: Reproduced from Lamprecht (1980) Copyright © 1980, Academic Press.

changes in experiments using "indirect calorimetry" with a Warburg apparatus. However, this procedure is subject to experimental error, although a first estimation of RQs can be performed. Under fully aerobic conditions, the RQ values for carbohydrates (1.00), fats (0.707), and proteins (0.801), were measured. Since the heat of combustion of these compounds were known ($17.2 \, \text{kJ} \, \text{g}^{-1}$ for carbohydrates, $38.9 \, \text{kJ} \, \text{g}^{-1}$ for fats and $22.6 \, \text{kJ} \, \text{g}^{-1}$ for proteins), the enthalpy changes could be readily calculated (reviewed by Lamprecht (1980)). At present, with more sophisticated equipment for both calorimetric and RQ measurements, indirect calorimetry was found to give only a qualitative assessment of enthalpy changes associated with metabolism (Lamprecht, 1980).

As heat production is the result of the whole metabolic machinery, nowadays calorimetry is used in combination with other methods and these enable a thorough interpretation of the energetics of growth and product formation in the analysis of metabolic networks (Larsson *et al.*, 1993).

Heat of combustion

A useful thermodynamic state function is the heat of combustion. This is defined as the amount of heat released during the reaction of a substance with oxygen to form oxidized products such as CO_2, H_2O, and N_2, at standard conditions of temperature and pressure, i.e. $25°C$, 1 atm.

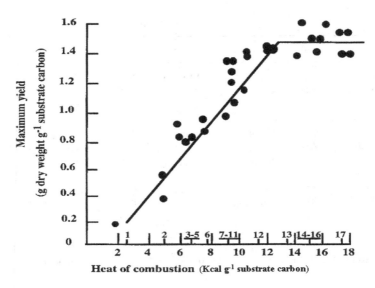

Figure 2.3.   Maximum growth yield relationship with enthalpy of combustion of various organic compounds supporting microbial growth. The maximum growth yields observed in a wide range of organic substrates for various microorganisms growing in batch or continuous cultures are plotted against the heat of combustion as a measure of the energy content of the compound and directly related to its degree of reduction (Table 2.2). 1, oxalate; 2, formate; 3, citrate; 4, malate; 5, fumarate; 6, succinate; 7, acetate; 8, benzoate; 9, glucose; 10, phenylacetic acid; 11, mannitol; 12, glycerol; 13, ethanol; 14, propane; 15, methanol; 16, ethane; 17, methane.
Source: Reprinted from Linton and Stephenson (1978) ©copyright 1978, with permission from Elsevier Science.

The standard heat of combustion is used to estimate the reaction enthalpy changes, as follows:

$$\Delta H_c^o = \sum_{\text{reactants}} n\Delta h_c^o - \sum_{\text{products}} n\Delta h_c^o. \qquad (2.29)$$

Indeed, the growth of a microorganism is related to its physiology and bioenergetics. The growth yield of microorganisms on a given substrate, and the heat of combustion of that substrate were found to be linearly related (Fig. 2.3). The heat of combustion was interpreted as a measure of the carbon and energy content of an organic substrate. Figure 2.3 shows that in the range of 2.5 to 11.0 kcal g^{-1} substrate-C, the enthalpy of combustion and the growth yield (ranging from 0.2 to 1.4 g dw g^{-1} substrate-C) display a linear correlation. At enthalpies of combustion higher than 11.0 kcal g^{-1} substrate-C, growth is no longer energy-limited, and an upper limit in the relationship between Y_{XS} and ΔH_c^o is attained. The maximal Y_{XS} of 1.43 g dw g^{-1} substrate-C, was interpreted as the achievement of a stoichiometric limit. This relationship is also discussed in the context of the work of Sauer *et al.* (1998) as applied to the growth and metabolite production by *Bacillus subtilis*, and analyzed with a stoichiometric model (see Chap. 4, section: *A comparison between different methods of metabolic flux analysis*).

Stouthamer and van Verseveld (1985) have redrawn the data of Linton and Stephenson (1978) on the bases of the degree of reduction of the growth substrates. They introduced the concept that during growth of organisms, adaptation of the energy generation mechanisms occur according to ATP demand by biomass synthetic processes. Such adaptation can be accomplished through changes in the P:O ratio (deletion of one or more phosphorylation sites in the respiratory chain) and occurrence of energy spilling mechanisms (Stouthamer and van Verseveld, 1985). This explains why the growth of *Pseudomonas denitrificans* with mannitol or methanol is energy-limited.

Table 2.2 shows the proportionality between the enthalpy and free energy of combustion of organic compounds (expressed on a carbon basis) and their degree of reduction. Among the organic compounds analyzed, are a number of carbon and energy substrates able to sustain microbial growth as well as compounds that take part of the macromolecular composition of cells, e.g. amino acids. The abovementioned proportionality is evidenced through the constancy in the ratio between the heat of combustion over the degree of reduction, with an average proportionality constant of 115 kJ C-mole^{-1} (Table 2.2). The data shown in Table 2.2, allow writing the following formula to compute the molar heat of combustion of any organic compound:

$$\Delta h_c^o = -q\gamma x_C, \tag{2.30}$$

where q is the heat evolved per mole of available electrons transferred to oxygen during combustion; γ, the degree of reduction of the compound with respect to N_2, and X_C, the number of C atoms in the molecular formula.

Usually, biotechnological processes involve microbial growth. The growth of a microorganism results from a complex set of chemical reactions as well as signalling networks (see Fig. 1.10). The energy or heat evolved by microbial growth may be computed through the heat of combustion of the various chemicals involved, carbon and nitrogen substrates or biomass, as follows:

$$C_wH_xO_yN_z + aO_2 + bH_gO_hN_i \Rightarrow cCH_\alpha O_\beta N_\delta + dCO_2 + eH_2O. \quad (2.31)$$

Growth can occur either under aerobic or anaerobic conditions depending on the electron acceptor used. Equation (2.29) corresponds to growth with oxygen as electron acceptor, while Eq. (2.32) represents growth with endogenous organic compounds as electron acceptors (anaerobic growth):

$$C_wH_xO_yN_z + b\,H_gO_hN_i$$
$$\Rightarrow cCH_\alpha O_\beta N_\delta + dCO_2 + eH_2O + fC_jH_kO_lN_m. \quad (2.32)$$

In all cases, the heat of growth may be calculated from the following expression:

$$\Delta H_{rxn}^o = n(\Delta h_c^o)_{\text{substrate}} + n(\Delta h_c^o)_{\text{NH}_3} - n(\Delta h_c^o)_{\text{biomass}} - n(\Delta h_c^o)_{\text{product}}$$
$$(2.33)$$

In microbiological processes, it is usual that only the heat of reactions, phase change, and shaft work, are employed to calculate the energy balance, whereas other energy forms are frequently negligible.

So far, only the general thermodynamic concepts have been discussed. In the next section we open the "black box."

An energetic view of microbial metabolism

The main aim of this section is to visualize the sinks of energy during microbial growth. The energy demands during microbial growth are associated with:

(1) Synthesis of precursors of cellular material.
(2) Polymerization and acquisition of secondary and tertiary structure of proteins (folding), DNA and RNA.
(3) Formation of supramolecular assemblies (enzyme–enzyme complexes, membranes, ribosomes).
(4) Assembly of cell organelles and the whole cell.

Table 2.10 Efficiency of microbial growth under aerobic conditions on substrate with different degrees of reduction

Organism	Substrate	μ^* kJ C-mol^{-1}	γ	J_b/J_s	η_{th}^R	η_{th}^{Wff}
Ps. oxalaticus	oxalate	-337	1	0.07	24	20
	formate	-335	2	0.18	34	20
Pseudomonas sp.				0.18	34	20
Ps. denitrificans	citrate	-195	3	0.38	54	26
A. aerogenes				0.34	49	23
Ps. denitrificans	malate	-212	3	0.37	52	24
Ps. fluorescens				0.33	46	19
Ps. denitrificans	fumarate	-151	3	0.37	53	25
Ps. denitrificans	succinate	-173	3.5	0.39	49	16
Pseudomonas sp.				0.41	52	19
A. aerogenes	Lactate	-173	4	0.32	34	3
Ps. fluorescens				0.37	40	5
Pseudomonas sp.	acetate	-186	4	0.44	49	9
C. utilis				0.42	47	9
Ps. fluorescens				0.32	36	6
C. tropicalis				0.36	40	6
S. cerevisiae	glucose	-153	4	0.59	59	0
S. cerevisiae				0.57	57	0
E. coli				0.62	62	0
P. chrysogenum				0.54	54	0
A. acogen	glycerol	-163	4.67	0.66	57	-26
C. tropicalis	ethanol	-91	6	0.61	44	-44
C. boidinii				0.61	44	-44
Candida utilis				0.61	44	-44
Ps. fluorescens				0.43	23	-35
C. utilis				0.55	39	-36
C. brassicae				0.64	46	-50
C. boidinii	methanol	176	6	0.52	36	-33
Klebsiella sp.				0.47	32	-28
M. methanolica				0.6	41	-48
Candida N-17				0.46	31	-28
H. polymorpha				0.45	31	-25
Pseudomonas C				0.67	46	-64
Pseudomonas EN				0.67	46	-64
Torulopsis				0.7	48	-73
Methylomonas sp.				0.5	34	-31
M. methanolica				0.64	44	-56
C. tropicalis	hexadecane	5	6.13	0.56	41	-34

(*Continued*)

Table 2.10   (*Continued*)

Organism	Substrate	μ^* kJ C-mol^{-1}	γ	J_b/J_s	η_{th}^R	η_{th}^{Wff}
C. lipolytica	dodecane	4	6.17	0.41	30	−19
Job 5	propane	−8	6.67	0.71	48	−79
	ethane	−16	7	0.71	46	−82
M. capsulatus	methane	−51	8	0.63	37	−70
M. methanooxidans				0.68	40	−88

The values of standard Gibbs free energy of the substrate, its degree of reduction, the growth yield of the indicated organism on that substrate (fifth column) and the efficiencies in column (η_{th}^R) were taken from Westerhoff *et al.* (1982). The η_{th}^{Wff} was calculated with the following formula:

$$\eta_{th}^{Wff} = \frac{\eta_{th}^R - (J_b/J_s)}{1 - (J_b/J_s)}$$

Source: Reproduced from Westerhoff *et al.* (1982) ©copyright 1982, with permission from Elsevier Science.

Other energy-demanding processes taking place during cell growth, but not directly associated with biomass synthesis, are:

(1) Maintenance of cell physiological status (intracellular pH, ionic balance, osmotic regulation).
(2) Transport of substances to and from the cell.

The energetic yield of a microbial process may be directly calculated from the known biochemistry of catabolic pathways (e.g. glycolysis occurring through the Embden–Meyerhoff or Entner–Doudoroff pathways, and the efficiency of oxidative phosphorylation through the P:O ratio).

A useful quantity regarding the energetics of microbial growth is the biomass yield based on ATP, Y_{ATP}, that has been extensively analyzed by Stouthamer (1979). This can be exactly calculated only during anaerobic growth, because the efficiency and branching of the aerobic respiratory chain are difficult to assess experimentally (Stouthamer and van Verseveld, 1985). Y_{ATP} values are influenced by a number of factors, e.g. the presence of exogenous electron acceptor, the carbon source and the complexity of the medium, the nature of the nitrogen source, and the chemical composition of the organism under specific conditions. Regarding the latter it has been found that organisms exhibiting large differences in macromolecular compositions showed little variance in Y_{ATP}^{max}, e.g. 28.8 against 25 (g cells mol^{-1} ATP) for *E. coli* and *Aerobacter aerogenes*, respectively. On the other hand the nature of the carbon or the nitrogen sources and especially the metabolic pathway through which they are assimilated exerted a strong influence on the ATP yield (Stouthamer, 1979).

It has been calculated that only between 15% and 25% of the total energy demand was required for the synthesis of macromolecular precursors (Lamprecht, 1980). A large proportion of total energy expenditure is required to maintain cellular organization. This implies that multicellular organisms with a higher degree of complexity require even a larger energy supply to maintain that organization. Calorimetric determinations have revealed that yeast cells release 157-fold higher amounts of heat than a human; the comparison made on a nitrogen content basis. In turn, a horse releases three times as much heat as a human (quoted in Lamprecht, 1980).

Bioenergetics regard cellular metabolism as an energy-transducing network that couples the release of energy from the oxidation of organic compounds to the synthesis of cellular biomass and other energy-demanding processes. Within this framework, a series of thermodynamic functions have been defined to understand cellular optimization principles as a function of fitness (Westerhoff and Van Dam, 1987). These authors analyzed microbial growth (e.g. bacterial, fungal) on various substrates. Thermodynamic efficiency was calculated according to two different definitions (Table 2.10) (Roels, 1983; Westerhoff and Van Dam, 1987). According to Westerhoff and van Dam (1987), thermodynamic efficiency becomes negative when the energy content of the substrate is so high that additional energy provided by catabolism (ATP and reducing equivalents) contribute to fuelling the microbial system just to make it run faster (Westerhoff *et al.*, 1982). Reasoning in terms of steps I and II of anabolism (see section: *Anabolic fluxes*), a substrate with a high energy content generates in the pathways involved in step I a surplus of ATP sufficient to fuel step II, together with all other energy-demanding processes such as polymerization or assembly of cellular structures. Efficiencies lower than zero are obtained with substrates more reduced than biomass. When microbial growth occurs on C substrates with a similar reduction degree as biomass, e.g. glucose, lactate or acetate, the thermodynamic efficiency of growth is almost zero. Substrates that are more oxidized than biomass, e.g. formate, oxalate, sustain maximal efficiency values of 25% (Table 2.10). In fact, the theoretical maximum of thermodynamic efficiency (i.e. 100%) is never attained, since this would imply that there are no flows through the system, i.e. the growth rate would be null in such a case (Westerhoff and Van Dam, 1987).

Chapter 3

Cell Growth and Metabolite Production: Basic Concepts

Microbial Growth under Steady and "Balanced" Conditions

Microbial populations grow in geometric progression. The dynamic states of cultures of microorganisms are essentially of two sorts: transient or steady, irrespective of their stage of growth or quiescence. The main difference between transient and steady state is their time-dependence (see Chap. 6).

Growth is balanced when the specific rate of change of all metabolic variables (concentration or total mass) is constant (Barford *et al.*, 1982):

$$\frac{1}{x_{iAV}} \frac{\Delta x_i}{\Delta t} = \text{constant}, \tag{3.1}$$

where Δx_i is the change in x_i during time Δt; x_{iAV} is the average value of x_i during time Δt.

A steady state is a sort of balanced growth, but with the further requirement that the overall rate of change of any metabolic variable or biomass be zero:

$$\frac{dx_i}{dt} = 0. \tag{3.2}$$

Analysis of a growing population, at any time, reveals that on average, all cells process the available nutrients and produce new biomass and increase in cell size at similar rates. By definition, the time taken to go from one generation $(n - 1)$ to the next generation (n) is constant under stable environmental conditions, and so the rate of increase in cell number in the population increases with time. In fact, the population number increases exponentially, and the rate of acceleration in the population size is governed by two major factors. First, the rate is controlled by the intrinsic nature of the organism that is its basic genetic structure and potential, which largely determine its physiological characteristics. Second, the rate of population increase is substantially modified by extrinsic factors due to the organism's growth environment (Slater, 1985). So in "rich" environments containing many preformed cellular components, such as amino acids and nucleosides, the rate of increase is more rapid than in a "poor" environment in which the growing organism has to synthesize all its cellular requirements, devoting proportionally much more of the available resources and energy to these activities. Thus the rate of cell number increase with time is a variable dictated by the type of organism and the nature of its growth environment. Under constant environmental conditions the time taken to complete each generation and double the size (the growth rate) is a characteristic constant known as the culture doubling time t_d.

The rate of change of the population size, or more conveniently the rate of growth, is directly proportional to the initial population size,

$$\frac{dx_i}{dt} = \mu x_o. \tag{3.3}$$

In Eq. (3.3), μ, the specific growth rate is a measure of the number of new individuals produced by a given number of existing individuals in a fixed period of growth time. If the rate of growth is optimized, μ, becomes the maximum specific growth rate μ_{max}. In many closed (batch) cultures, the ideal conditions are achieved and growth proceeds at μ_{max}.

By integration from $t = 0$ to $t = t$ Eq. (3.3) has the solution:

$$x_i = x_o e^{\mu t}, \tag{3.4}$$

that describes an exponential curve, which in linear form reads:

$$\ln x_t = \ln x_o + \mu t, \tag{3.5}$$

its slope is:

$$\mu = \frac{0.693}{t_d}. \tag{3.6}$$

Thus, the specific growth rate is inversely proportional to the culture doubling time.

The basic growth (Eq. (3.4)) may be derived either by considering populations as numbers of individual cells, or rates of increase as increases in the number of cells in the population (see Slater (1985)).

Unlimited quantities of all growth resources might exist only for short periods of time (e.g. early stages of a batch culture), but in most natural habitats, resource limitation is the rule rather than the exception. Thus, growth limitation by substrate depletion (of the growth limiting substrate) restricts the final size of the population, and indeed the two parameters may be shown to be proportional to each other:

$$xf \propto S_r, \tag{3.7}$$

where S_r is the initial concentration of the growth-limiting substrate, and xf is the final population size. Thus:

$$xf = YS_r, \tag{3.8}$$

where Y is a proportionality constant known as the observed growth yield. It is defined as that weight of new biomass produced as a result of utilizing unit amount of the growth-limiting substrate. The yield term indicates how much of the available substrate is used for new biomass production plus energy generation to drive biosynthetic reactions (see Chap. 2). Table 3.1 shows the biomass growth yield on the bases of glucose or ATP for several organisms growing anaerobically (Stouthamer, 1979). That a relationship exists between the amount of growth of a microorganism and the amount of ATP that could be obtained from the energy source in the medium was first suggested for anaerobic cultures (Bauchop and Elsden, 1960). However, for aerobic growth, this relationship has turned out to be more complex than the originally suggested direct proportionality (Stouthamer, 1979; Stouthamer and van Verseveld, 1985). Thus, both theoretical and experimental data show that Y_{ATP} is not a biological constant (Table 3.1), and that the ATP requirement for biomass formation depends on the carbon substrate, the efficiency of oxidative phosphorylation, the anabolic and catabolic pathways utilized (see Table 4.1), the presence of preformed monomers, and the nitrogen source (Stouthamer, 1979; Stouthamer and van Verseveld, 1985). Moreover, measured values are mostly 50%–60% of the theoretical calculated

Table 3.1 Y_{ATP} for batch cultures of a number of microorganisms growing anaerobically in various media on glucose

Organism	$Y_{X/C}$ (g dry weight mol^{-1} substrate)	Y_{ATP} (gmol gmol^{-1} substrate)
Streptocoecus faecalis	20.0–37.5	10.9
Streptococcus agalactiae	20.8	9.3
Streptococcus pyogenes	25.5	9.8
Lactobacillus plantarum	20.4	10.2
Lactobacillus casei	42.9	20.9
Bifidobacterium bifidum	37.4	13.1
Saccharomyces cerevisiae	18.8–22.3	10.2
Saccharomyces rosei	22.0-24.6	11.6
Zymomonas mobilis	8.5	8.5
	6.5	6.5
	4.7	4.7
Zymomonas anaerobia	5.9	5.9
Sarciiza ventriculi	30.5	11.7
Aerobacter aerogenes	26.1	10.2
	47	18
	69.5	28.5
Aerobacter cloacae	17.7–27.1	11.9
Escherichia coli	25.8	11.2
Ruminococcus flavefaciens	29.1	10.6
Proteus mirabilis	14	5.5
	38.3	12.6
	48.5	18.6
Actinomyces israeli	24.7	12.3
Clostridium perfringens	45	14.6
Streptococcus diacetilactis	35.2	15.6
	43.8	21.5
Streptococcus cremoris	31.4	13.9
	38.5	18.9

Source: Reproduced from Stouthamer (1979).

ones. Several possible explanations have been offered to understand this phenomenon (Stouthamer and van Verseveld, 1985; Tempest and Neijssel, 1984; Verdoni *et al.*, 1990).

Assimilatory and dissimilatory flows are needed to incorporate carbon substrates into biomass. The assimilatory flow is defined as the substrate flux directed toward biomass synthesis (anabolism), whereas the dissimilatory flow is the amount of substrate needed to release energy for other purposes such as transport and biosynthesis (production of new biomass). Futile cycling reactions may also occur (i.e. those in which metabolic energy is dissipated without performing useful biological work) (Gommers *et al.*, 1988). Clearly, the greater the amount of energy generation which is diverted for biosynthetic purposes, the less the amount of available substrate which has to be dissimilated, and the higher the yield (see Table 4.1).

During the growth and decline phases of batch culture, the specific growth rate of cells is dependent on the concentration of nutrients in the medium. Often, a single substrate exerts a dominant influence on rate of growth; this component is known as the growth-limiting substrate. The growth-limiting substrate is often the carbon or nitrogen source, although in some cases it is oxygen or another oxidant such as nitrate. Monod was the first to establish the relationship between growth-limiting substrate concentration and specific growth rate by recognizing that it follows a rectangular hyperbola and closely resembles that between the velocity of an enzyme-catalyzed reaction and substrate concentration. Monod deduced the following expression for microbial cultures:

$$\mu = \frac{\mu_{max}[S]}{K_S + [S]}. \tag{3.9}$$

In Eq. (3.9), $[S]$ is the concentration of growth-limiting substrate, and K_S is the saturation constant. The saturation constant is a measure of the affinity the organism has for the growth-limiting substrate. Typical values of K_S are shown in Table 3.2; they are very small, of the order of milligram per liter for carbohydrate and microgram per liter for other compounds, such as amino acids (Doran, 1995).

The level of the growth-limiting substrate in culture media is normally much greater than K_S (usually $[S] > 10\,K_S$). This explains why μ remains constant and equal to μ_{max} in batch culture until the medium is virtually exhausted of substrate (see below and Chap. 6 for the treatment of this subject in open, continuous cultures).

From Eqs. (3.7) and (3.8) it holds that in a small time interval, δt, the population increases in size by a small amount, δx, as a consequence of using a small amount of the growth-limiting substrate $-\delta s$. Thus:

$$-\frac{dx}{ds} = Y, \tag{3.10}$$

Table 3.2   Effects of substrate concentration on growth rate. K_s values for several organisms

Microorganism (genus)	Limiting Substrate	Ks $(mg\ l^{-1})$
Saccharomyces	Glucose	25
Escherichia	Glucose	4.0
	Lactose	20
	Phosphate	1.6
Aspergillus	Glucose	5.0
Candida	Glycerol	4.5
	Oxygen	0.042–0.45
Pseudomonas	Methanol	0.7
	Methane	0.4
Klebsiella	Carbon dioxide	0.4
	Magnesium	0.56
	Potassium	0.39
	Sulphate	2.7
Hansenula	Methanol	120.0
	Ribose	3.0
Cryptococcus	Thiamine	1.4×10^{-7}

Source: Data were taken from Pirt (1975) and Wang *et al.* (1979).

which is the rate of change of biomass with respect to substrate used, and it depends on the observed growth yield. Hence the rate of change of the growth-limiting substrate may be calculated by substituting Eq. (3.3) for dx and rearranging:

$$-\frac{d[S]}{dt} = \frac{\mu x}{Y},$$

(3.11)

the negative sign indicates a decrease in substrate concentration with time. The constants μ and Y may be replaced by another constant q, the specific metabolic rate:

$$-\frac{d[S]}{dt} = qx.$$

(3.12)

The specific metabolic rate defines the rate of uptake of the growth-limiting substrate by a unit of amount of biomass per unit time, and can vary widely, since μ and Y are subject to environmental variation.

In heterogeneous systems the rates of reaction and substrate mass transfer are not independent. In this case the rate of mass transfer depends on the concentration gradient established in the system which in turn depends on the rate of substrate depletion by reaction. Under those conditions, the rate of substrate uptake and the affinity constant of the substrate K_S by the cell become very important.

Considering that the rate of substrate S depletion by a microorganism immobilized in a spherical particle is

$$r_s = r_s^{max} V_p \frac{[S]}{K_S + [S]}, \qquad (3.13)$$

where r_s are moles of S consumed per unit time; K_s, affinity constant; V_p, particle volume; r_s^{max}, maximal rate of S consumption inside the particle expressed in terms of S concentration. When the rate of transport of S to the particle is greater than r_s the process is kinetically controlled. Diffusional control occurs when the rate of transport of S to the particle is lower than r_s. The metabolism and physiology of the microorganism become limiting when the process is kinetically controlled. In several cases, compelling evidence shows that the rate of substrate uptake by the cell is a main rate-controlling step of yeast catabolism (Cortassa and Aon, 1994a; Cortassa and Aon, 1997; Cortassa and Aon, 1998).

Microbial Energetics under Steady-State Conditions

Monod defined the macroscopic yield of biomass, or observed growth yield (see Eq. (3.8)) as the ratio of the biomass produced to substrate consumed. He described the dependence of the growth rate on the concentration of the growth-limiting substrate (see Eq. (3.9)). Following the introduction of continuous cultivation techniques, it was shown that in carbon-limited continuous cultures the growth yield, Y_s, was not constant but decreased as the dilution rate, D (=growth rate), decreased (see Slater (1985) and refs. therein). This effect was attributed to what he called the *endogenous metabolism*. Thus, during growth the consumption of the energy source is partly growth-dependent and partly growth-independent. Assuming the latter, the following equation can be derived that relates the growth yield and specific growth rate (Pirt, 1975; Stouthamer, 1979):

$$\frac{1}{Y_{xs}} = \frac{1}{Y_{xs}^{max}} + \frac{m_S}{\mu}, \tag{3.14}$$

where Y_{xs} is the yield of biomass, x, on substrate, s; m_s, the maintenance coefficient (mol of substrate per g dry weight per hr), and Y_{xs}^{max} is the yield biomass after correction for energy of maintenance (in g dry weight per mol of substrate). In chemostat cultures, the maximum growth yields for substrate (Y_{xs}^{max}) and oxygen (Y_{O2}^{max}) may be obtained from the plots of q_s and q_{O2} versus D by the following equation (Stouthamer, 1979; Verdoni *et al.*, 1992):

$$q_S = \frac{1}{Y_{xs}^{max}} D + m_S. \tag{3.15}$$

A similar reasoning may be followed for the determination of the maximum growth yield based on oxygen (Y_{O2}^{max}) or ATP (Y_{ATP}^{max}) from a plot of q_{O2} or q_{ATP} versus D (Stouthamer, 1979; Verdoni *et al.*, 1992).

As an example, the specific rate of ATP production, q_{ATP} (in mmol of ATP h^{-1} g [dry weight]$^{-1}$) in oxygen-limited chemostat cultures of *P. mendocina* growing on glucose, was calculated as follows (Verdoni *et al.*, 1992):

$$q_{ATP} = q_{glc}(1 - \beta) + q_{AA} + q_{O2}3(P/O) + 3q_{alg}, \tag{3.16}$$

where q_{AA} and q_{alg} are the specific rates of acetate and alginate production, respectively; the number of ATP molecules formed by substrate phosphorylation during the complete oxidation of the substrate is taken into account by q_{glc} (one ATP) and q_{AA} (one ATP). The fermentation of glucose to acetic acid by the Entner–Doudoroff pathway that operates in *P. mendocina* yields 2 mol of ATP per mol of glucose (Lessie and Phibbs, 1984); β is the part of the substrate that is assimilated. The value of β may be calculated as the ratio between Y_{glc}^{max} values measured under oxygen-limited conditions ($Y_{glc}^{max} = 70$) and the molecular weight of *P. mendocina* according to its chemical composition (molecular weight $= 152.4$; see Chap. 2) (Verdoni *et al.*, 1992).

The coefficient 3 that affects the q_{alg} term represents the net yield in ATP as a result of alginate synthesis. It is estimated that for alginate, 3 ATP equivalents are utilized for each uronic acid monomer incorporated

(Jarman and Pace, 1984; Verdoni *et al.*, 1992). At P:O ratios of 3, the alginate biosynthetic pathway becomes net ATP yielding, the 3 ATP equivalents required per monomer polymerized being supplied by the oxidation of the 2 NAD(P)H (6 ATP produced when oxidized via oxidative phosphorylation) generated from uronic acid synthesis (Jarman and Pace, 1984; Verdoni *et al.*, 1992).

Growth Kinetics under Steady-State Conditions

Open growth systems differ from closed ones in that there is a continuous input of growth substrates and removal of waste products, cells and unused substrates (see Chap. 2). These systems, known as continuous-flow cultures, enable the exponential growth phase to be prolonged indefinitely, establishing steady state conditions. Continuous-flow cultures present additional advantages: specific growth rate may be directly set by the experimenter, substrate-limited growth may be established, submaximal growth rates can be imposed, and biomass concentration may be set independently of the growth rate (Pirt, 1975; Slater, 1985).

The most widely used continuous culture system is the chemostat, characterized by growth control through a growth-limiting substrate. The use of chemostat or continuous cultures to study a microorganism at the steady state provides a rigorous experimental approach for the quantitative evaluation of microorganism's physiology and metabolism (see Chaps. 1, 4, and 5). Besides, chemostat cultures allow the definition of the phase of behavior which suits a purpose we may decide upon, e.g. maximum substrate consumption and output fluxes of metabolic by-products of interest (see Chap. 4). The behavior of the steady-state fluxes as a function of a parameter in the chemostat is, conceptually, the same to that described by a bifurcation diagram (Aon and Cortassa, 1997; see Chap. 5).

The dilution rate

In a chemostat, the concentration of the growth-limiting substrate clearly depends on the rate at which the organisms use it and on its rate of supply. The growth-limiting substrate concentration depends on a ratio, known as the dilution rate, D, of the flow rate F through the system and the volume of the culture, V, in the vessel: $D = F/V$. The dilution rate has units of reciprocal time (h^{-1}), and is a measure of the number of volume changes achieved in unit time.

The dilution rate and biomass concentration

If it is assumed for a moment that the organisms are nongrowing, then the rate of change of the organism concentration in a system with D is given by

$$\frac{dx}{dt} = -Dx, \tag{3.17}$$

and by integration:

$$x_i = x_o e^{-Dt}. \tag{3.18}$$

Equation (3.18) describes exponential decay and constitutes a measure of the culture washout.

In a growing culture, an additional factor influencing the rate of change of biomass concentration is a growth term, such that

$$\frac{dx}{dt} = \mu x - Dx, \tag{3.19}$$

$$\frac{dx}{dt} = [\mu - D]\,x, \tag{3.20}$$

$$\frac{dx}{dt} = \left[\frac{\mu_{max}[S]}{K_S + [S]} - D\right]x. \tag{3.21}$$

Equation (3.21) predicts three general cases for the overall rate of change of biomass concentration (Pirt, 1975; Slater, 1985) (Fig. 3.1, top panel). (i) If $\mu > D, dx/dt$ is positive; the biomass concentration increases and the rate of biomass production is greater than the rate of culture washout (Fig. 3.1, top panel, curve 3). (ii) If $\mu < D, dx/dt$ is negative; the biomass concentration declines and the growth rate is less than the washout rate (Fig. 3.1, top panel, curve 1). (iii) If $\mu = D, dx/dt$ is zero and the rate of biomass production balances the rate of culture washout. Under these conditions the culture is said to be in a steady state (Fig. 3.1, top panel, curve 2). This is the preferred stable state and in time all chemostat cultures will reach a steady state provided that $D < \mu_{max}$ and the environmental conditions are kept constant.

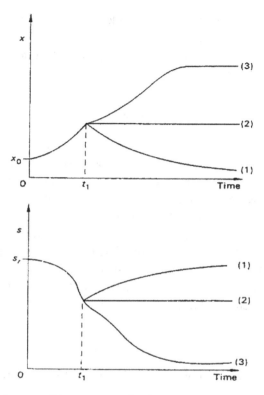

Figure 3.1. The three possible outcomes of a chemostat culture in which growth rate of the biomass (x) is limited by the concentration of the growth-limiting nutrient (s). The flow of medium containing growth-limiting nutrient at concentration s, is started at time, t_1. The different cases are: (1) rate of wash out of biomass exceeds maximum growth rate; (2) rate of wash out of biomass = maximum growth rate; (3) initial rate of wash out is less than maximum growth rate of biomass.

Source: Reproduced from Pirt (1975).

The Dilution Rate and the Growth-limiting Substrate Concentration

A similar balanced equation may be derived to describe the change of growth-limiting substrate, such that

$$\frac{d[S]}{dt} = D[S_r] - D[S] - \frac{\mu x}{Y}, \tag{3.22}$$

$$\frac{d[S]}{dt} = D([S_r] - [S]) - \frac{\mu x}{Y}. \tag{3.23}$$

The first two terms are rates dependent on D, while the growth term was derived previously (Eq. (11)). The two terms ($[Sr]$–$[S]$) and x/Y are equivalent, and both measure the proportion of substrate used for growth, Sg, rendering equation:

$$\frac{d[S]}{dt} = D[S_g] - \mu x_g. \tag{3.24}$$

Three general cases may be considered (Pirt, 1975; Slater, 1985) (Fig. 3.1, bottom panel). (i) If $D > \mu$, $d[S]/dt$ is positive; the growth-limiting substrate concentration increases and the rate of supply of growth-limiting substrate is less than the rate of use by x (Fig. 3.1, bottom panel, curve 1). (ii) If $D < \mu$, $d[S]/dt$ is negative, and the overall growth-limiting substrate concentration declines (Fig. 3.1, bottom panel, curve 3). (iii) If $D = \mu$, $d[S]/dt$ is zero and the growth-limiting substrate concentration in the culture vessel reaches a steady state (Fig. 3.1, bottom panel, curve 2). This occurs at the same time as the value for $dx/dt = 0$ (Fig. 3.1, top panel).

Biomass and Growth-limiting Substrate Concentration at Steady State

Chemostat culture systems establish steady-state conditions when $d[S]/dt = 0$ and $dx/dt = 0$. These values are attained for unique values of x and $[S]$ for a given value of D and constant environmental conditions (Fig. 3.2).

By substituting in Eq. (3.21) and then rearranging, Eq. (3.27) is obtained as follows:

$$0 = \left[\frac{\mu_{\max}[S]}{K_S + [S]} - D \right] x, \tag{3.25}$$

$$D = \frac{\mu_{\max}[S]}{K_S + [S]}, \tag{3.26}$$

$$[\bar{S}] = \frac{K_S D}{\mu_{\max} - D}. \tag{3.27}$$

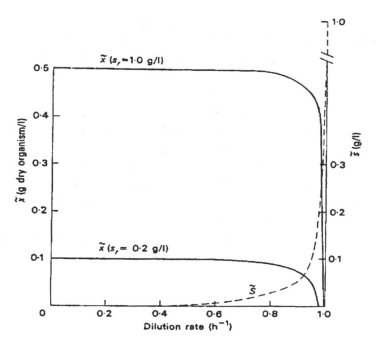

Figure 3.2. Steady-state values of biomass (x) and growth-limiting substrate (s) concentrations in a chemostat. A plot of steady-state values of x and s against dilution rate, D, with typical parameter values ($\mu_m = 1.0\,h^{-1}$; $K_s = 0.005\,g/l$; $Y = 0.5$) is shown. *Source*: Reproduced from Pirt (1975).

Similarly Eq. (3.23) leads to Eq. (3.29).

$$0 = D([S_r] - [\bar{S}]) - \frac{\mu \bar{x}}{Y}, \tag{3.28}$$

$$\bar{x} = Y([S_r] - [\bar{S}]). \tag{3.29}$$

Substituting Eq. (3.27) into Eq. (3.29) gives Eq. (3.30).

$$\bar{x} = Y\left([S_r] - \frac{K_S D}{\mu_{max} - D}\right). \tag{3.30}$$

Equations (3.27) and (3.30) enable the growth-limiting substrate and steady-state biomass concentrations, respectively, to be calculated provided that three basic growth parameters, Ks, μ_{max}, and Y, are known. Since for a given organism under constant conditions these parameters are constant, then the unused growth-limiting substrate concentration in the culture vessel depends solely on the imposed dilution rate: it is even independent of the initial substrate concentration S_r. On the other hand, the biomass concentration depends on D and S_r. The relationships between x, $[S]$, the organism constants and D are demonstrated in Fig. 3.2.

In a batch culture, the biomass concentration will increase rapidly and the maximal amount (as determined by the engineering parameters of the fermentor) will be reached within a few hours. If we start with a lower concentration of substrate and start to feed further substrate when it becomes exhausted, it is possible to approach the maximal amount of biomass in the fermentor slowly (Fig. 3.3). Therefore, in a fed-batch culture, the capacity of the fermentor is utilized to a large extent for a much longer period than in a batch culture. This is the main reason why the processes for industrial fermentations usually use fed-batch cultures (Stouthamer and Van Verseveld, 1987).

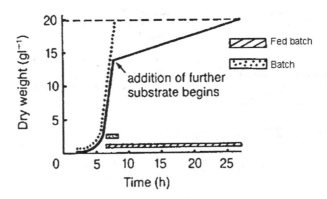

Figure 3.3.   Growth of a microorganism in a batch and fed-batch culture. At $t = 0$, the culture was inoculated to a biomass concentration of 0.01 g l^{-1}. The division time of the organism was supposed to be 45 min. It was assumed that due to restricting engineering parameters the biomass concentration in the fermentor cannot exceed 20 g l^{-1}, since at higher concentrations the capacity for oxygen transfer of the fermentor is insufficient. Already after 8.1 h this limit is reached in the batch culture. If the substrate concentration is insufficient to reach this amount of biomass and a feed of substrate is started after the substrate has been fully consumed (indicated by an arrow) the limits of the fermentor capacity can be gradually approached. The dashed line represents biomass limit due to O$_2$ transfer; dotted line represents batch culture; thick line represents fed-batch culture. Productive times in batch or fed-batch cultures are indicated by horizontal bars.
Source: Reprinted from Stouthamer and van Verseveld (1987).

In addition to balance equations, kinetic ones are needed to describe microbial production processes. The linear equation for substrate consumption with product formation is the most important. The total rate at which substrate is used is the sum of the rates at which it is used for maintenance purposes, biomass and product formation (Stouthamer and Van Verseveld, 1987).

$$r_s = m_s x_t + \frac{1}{Y_{xsm}} r_x + \frac{1}{Y_{psm}} r_p, \tag{3.31}$$

where r_s, r_x, and r_p are, respectively, the rates of substrate consumption, biomass, and product formation, and m_s and x_t are the maintenance coefficient and biomass concentration; Y_{xsm} and Y_{psm} are, respectively, the maximal yield of biomass and product on substrate.

The parameters of Eq. (3.31) are of great importance because they can be used:

- in mathematical models to optimize both the conversion of substrate into product, and the use of the bioreactor at full capacity;
- to analyze substrate-related production costs when a new process is being developed;
- to evaluate the yield of products of recombinant strains obtained through DNA technology along with the suitability of a host for producing a certain product.

Metabolic Fluxes during Balanced and Steady-State Growth

The regulation of the degree of coupling between catabolic and anabolic fluxes, in turn, modulates the amount of flux redirection toward microbial products. Microbial products may be divided into two groups: (i) fermentation products, and (ii) proteins and polysaccharides. In the first group, redirection of fluxes toward fermentation products may result from alteration in the redox and phosphorylation potentials (Aon *et al.*, 1991; Cortassa and Aon, 1994b; Fuhrer and Sauer, 2009; Verdoni *et al.*, 1992). During transitions from aerobic to oxygen-limited or anaerobic conditions, the correct balance of both, redox (NADH/NAD; NADPH/NADP) and phosphorylation

Table 3.3   Classification of low-molecular-weight fermentation products

Class of metabolite	Examples
Products directly associated with generation of energy in the cell	Ethanol, acetic acid, gluconic acid, acetone, butanol, lactic acid, other products of anaerobic fermentation
Products indirectly associated with energy generation	Amino acids and their products, citric acid, nucleotides
Products for which there is no clear direct or indirect coupling to energy generation	Penicillin, streptomycin, vitamins

Source: Reproduced from Doran (1995).

(ATP/ADP) couples (Anderson and Dawes, 1990; Kell *et al.*, 1989; Senior *et al.*, 1972) are germane to the question of the formation of fermentation products.

Fermentation products can be classified on the basis of their relationship with product synthesis and energy generation in the cell (Table 3.3) (Doran, 1995; Stouthamer and van Verseveld, 1985). The first category corresponds to end- or by-products of energy metabolism synthesized in pathways which produce ATP. Those of the second class are partly linked to energy generation but require additional energy for synthesis. The third class involves the production of antibiotics or vitamins far-removed from central energy metabolism.

The rate of product formation in cell culture can be expressed as a function of biomass concentration:

$$r_p = q_p x, \tag{3.32}$$

where r_p is the volumetric rate of product formation, x, biomass concentration, and q_p, the specific rate of product formation. Here q_p is not necessarily constant during batch culture (Doran, 1995). Depending on whether the product is linked to energy metabolism or not, equations for q_p as a function of growth rate and other metabolic parameters can be developed. For products formed in pathways which generate ATP (Table 3.3), the rate of production is related to cellular energy demand. Thus, as growth constitutes the major energy-requiring function of cells, product has to be formed whenever there is growth, so long as its production is coupled to energy metabolism. Under these conditions, kinetic expressions for product formation must account for growth-associated and maintenance-associated production:

$$r_p = Y_{px}r_x + m_px, \tag{3.33}$$

where r_x is the volumetric rate of biomass formation, Y_{px}, the theoretical or true yield of product from biomass, m_p, the specific rate of product formation due to maintenance, and x, biomass concentration. Equation (3.33) states that the rate of product formation depends partly on rate of growth but also partly on cell concentration. Taking r_x as equal to μx, then:

$$r_p = (Y_{px}\mu + m_p)x. \tag{3.34}$$

Comparison of Eqs. (3.32) and (3.34) shows that, for products coupled to energy metabolism, q_p is equal to a combination of growth-associated and nongrowth-associated terms:

$$q_p = Y_{px}\mu + m_p. \tag{3.35}$$

Similar expressions have been derived for the production of extracellular enzymes by *Bacillus*. In this case, a linear relationship between r_p and r_x has been demonstrated (Stouthamer and van Verseveld, 1987):

$$r_p = ax_t + br_x, \tag{3.36}$$

in which a and b are constants. The amount of product formed depends both on the amount of biomass and on its rate of increase. The amount of biomass, x_t, and the growth rate r_x will be dictated by the culture system used.

Growth as a Balance of Fluxes

Any dynamic view of cell growth and proliferation, must consider growth as a dynamically coordinated, dissipative balance of fluxes, i.e. the result of fluxes of the different materials consumed and the interactions between those fluxes (see Chap. 5).

Growth as a result of the dynamic balance between synthesis and degradation of macromolecular components, e.g. DNA, RNA, proteins, and polysaccharides, produces the typical growth curve shown by microorganisms in batch culture. Synthetic process reactions decrease their rates with time, while degradative ones increase their rates, thereby achieving a pseudo-steady-state level which will persist for some time and then decline (Aon and Cortassa, 1995). In Fig. 3.4, the results of a numerical

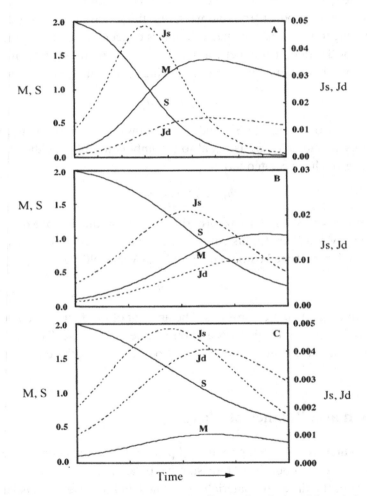

Figure 3.4. Cell growth as a result of a dynamic balance of synthetic and degradative fluxes. A simple model consisting of the autocatalytic synthesis of a macromolecular component (M) from a substrate (S). M is degraded according to a first-order kinetic law:

$$S + ATP + NADPH \rightarrow M \rightarrow$$

simulation of a typical growth curve of a batch culture of microorganisms at decreasing contents (from A to C) of available redox equivalents and high free-energy transfer bonds (NAD(P)H, ATP) are shown. Growth measured as the net accumulation of macromolecules, M, results from the dynamic balance of synthetic, J_s, and degradative, J_d, fluxes from catabolized substrate, S. As shown in Fig. 3.4, J_s decreases with time after reaching a maximum, whereas J_d increases until reaching a plateau. In all cases, the intersection of J_s and J_d corresponds to the maximal accumulation of M. A pseudo-steady-state level of M is achieved for some time and then decline; the latter corresponding to lower J_s than J_d fluxes. Figure 3.4 also shows that the pseudo-steady-state level of M also depends on the magnitude of fluxes which in turn are influenced by ATP and NAD(P)H contents.

The analysis of the dynamics of cellular processes in the light of the concepts of dynamics and thermodynamics reveals that flux redirection at the cellular level may be accomplished taking advantage of the intrinsic dynamic properties of cellular metabolism. (Aon and Cortassa, 1995) (see Chaps. 4–6). Large dissipation rates through a metabolic path or branch flux redirection may take place because of kinetic limitations (see below).

←——————————————————————————————

Figure 3.4. (*Continued*) The model is mathematically expressed by two differential equations:

$$\frac{dM}{dt} = J_s - J_d$$

$$\frac{dS}{dt} = -J_s$$

with synthetic: $J_s = k_1 \cdot S \cdot \text{ATP} \cdot \text{NADPH}$; and degradative fluxes: $J_d = k_2 \cdot M \cdot k_1$ and k_2 are the rate constants of synthesis and degradation of macromolecules (M), respectively. ATP and NADPH contents are parameters, i.e. they are assumed constant for simplicity. The parameter values were: $k_1 = 0.1\,(\text{mM}^{-2}\text{h}^{-1}); k_2 = 0.01\,(\text{h}^{-1}); S_0$ (the initial concentration of substrate $= 2.0$; ATP $= 1.0$. According to the units employed in the model, S, ATP, NADPH and M are in mM concentration units. From (A) to (C) the varying parameter is NADPH: (A) 0.5; (B) 0.25; (C) 0.1; indicating decreasing redox equivalent contents. Notice that when the flux of synthesis (J_s) intercepts the flux of degradation (J_d) the accumulation of M is maximal; thus, when $J_s > J_d$ the flux balance favors the synthesis of M and consequently it accumulates. On the contrary, when $J_s < J_d$ the flux balance favors the degradation of M which begins to decline. The time as well the amount reached by M depends on the metabolic status, e.g. available redox equivalents.
Source: Reprinted from Aon and Cortassa (1995).

Kinetic limitation occurs when a metabolic pathway has attained the maximum possible rate of dissipation. The rate of dissipation suggests that at fixed ΔG_i then fixed stoichiometry the maximum flux, J_i, may be set by the maximal through-put velocity sustained in that path (see Eq. (3.37) below). Since the maximum velocity depends on enzyme concentrations at rate-controlling steps, then the higher the enzyme concentrations, the higher the flux.

The Flux Coordination Hypothesis

In order to conceptualize the cell, we have introduced the Flux Coordination Hypothesis (FCH). This hypothesis emphasizes the regulation of the degree of coupling between catabolic and anabolic fluxes as a regulatory mechanism of the growth rate. FCH stresses the fact that cells exhibit global regulatory mechanisms of flux balance associated mainly with the rate of energy dissipation (Aon and Cortassa, 1997). Overall, the rate of energy dissipation, σ, by a cell is given by the product between fluxes, Ji, and their conjugated forces, ΔGi, as follows:

$$\sigma = J_i \Delta G_i. \tag{3.37}$$

Taking into account the fact that the free-energy difference, ΔGi, in every ith enzyme-catalyzed reaction of either catabolism or anabolism is fixed by the nature of reactants and products, the effective flux to be sustained will depend on the genetic makeup, as well as the availability of effectors according to environmental conditions (Aon and Cortassa, 1997). Thus, cells may redirect fluxes, modifying in this way the effective stoichiometry of metabolic pathways, especially when facing unfavorable environmental challenges (Fig. 3.5). Under these conditions cells uncouple catabolic and anabolic fluxes, becoming able to excrete products, some of them of commercial value. We may take advantage of these global regulatory mechanisms exhibited by cells for engineering them with the aim of obtaining specific products (see Chap. 10).

An imbalance between catabolic and anabolic fluxes may be associated with the onset of cell cycle arrest through growth limitation (Fig. 3.5). According to FCH, some cellular mechanism for metabolic fluxes

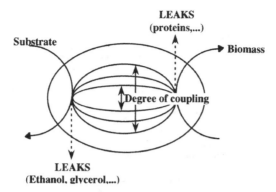

Figure 3.5   The Flux Coordination Hypothesis. The degree of coupling between catabolic and anabolic fluxes as a phenomenon involved in the regulation of microbial proliferation and metabolic flux redirection. The hypothesis postulates that for higher degrees of coupling, i.e. longest arrow, the more coupled anabolic and catabolic fluxes will be, and that would correspond to lower leaks (dashed arrows). Under those conditions, cells will continue their mitotic cycling. Cells will leave the mitotic cycle whenever they are challenged by an unfavorable environment, or when environmental changes, e.g. oxygen and carbon source, induce a particular metabolic and energetic status that leads to a differential gene expression which in turn induces metabolic flux redirection and lower growth rates.
Source: Reprinted from Aon and Cortassa (1995).

redirection toward product (either catabolic, organic acids, or anabolic, polysaccharides) or biomass (macromolecules) formation should be associated with the onset of cell division arrest (see Chaps. 4 and 10). This is indeed very relevant since through regulation of the growth rate, an effective change is attained in the duration of the cell cycle, and thus of gene expression. In this way, the cell cycle becomes a potential (supra)regulatory mechanism of gene expression (Aon and Cortassa, 1995; Aon and Cortassa, 1997; Gubb, 1993). We have shown that, independently of the way the growth rate was varied, the time taken to stay in the G1 phase of the cell cycle decreases exponentially with growth rate (Fig. 3.6).

This is in agreement with the fact that G1 phase of the cell cycle is the most sensitive to environmental conditions, and is coincident with the main molecular components acting at Start (Aon and Cortassa, 1998; Forsburg and Nurse, 1991; Hartwell, 1991; Reed, 1992).

In recent work, the physiological response of chemostat cultures of *S. cerevisiae* to nutritional limitations (C, N, P, S) was analyzed with a Systems Biology approach (Gutteridge *et al.*, 2010). The transcriptome,

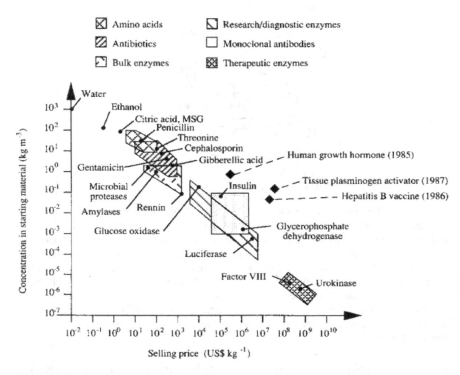

Figure 3.6 Relationship between selling price and concentration before downstream processing for several fermentation products.
Source: Reproduced from Doran (1995).

proteome, and metabolome of yeast were analyzed in each condition comparing excess versus nutrient-limitation as a function of three growth rates ($0.07, 0.1, 0.2\,h^{-1}$). Carbon limitation exhibited the largest effect on transcriptional patterns due to the release of glucose repression on many genes (see Chap. 4), and the activation of respiratory metabolism, as indicated by large increases in the expression of genes related to mitochondrial functions (Gutteridge *et al.*, 2010). Under C-limitation the expression of mitochondrial and respiratory genes increases with growth rate, as well as genes related with synthesis and degradation of carbohydrate storage molecules and ribosomal biogenesis. A relatively smaller subset of genes is affected under N-, P-, and S-limitations when compared to carbon. The results of the studies above are in agreement with the idea introduced long time ago by Beck and von Meyenburg (1968), who proposed that enzyme synthesis is a function of the rate of glucose consumption and not of glucose concentration per se. This early proposal was later confirmed by several authors (reviewed in Aon and Cortassa (1997)).

Flux redirection may be viewed as a bifurcation in the phase portrait dynamics of metabolic flux behavior (Abraham and Shaw, 1987; Aon and Cortassa, 1997; Bailey *et al.*, 1987; Kauffman, 1989; Lloyd *et al.*, 2001). Which are those nonlinear mechanisms that at the cellular level give rise to such bifurcations in the dynamic behavior of metabolic fluxes? The stoichiometry of ATP production by glycolysis is itself autocatalytic (Cortassa *et al.*, 1991). This autocatalytic feedback loop gives rise to complex behavior in addition to stable, asymptotic steady states, and oscillatory dynamics (Aon *et al.*, 1991; Cortassa and Aon, 1994b). Other various mechanisms of nonlinear behavior as a source of emergent behavior, such as metabolic flux redirection, are described in Chaps. 1, 5, 6, and 10 of this book.

Flux imbalance and cell cycle arrest

During arrest of cell proliferation, flux redirection was found to be associated with carbon and energetic uncoupling (Aon *et al.*, 1995; Monaco *et al.*, 1995). According to FCH, when cells are challenged by an unfavorable environmental condition (e.g. temperature), distinct metabolic and energetic requirements are induced leading to differential gene expression, metabolic flux redirection, that in turn induce lower growth rates (Aon and Cortassa, 1997). Indeed, cell arrest of proliferation at 37°C (restrictive temperature) in the cell division cycle (cdc) mutants of *S. cerevisiae cdc28, cdc35, cdc19, cdc21*, and *cdc17*, was correlated with carbon and energy uncoupling (Aon and Cortassa, 1995; Aon *et al.*, 1995; Monaco, 1996). Apparently, the flux imbalance associated with cell division arrest was more pronounced for carbon assimilation than for energy generation. At 37°C, cdc mutants diverted to biomass synthesis only 3% and around 20% of the fluxes of carbon consumed and ATP obtained by catabolism, respectively, compared with 50% and 30% in the wild type strain A364A. At 37°C, cdc mutants directed 60% of the carbon to ethanol production. *Cdc28* and *cdc35* showed decreased mitochondrial biogenesis as well as impaired respiration (Genta *et al.*, 1995). When the anabolic fluxes sustained by *cdc28, cdc35*, and *cdc21* were analyzed through incorporation of

Table 3.4　Average specific incorporation of radioactive precursors into macromolecules during 8 h at the restrictive temperature in wild type *S. cerevisiae* cells and mutants *cdc21* and *cdc35*

Strain	Specific incorporation into macromolecules (cpm mg^{-1} dw h^{-1})		
	Proteins	RNA	DNA
WT	6592	1613	908
cdc21	7883	1997	1001
cdc35	5657	1551	719

Source: Reproduced from Monaco (1996).

Table 3.5　Chemical composition of strains WTA364A, *cdc21* and *cdc35* at the restrictive temperature in 1% glucose and minimal medium

Strain	RNA (%)	Carbohydrates (%)	Proteins (%)	Lipids (%)
WT				
t0	13.08 ± 0.99	22.2 ± 1.8	43.65 ± 1.9	17.07 ± 4.6
t4	10.70 ± 2.1	30.55 ± 1.2	42.1 ± 3.1	13.65 ± 6.4
t8	8.73 ± 1.9	25.85 ± 2.9	40.75 ± 1.7	15.8 ± 6.5
cdc21				
t0	13.05 ± 1.6	29.95 ± 1.7	47.25 ± 1.7	10.1 ± 5
t4	9.37 ± 1.7	29.5 ± 0.8	47.75 ± 1.1	11.37 ± 3.6
t8	7.8 ± 0.95	27.25 ± 3.2	45.2 ± 1.7	16.25 ± 5.85
cdc35				
t0	12.15 ± 2.0	35.7 ± 1.1	34.9 ± 2.3	13.25 ± 5.4
t4	10.3 ± 1.9	39.95 ± 1.5	31.6 ± 1.9	14.15 ± 7
t8	7.88 ± 1.7	34.75 ± 3.9	37.05 ± 1.2	16.25 ± 6.8

Source: Reproduced from Monaco (1996).

radioactive precursors into macromolecules (DNA, RNA, proteins), similar specific rates of incorporation were obtained in cdc mutants with respect to the wild type at the restrictive temperature (Table 3.4). The chemical composition of *cdc35* and *cdc21* mutants at either permissive (25°C) or restrictive temperatures did not differ significantly with respect to the wild type (Table 3.5) (Monaco, 1996; Monaco and Aon, unpublished results). When analyzed together, the results obtained are in agreement with the idea that catabolism is the major contributor to the carbon and energetic imbalance associated with the arrest of cell proliferation in cdc mutants.

Within the large block of catabolism, ethanolic fermentation was one of the major pathways activated and responsible of the carbon and energetic uncoupling exhibited by the cdc mutants during cell proliferation arrest (Aon *et al.*, 1995).

Further experimental evidence suggested that the increased fermentative ability, and the drastic carbon uncoupling, must be somehow linked to glucose-induced catabolite repression (Monaco *et al.*, 1995; see section: *Catabolite repression and cell cycle regulation in yeast*).

Redirecting Central Metabolic Pathways under Kinetic or Thermodynamic Control

In metabolism, central catabolic pathways (glycolysis, pentose phosphate pathway, tricarboxylic acid cycle, and oxidative phosphorylation) provide the key intermediary metabolites precursors of monomer synthesis (amino acids, lipids, sugars, and nucleotides) for the main macromolecular components (proteins, lipids, polysaccharides, nucleic acids) (see Fig. 2.2). The production rate and yield of a metabolite are ultimately limited by the ability of cells to channel the carbon flux from central catabolic pathways to main anabolic routes leading to biomass synthesis (Celinska, 2010; Cortassa *et al.*, 1995; Dong *et al.*, 2011; Liao *et al.*, 1996; Vaseghi *et al.*, 1999).

Several criteria may be adopted when deciding the feasibility of modifying a cell or microorganism through DNA recombinant technology: (i) how far are we below the theoretical yield; (ii) if we are close to the theoretical yield (say 90%–95%), then we must consider whether a 5% to 10% increase in the production of the desired metabolite is economically worthwhile. This decision depends on the commercial value of the product and its concentration before downstream processing. The higher the starting concentration the less costly is the final product (Fig. 3.6).

Once it has been decided to genetically modify the organism, a strategy must be designed. The first step of this strategy is to decide which steps of the metabolic pathway are appropriate. Thus, an important question in MCE is how the fluxes toward key precursors of cell biomass are controlled (Morandini, 2009). This is a very important topic beyond the peculiarities established by the production of a desired metabolite or molecule since it points out toward general constraints of biological performance.

A large experience in the physiological behavior of microorganisms shows that their ability to channel carbon flux from central catabolic pathways to main anabolic routes is either thermodynamically or kinetically controlled. This distinction is timely topic when the performance of a microorganism is aimed to be improved through MCE or optimization of environmental conditions. In the following, we focus to show that distinguishing about kinetic or thermodynamic control of metabolic flux under a particular growth condition in which a cell (e.g. a unicellular microorganism) is searched to be improved, is critical for two main reasons: (i) the decision about the strategy of engineering by acting on metabolism or optimization of culture conditions; and (ii) the evaluation of cellular performance, either genetically engineered or not, under a defined condition.

Thermodynamic or Kinetic Control of Flux under Steady-State Conditions

In terms of enzyme kinetics, a steady-state flux of a linear metabolic pathway may be visualized by plotting individual enzyme activities as a function of substrate concentration (Fig. 3.7). At the steady state, all activities are set at the same value. In the case none of the enzymes are operating at substrate saturation (i.e. at V_{max}), the flux through the pathway may increase just by an increase in the amount of substrate (Fig. 3.7(A)), namely through an increase in the chemical potential of the reaction. In this case, the flux is thermodynamically controlled, and further optimization can be achieved by manipulating the environment since the biological limits have not been attained yet.

Figure 3.7(B) depicts the situation for which the enzymatic activities of the pathway are operating near their maximal rates, unlike the case described in panel (A) (Fig 3.7). At plateau levels, the flux through the pathway is rather kinetically limited than by substrate. In terms of MCA, the saturated enzymes have a large flux control coefficient; thus, an increase in flux can be achieved via simultaneous increases in the amounts of the flux-controlling enzymes (see Chap. 5). An alternative strategy will be to increase all the

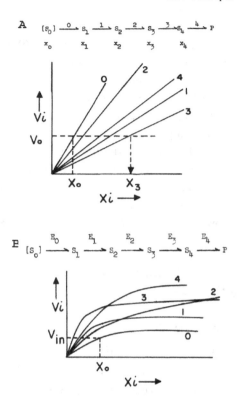

Figure 3.7 Initial rate of a hypothetical metabolic pathway as a function of substrate concentration. Each enzyme-catalyzed step of the linear metabolic pathway at the top of each panel exhibits a dependence on substrate concentration, X, as indicated in the diagrams. (A) All enzymes display linear kinetics in the physiological range of variation; at the steady state, the rate of each enzyme is equal to V_O and the level of each substrate is given by the intersection between enzyme rate, V_i, and the V_O line. (B) The enzymes of the linear pathway operate near saturation and the enzyme activity E_O corresponds to the rate-controlling step. As in panel (A), the steady-state flux is set at the V_{in} value as well as all enzyme rates (i.e. steady-state condition); the substrate concentrations attained are given by the intersection between the corresponding rate curve and the V_{in} line.
Source: Adapted from Higgins (1965).

enzyme levels corresponding to the elementary flux mode that gives the largest yield of the desired product (see Chap. 5).

An analogy may further help to clarify thermodynamic or kinetic control. The metabolite flow being sustained through metabolic pathways in a cell would be analogous to a river's basin whose depth determines the extent of the maximal flux that the river is able to sustain. Under thermodynamic limitation, the basin is large with respect to the water provision; thus, the cell is able to sustain even larger fluxes. Under those conditions, the process can be further optimized by manipulating the environment, as the biological limits have not been attained. On the other hand, in the case of kinetic limitation, the improvement of the flux will require manipulation of the metabolic pathway through genetic engineering.

The point at which the thermodynamic limitation is achieved can be assessed through the yield, which is in fact, a thermodynamic parameter (Heijnen and Van Dijken, 1992; Roels, 1983) (see Chap. 2). The latter notion points out the importance of a correct evaluation of biomass yields during microbial growth (Cortassa *et al.*, 1995; Verduyn *et al.*, 1992).

Kinetic and Thermodynamic Limitations in Microbial Systems: Case Studies

Saccharomyces cerevisiae

As an example let us discuss the case of the yeast *S. cerevisiae* growing in aerobic, glucose- or nitrogen-limited chemostat cultures. The maximal theoretical anabolic flux necessary for yeast cells growing in the respiratory regime of continuous cultures can be calculated. It has been found that during solely respiratory glucose breakdown, the flux through anabolic pathways (i.e. that directed toward the synthesis of cell biomass) is thermodynamically controlled (Aon and Cortassa, 2001). This means that if we add more substrate, the anabolic flux could still increase, or in other words the pools of the key precursors (e.g. hexose-6-P, triose-3-P, ribose-5-P, erythrose-4-P, and 3-P-glycerate) of macromolecules (lipids, proteins, polysaccharides, and nucleic acids) are not saturated with respect to the maximal flux attainable under such conditions. On the contrary, in the so-called respiro-fermentative regime of yeast physiology in chemostat cultures, the fluxes through central anabolic pathways are kinetically controlled. The accumulation of substrate (e.g. glucose) and of catabolic by-products (e.g. ethanol, glycerol) in the extracellular medium provides evidence for anabolic flux saturation. Stated otherwise, the amount of intermediates necessary for growth are far in excess of needs compared with the possible maximal rate of utilization of such intermediates. Under these conditions, catabolism uncouples from anabolism, and catabolic or anabolic products start to accumulate either in the intra- or extra-cellular milieu (kinetic limitation).

In the thermodynamically limited regime (i.e. at low D), the biomass appeared to be more sensitive to N-limitation (i.e. an anabolic limitation) than at high D where the limitation was mainly kinetic. Figure 3.8 shows the carbon fate observed in the respiratory regime or in the respiro-fermentative mode of glucose breakdown either in ammonia- or glutamate-fed cultures for which the carbon recovery was complete (Aon and Cortassa, 2001). Most carbon was recovered as cell biomass under C-limitation in the respiratory regime either with glucose (54%) or glucose plus glutamate (63%) (Fig. 3.8(A), (E)). On the contrary, carbon mostly evolved as CO_2 under N-limitation; this trend was more drastic in the presence of mixed carbon substrates (glucose plus glutamate) (Fig. 3.8, compare pies (C) and (G)). Under N-limited conditions, acetate and glycerol were excreted along with some ethanol, although q_{EtOH} was low enough (< 1.0 mmol h^{-1} g^{-1} dw) to be considered as a respiratory regime (Aon and Cortassa, 2001). In the presence of ammonia, during the respiro-fermentative regime, (i.e. high D), large differences in carbon distribution were observed between C- and N-limited cultures, especially with glucose as the C-source. In fact, cell biomass decreased from 34% to 16 % under N-limitation with respect to C-limitation with a two-fold increase in ethanol recovery, i.e. from 27% to 52% (Fig. 3.8(B), (D)). Concomitantly, carbon redistributed toward acetate and glycerol (3%), and CO_2 decreased from 39% to 29%. Interestingly, the fate of carbon was similar under C- or N-limitation despite the occurrence of a mixed-substrate utilization (glucose plus glutamate) (Fig. 3.8(F), (H)). Apparently, glutamate was able to provide precursors for biomass synthesis allowing more glucose to be directed toward ethanolic fermentation. However, similar carbon distributions were obtained from very different patterns of metabolic flux (Aon and Cortassa, 2001). A quite different picture was observed in N-limited cultures, both during purely respiratory and respiro-fermentative modes (Aon and Cortassa, 2001). From the total glucose consumption flux, only 26% was directed to biomass synthesis at D_c, whereas at high D this percentage decreased to only 15% of the total flux. Due to the excess carbon present in N-limited cultures, the contribution of glutamate to biomass precursors was less important and limited to

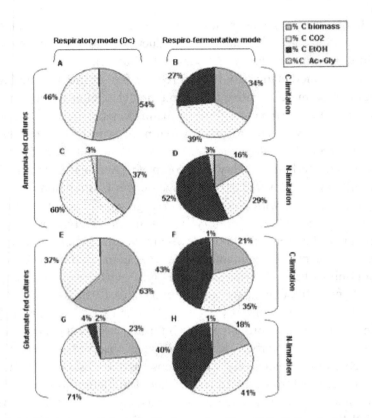

Figure 3.8   Steady-state carbon distribution under C- or N-limitation in the respiratory or respiro-fermentative regimes of glucose breakdown. The pies indicate the fate of carbon from glucose (panels (A)–(D)) or from glucose plus glutamate (panels (E)–(H)) at the critical dilution rate, D_c (panels A, C, E, G, respiratory metabolism), and at a high dilution rate, D (0.34–0.36 h^{-1}) (panels B, D, F, H, respiro-fermentative metabolism) from chemostat cultures run under C-limitation (A, B, E, F) or under N-limited conditions (C, D, G, H). In panels (A)–(D), the N-source was ammonia, while glutamate was the N-source in panels (E)–(H). When the percentage carbon recovered as a given compound was 1% or over, the value is indicated near its "portion" in the pie.
Source: Reproduced from Aon and Cortassa (2001) ©copyright 2001 Academic Press, with permission.

the provision of carbon through α ketoglutarate- and oxalacetate-derived C compounds (Aon and Cortassa, 2001).

A threshold value of glucose consumption rate, q_{Glc}, was determined at the critical dilution rate, D_c, just before the onset of the respiro-fermentative regime (Aon and Cortassa, 2001). In fact, the glucose consumption rate at D_c was the same, independent of the nature of the nutrient limitation, i.e. carbon or nitrogen, and only slightly dependent upon the quality of the N-source. The latter may be indicating a threshold for glucose consumption at which yeast cells are no longer able to process glucose solely through the oxidative pathway. It also denotes the attainment of a biological limit upon which a redirection of the catabolic carbon flux through ethanolic fermentation is triggered.

The interaction between carbon and nitrogen metabolism determine the level of anabolic flux since both C- and N-sources supply intermediates for biosynthesis. The role of the nitrogen metabolism in the triggering of ethanol production would be realized through setting the anabolic flux and in turn the biomass level in the chemostats. These biomass levels determine when the threshold glucose consumption rate is achieved after which ethanolic fermentation is triggered.

Catabolite repression and cell cycle regulation in yeast

Catabolite repression has to be considered within the context of the cell division cycle: this is well illustrated in experiments with yeasts. The eukaryotic cell cycle comprises two main stages, mitosis (M) and DNA synthesis (S), separated by two gap periods, G1 and G2 (Lloyd, 1998; Lloyd *et al.*, 1982). Genetic and molecular biological approaches have contributed a great deal to the understanding of cell cycle regulation (Hartwell, 1991; Reed, 1992; Surana *et al.*, 1991). Particularly important has been the discovery that the *cdc28* gene product of *S. cerevisiae*, the *cdc2* gene product of *Schizosaccharomyces pombe*, and the maturating promoting activity of *Xenopus* were all related serine-threonine protein kinases (Hartwell, 1991). Furthermore, it has also been shown after transfer genetic experiments that the kinases of *Schizo. pombe*, *S. cerevisiae*, and humans are functionally homologous (Forsburg and Nurse, 1991). Therefore, all the evidence available suggests that the mechanisms of cell cycle regulation are likely to be shared even in distantly related species.

As shown in Fig. 3.9, further increases in growth rate trigger an exponential shortening of the G1 phase in yeast grown under batch (Fig. 3.9(A)) or chemostat (Fig. 3.9(B)) culture conditions. Thus there is a linear relationship

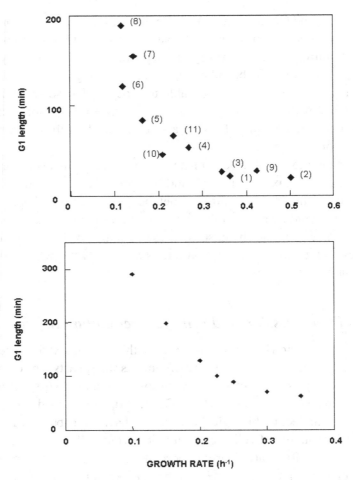

Figure 3.9 G1 duration as a function of the growth rate shown by the WT CEN.PK122 and catabolite (de)repression mutants *snf1*, *snf4* and *mig1* in batch cultures (A) and aerobic, glucose-limited, chemostat cultures (B). (A) The values of G1 lengths presented correspond to logarithmically growing batch cultures of the strains on 4% glucose, 3% ethanol or 4% glycerol as described in Aon and Cortassa (1999). The numbers correspond to: *mig1*, 4% glucose, synchronized cultures with mating pheromone (1); WT, 4% glucose, synchronized cultures with mating pheromone (2); WT, 4% glucose (3); *mig* 1, 4% glucose (4); *mig1*, 3% ethanol (5); WT, 3% ethanol (6); *mig1*, 4% glycerol (7); WT, 4% glycerol (8); *snf4*, 4% glucose, synchronized cultures with mating pheromone (9); *snf4*, 4% glucose (10); *snf1*, 4% glucose (11). (B) represents the data of G1 length obtained from aerobic, glucose-limited chemostat cultures run under the conditions described in Cortassa and Aon (1998) and Aon and Cortassa (1998). After attaining the steady state, judged through the constancy in

between the temporal duration of G1 and the doubling time. Interestingly, the specific rate of production of homologous proteins also depends on the growth rate (Giuseppin *et al.*, 1993; Hensing, 1995; Park *et al.*, 1994; Park *et al.*, 1995; Rouwenhorst *et al.*, 1988). Under the same conditions, the combined length of the cell cycle periods S+G2+M only slightly declines with the growth rate, especially at low growth rates. The sort of limitation to which the cells are subjected under continuous culture appears also to influence the cell cycle distribution of the population. For instance, the response of the cells to nitrogen limitation in continuous cultures was to stay longer in G1 (Aon and Cortassa, 1995).

Carbon catabolite repression regulates several genes that code for enzymes of the Embden–Meyerhoff, TCA cycle, gluconeogenesis, and OxPhos pathways, the expression of which is controlled by several *cis-* or *trans-*acting regulatory genes (see Fig. 4.10) (Aon and Cortassa, 1995; Entian and Zimmermann, 1982; Gancedo, 1992). Some experimental evidence showed that invertase, respiratory enzymes, and FBPase, are catabolite repressed in cdc mutants of *S. cerevisiae* (Monaco *et al.*, 1995). Furthermore, glucose repression appears to be related with the carbon and energetic uncoupling exhibited by cdc mutants at the restrictive temperature, concomitantly with the arrest of cell proliferation, increased ethanolic fermentation, and decreased mitochondrial biogenesis (Aon and Cortassa, 1995; Aon *et al.*, 1995; Genta *et al.*, 1995; Monaco *et al.*, 1995). Thus, cell cycle-related gene expression seems to be affected by catabolite repression.

When the same problem was approached from catabolite repression-related genes, similar results were observed. We focused our studies on the effect of catabolite repression-related genes on yeast cell cycle, through disruption of *SNF1, SNF4,* and *MIG1.* The two former genes code for

←—————————————————————————————

Figure 3.9. (*Continued*)   biomass concentration in the reactor vessel and in the concentration of O_2 and CO_2 in the exhaust gas, samples were taken for staining with propidium iodide and subsequent flow cytometry analysis (see Aon and Cortassa, 1998). The growth rate shown in panel (B) is fixed by the experimenter through the dilution rate, D, and not by the strain or the carbon source as shown in panel (A).
Source: Reproduced from Aon and Cortassa (1999) ©copyright Springer-Verlag, with permission.

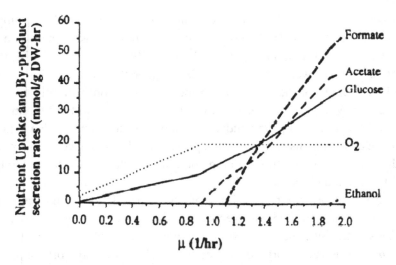

Figure 3.10 Optimal aerobic growth and secretion of by-products predicted by a flux balance model for *E. coli* during glucose-limited growth.
Source: Reprinted with permission from *Nature*. Varma and Palsson (1994) ©copyright (1994) Macmillan Magazines Limited.

proteins involved in catabolite derepression and the product of the latter has been shown to be involved in glucose repression (see Fig. 4.10) (Celenza and Carlson, 1986; Entian and Zimmermann, 1982; Gancedo, 1992; Schuller and Entian, 1987; Zimmermann *et al.*, 1977). It was shown that the onset of fermentative metabolism as well as cell cycle behavior either at cell population or molecular levels in aerobic glucose-limited chemostat cultures are dependent upon properties linked to the genes studied, known to be involved in the regulation of catabolite (de)repression (see Figs. 4.11 and 4.12) (Aon and Cortassa, 1998). The *snf4*, and to a lesser extent, the *snf1* mutants exhibited not only effects at the level of the G1 length but also at the differential cell cycle length exhibited by parent and daughter cells in chemostat cultures (Aon and Cortassa, 1998; Aon and Cortassa, 1999). In batch cultures, *snf1* and *snf4* disruptants accumulated to an extent of 70% or 80% on ethanol, whereas cells arrested randomly when transferred to glycerol, i.e. the percentage of cells in G1 did not evolve after 24 h of exposure to glycerol (Aon and Cortassa, 1998). The levels of expression of a *CDC28-lacZ* fusion gene in *snf1* and *snf4* mutants were 2- to 3-fold lower than in the wild

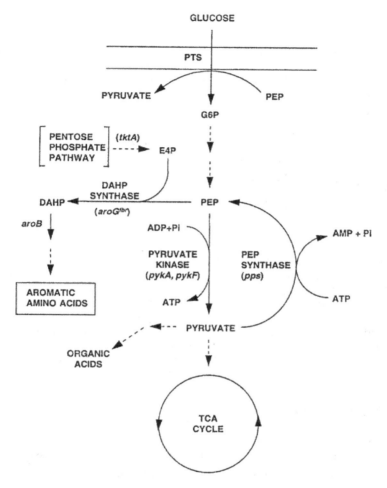

Figure 3.11   Metabolic pathways related to the formation and consumption of PEP, and the relationship of PEP to aromatic amino acid biosynthesis. Dashed lines represent multiple enzymatic steps.
Source: Reprinted with permission from Nature Publishing Group. Gosset *et al.* (1996) ©copyright 1996.

type, and consistently lower on ethanol or glycerol. These mutants are unable to grow on glycerol or ethanol because they do not express gluconeogenic enzymes (Cortassa and Aon, 1998; Entian and Zimmermann, 1982).

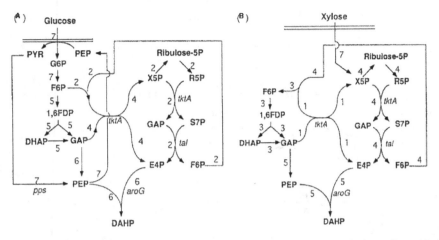

Figure 3.12 Optimal pathways and flux distribution involved in 3-deoxy-D-arabino-heptulosonate-7-phosphate (DHAP) synthesis from glucose (A) or xylose (B). The pathways that convert sugars to aromatic metabolites in *E. coli* has been examined by Liao *et al.* (1996). The first step after commitment to the aromatic pathway is the condensation between PEP and erythrose 4-phosphate (E4P) to form 3-deoxy-D-arabino-heptulosonate-7-phosphate (DHAP), that is catalyzed by DHAP synthase. To produce aromatic metabolites from glucose, carbon flow has to be effectively channelled through DHAP synthase. Typically, the carbon yield of aromatic metabolites from glucose is less than 30%.

Source: Liao *et al.* (1996) ©copyright 1996 John Wiley & Sons, Inc. Reprinted by permission of Wiley-Liss, Inc., a subsidiary of John Wiley & Sons, Inc.

Escherichia coli

This bacterium was selected for exploiting rich medium for rapid growth as *E. coli* can sustain very high maximal rates of gene expression of the components of the machinery for protein and RNA synthesis. (Jensen and Pedersen, 1990). In this growth mode, biosynthetic mechanisms are highly dependent on: (i) the accumulation of large pools of activated precursors (e.g. charged tRNAs complexed with GTP and elongation factor (EF)-Tu), and (ii) free catalytic components (ribosomes). Upon achievement of maximal rates of biosynthesis, *E. coli* becomes depleted of precursors for macromolecules synthesis, as the cell's ability to use the activated precursors and catalytic components exceeds the capacity of intermediary metabolism to provide these precursors (Jensen and Pedersen, 1990). By analogy with the growth condition in a chemostat, the bacterium is intrinsically able to grow faster than it is allowed to by the peristaltic pump feeding the vessel. This is a similar case as the one described above for yeast growing in chemostat culture in the respiratory regime (i.e. low growth rates). We stated that under those conditions the anabolic flux is thermodynamically controlled (in the sense described above, i.e. the Gibbs free energy contributed by the substrate is limiting growth), and are not saturated with respect to the key intermediary precursors of the biomass.

Jensen and Pedersen (1990) pointed out that the transcription and translation steps are the ones involved in the control of growth rate and macromolecular composition. Nevertheless, this view does not hold for very high growth rates in which the situation becomes the inverse of the one described. This will be the case for yeast growing at a high growth rate in chemostat culture where the pools of precursors are saturated, and the system is kinetically controlled. Apparently, the same happens for *E. coli* grown in glucose-limited chemostat cultures, i.e. by-product secretion occurs at a growth rate of $0.9\,h^{-1}$, at a maximal oxygen uptake rate of $20\,mmol\,O_2\,g^{-1}$ dry weight h^{-1} (Varma *et al.*, 1993) (Fig. 3.10). Acetate is the first by-product secreted as oxygen becomes limiting, followed by formate, and then by ethanol at even higher growth rates (Varma *et al.*, 1993). *E. coli* excretes 10%–30% of carbon flux from glucose as acetate under aerobiosis (Holms, 1986).

Flux redirection may also arise when either phosphorylation or redox potentials become unbalanced, e.g. due to restriction of electron acceptors in aerobic bacteria (Verdoni *et al.*, 1990; Verdoni *et al.*, 1992).

Metabolic Design of Cells as Catalysts

Increasing carbon flow to aromatic biosynthesis in Escherichia coli

Based on the stoichiometric models of metabolism that constitute the basis of MFA, the first step is to identify possible pathways leading to the production of the metabolite(s) of interest. The main aim of this analysis is then to identify all stoichiometrically plausible pathways, determine flux distributions, and calculate theoretical yields.

Figure 3.11 shows the central metabolic pathways related to the formation and consumption of PEP, and the pathways leading to aromatic amino acid biosynthesis. (Gosset *et al.*, 1996). As can be seen in Fig. 3.11, PEP is used both in the aromatic biosynthesis and in the transport of glucose via a phosphotransferase system (PTS). To increase the intracellular levels of PEP may therefore help to divert the carbon flux to aromatic amino acid synthesis. After PEP is converted to pyruvate during glucose transport, pyruvate is not recycled to PEP under glycolytic conditions. Pyruvate recycling to PEP, or non-PTS sugars (e.g. xylose), may be used to avoid wastage of pyruvate (Liao *et al.*, 1996). Another possibility for increasing the intracellular concentration of PEP is to inactivate one or both genes coding for pyruvate kinase in *E. coli* (*pykA, pykF*) (Gosset *et al.*, 1996). Finally, overexpression of transketolase coded by the *tktA* gene, in order to augment the availability of E4P, has been also tested.

The strain choice is an important step in the design of a strategy to obtain recombinants. Since the condensation of E4P and PEP to form DAHP is irreversible, and since DAHP has no other known function in the cell, accumulation of DAHP is a good indicator of carbon commitment to aromatic biosynthesis. An *aroB* mutant of *E. coli* that cannot metabolize DAHP further and excretes this metabolite to the medium has been used (Gosset *et al.*, 1996; Liao *et al.*, 1996).

Even wild-type *E. coli* strains with a functional *aroB* gene and modified with a plasmid which overproduces DAHP could excrete the precursor into the medium (Gosset *et al.*, 1996).

The recycling of pyruvate to PEP can be achieved by overexpression of PEP synthase (Pps) in the presence of glucose. The optimal flux distributions for the strategies in which pyruvate is recycled or non-PTS sugars were used, are shown in Fig. 3.12. These approaches allow 100% theoretical carbon yield, with 86% theoretical yield of DAHP from hexose, or 71% theoretical molar yield when non-PTS sugars are used, even without recycling pyruvate to PEP (Liao *et al.*, 1996).

In Table 3.6 it can be seen that the overexpression of *tktA* produces a significant increase in the flux directed to DAHP synthesis (Gosset *et al.*, 1996). Within the framework of MCA, it may be deduced that transketolase represents a rate-controlling step of the carbon flux committed to aromatic amino acid synthesis. The latter increase was verified both with control and PTS$^-$ glucose$^+$ strains (Flores *et al.*, 1996; Gosset *et al.*, 1996). Some improvement could also be obtained with Pps overexpression. The highest yield of DAHP was achieved with strains in a PTS-background, in which

Table 3.6 Increasing carbon flow to aromatic biosynthesis in *Escherichia coli*. DAHP production and doubling times

Strain		Relevant property	mmol DAH(P) g⁻¹ dry cell weight		Doubling time (h)	
			Mean[a]	s.d.[b]	Mean	s.d.
1	PB 103(pRW300)	Control	0.37	0.08	1.91	0.07
2	PB 103A(pRW300)	PykA⁻	0.54	0.16	1.85	0.11
3	PB103F(pP\W300)	pykF⁻	0.44	0.07	1.83	0.12
4	PB 103AF(pRW300)	pykA⁻ pykF⁻	1.24	0.08	1.91	0.11
5	PB103(pRW300, pCLtkt)	tktA⁺⁺	1.66	0.26	1.78	0.24
6	PB103A(pRW300, pCLtkt)	pykA⁻ tktA⁺⁺	1.81	0.20	1.75	0.21
7	PB103F(pRW300, pCLtkt)	pykF⁻ tktA⁺⁺	1.63	0.05	1.76	0.22
8	PB 103AF(pRW300, pCLtkt)	pykA⁻ pykF⁻ tktA⁺⁺	1.94	0.12	1.85	0.07
9	PB 103(pRW5)	Control	0.77	0.07	2.07	0.37
10	PB 103(pRW5tkt)	tktA⁺⁺	1.65	0.23	2.06	0.46
11	PB103(pRW5, pPS341)	pps⁺⁺	1.37	0.30	2.05	0.40
12	PB 103(pRW5tkt, pPS341)	pps⁺⁺ tktA⁺⁺	2.93	0.63	2.15	0.59
13	NF9(pRW300)	PTS⁻	0.58	0.24	1.88	0.12
14	NF9A(pRW300)	PTS⁻ pykA⁻	1.61	0.68	2.19	0.52
15	NF9F(pRW300)	PTS⁻ pykF⁻	0.26	0.12	1.95	0.33
16	NF9AF(pRW300)	PTS⁻ pykA⁻ pykF⁻	3.39	0.68	2.26	0.35
17	NF9(pRW300, pCLtkt)	PTS⁻ tktA⁺⁺	2.15	0.94	2.00	0.14
18	NF9A(pRW300, pCLtkt)	PTS⁻ pykA⁻ tktA⁺⁺	3.29	0.59	1.85	0.26
19	NF9F(pRW300. pCLtkt)	PTS⁻ pykF⁻ tktA⁺⁺	1.18	0.60	1.87	0.13
20	NF9AF(pRW300, pCLtkt)	PTS⁻ pykA⁻ pykF⁻ tktA⁺⁺	7.37	0.37	3.24	0.37

[a]Results shown are averages of three independent experiments. Within each experiment, strains were tested in duplicate or triplicate.
[b]Standard deviation.
Source: Reprinted with permission from Nature Publishing Group. Gosset *et al.* (1996) ©copyright 1996.

pykA and *pykF* were deleted, and *tktA* overexpressed. This latter result is another illustration of a shared control of the flux by several participating steps in aromatic amino acid synthesis.

The significance of each gene in the production of aromatics was also evaluated by Liao *et al.* (1996) in terms of the DAHP yield from the carbon source, either glucose or xylose. The host, an *aroB* strain that cannot metabolize DAHP further and excretes this metabolite to the medium, was found to produce very little DAHP without the overexpression of *aroG*. With *aroG* overexpression alone, this strain gave about 60% molar yield of DAHP from glucose (Liao *et al.*, 1996). This yield was greater than the theoretical one (43%), without recycling pyruvate to PEP.

Another pathway had to be invoked for pyruvate recycling since the yield obtained was higher than the expected one. The pyruvate recycling through the basal level of Pps could not explain the yield of 60% obtained, because knocking out the chromosomal *Pps* gene showed no effect on the DAHP yield. It was found that pyruvate recycling can be also mediated through the glyoxylate shunt and the Pck reaction; the optimal yield for this pathway being 64% (Liao *et al.*, 1996). When both *Tkt* and *Pps* were overexpressed along with *AroG*, the molar yield was as high as 94%.

Multipurpose engineered microbes

A multipurpose ME strategy of introducing several non-native pathways in order to design a microbe (*E. coli*) as a catalyst has been pursued (Steen *et al.*, 2010). Main aims in the engineering strategy were the conversion of nonfood raw materials (hemicelluloses) into fatty esters (biodiesel), fatty alcohols, and wax esters from simple sugars. These three output materials from the engineered *E. coli* were obtained from genetic manipulation of endogenous or introduction of foreign pathways (Fig. 3.13). Moreover, an ethanol pathway and the capacity of expressing and secreting xylanases for degrading hemicellulose were also engineered into *E. coli*.

Fatty acid methyl and ethyl esters (FAMEs and FAEEs, respectively) derived from the chemical transesterification of plant and animal oils constitute biodiesel, the primary renewable alternative to diesel. Thus, the authors aimed at increasing the flux through the *E. coli* fatty acid pathway in order to improve production of free fatty acids (FAs) and acyl-CoAs

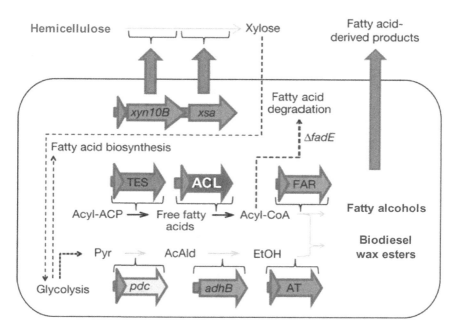

Figure 3.13 Various products were produced from non-native pathways (*xyn10B, xsa, ACL, AT*) including biodiesel, alcohols, and wax esters. Alcohols were produced directly from fatty acyl-CoAs by overexpressing fatty acyl-CoA reductases (FAR); the esters were produced by expressing an acyltransferase (AT) in conjunction with an alcohol-forming pathway; biodiesel was produced by introduction of an ethanol pathway (*pdc and adhB*) and wax esters were produced from the fatty alcohol pathway (FAR). Finally, expressing and secreting xylanases (*xyn10B and xsa*) allowed for the utilization of hemicellulose.
Source: Modified from Steen EJ, Kang Y, Bokinsky G, Hu Z, Schirmer A, McClure A, DelCardayre SB, Keasling JD. (2010) Microbial production of fatty-acid-derived fuels and chemicals from plant biomass. *Nature* **463**, 559–562.

by eliminating β-oxidation, and by overexpressing thioesterases (TES) and acyl-CoA ligases (ACL). Through pursuing FAs overproduction and deregulation of their synthesis, the authors diverted the pathway to fatty acyl-CoA, a key precursor of esters, alcohols, and other products. By introducing *Z. mobilis* genes for pyruvate decarboxylase (*pdc*) and alcohol dehydrogenase (*adhB*) the authors bypassed the need of feeding ethanol for producing FAs. The engineered *E. coli* produced FAEEs at 9.4% of the theoretical yield (674 mg l^{-1}).

Chapter 4

Engineering of Process Performance

Biochemical Rationale of Growth and Product Formation

In analyzing or modeling the behavior of the concentration of an intermediary metabolite, the first step is to evaluate its rate of change. This may be estimated as the sum of all fluxes producing the metabolite, $V_{production}$, minus the sum of all fluxes consuming it ($V_{consumption}$) or transporting it into and out of a given compartment ($V_{transport}$); each one of the fluxes being multiplied by the corresponding stoichiometric coefficient. The equation representing the rate of change of a metabolite concentration has the following general form:

$$\frac{d\text{M}}{dt} = +V_{production} \pm V_{transport} - V_{consumption}. \tag{4.1}$$

Metabolic flux quantifies the rate of conversion of precursors to products. Two techniques are used for flux determination: mass isotopomer analysis and extracellular metabolite balance models (Fiehn *et al.*, 2000; Raamsdonk *et al.*, 2001; Stephanopoulos, 2000; Yarmush and Berthiaume, 1997). Gas or liquid chromatography combined with increasingly powerful mass spectrometric techniques GC/MS or LC/MS together with application of nuclear magnetic resonance (NMR)-based methods enable quantitative assessment of metabolite pool sizes and *in vivo* metabolic fluxes (Doebbe *et al.*, 2010; Fiehn *et al.*, 2000, Klapa *et al.*, 2003a; Klapa *et al.*, 2003b; Matthew *et al.*, 2009).

Fluxes through anabolic pathways for precursor utilization for growth (formation of biomass) may be calculated by multiplication of stoichiometries by growth rate. In a steady-state continuous culture, growth rate is a constant equal to the dilution rate (see Chaps. 3 and 6). In batch cultures, it is directly the rate of increase in cellular dry weight if a condition of balanced growth can be demonstrated (Cortassa *et al.*, 1995). The demand of precursors for biomass synthesis is calculated from the biomass composition determined under each growth condition. This is the basic concept underlying metabolic fluxes analysis (MFA) estimation of anabolic fluxes, which then requires knowledge of metabolic pathways and their stoichiometry, biomass composition, and the growth rate.

When an insufficient number of fluxes can be measured experimentally, MFA may be performed on the basis of optimization principles (see below and Chap. 5). In Cortassa *et al.* (1995), the optimization principle based on a minimal energy dissipation was applied; this assumes that the function of glycolysis, the tricarboxylic acid (TCA) cycle and the respiratory chain is to generate enough ATP, and of the pentose phosphate (PP) pathway to produce adequate amounts of reducing equivalents to fulfill the biosynthetic demand. The addition of the carbon flux required to provide carbon intermediates (macromolecular precursors) to the catabolic flux (TCA cycle and PP pathways) gives the minimal flux of substrate consumption per gram of biomass dry weight. The reciprocal of this minimal flux multiplied by the growth rate renders the maximal yield of biomass on carbon substrate (Table 4.1). Following a similar procedure, the minimal catabolic flux of ATP required to accomplish the biosynthetic energy demand provides the flux of ATP formation. The maximal yield on ATP is given by the quotient of the growth rate to ATP flux.

Both yields, Y_{SX} and Y_{ATP}, are highly dependent on the P:O ratio (i.e. the number of moles of ATP synthesized by the mitochondrial ATP synthase per mole of electron pairs transported by the respiratory chain). This ratio gives a measure of the efficiency of oxidative phosphorylation (OxPhos) (see Chap. 2 section: *Catabolic fluxes*). As the criterion for calculating the catabolic flux is to minimize the ATP flux, the efficiency of OxPhos will have a large influence on such a calculation. Additionally, further oxidation in the respiratory chain of the redox potential generated during synthesis of precursors of macromolecules will also contribute to the ATP generation thereby influencing maximal values of Y_{SX} and Y_{ATP}.

As an example, Table 4.1 shows both yields, Y_{SX} and Y_{ATP}, as a function of different P:O ratios. If as a function of a given P:O ratio, the ATP flux gave

Table 4.1 Metabolic fluxes and the equivalence between biochemical stoichiometries and physiological parameters for yeast cells growing on different carbon sources in minimal medium

| Carbon source | Growth rate[a] (1/h) | P/O ratio | Fluxes[b] | | | | Y_{ATP}^{max} [c] | theo Y_S [d] | exp Y_S [e] | Yield index[f] |
			PP pathway	oxidative catabolic	anabolic	total carbon				
Glucose	0.307 ± 0.005	g	0.049	0.90	2.2	3.2	24	95	28 ± 3	0.30
		1	0.10	0.59		2.9	36	103		0.27
		2	0.17	0.18		2.6	76	117		0.24
Glycerol	0.18 ± 0.02	g	0.0067	1.2	2.5	3.7	19	47	15 ± 2	0.33
		1	0.11	0.57		3.2	39	55		0.28
Pyruvate	0.15 ± 0.02	g	0.00	2.2	1.2	4.5	13	33	10 ± 2	0.32
		1	0.00	2.1		4.4	14	33		0.31
		2	0.12	1.3		3.7	14	40		0.26
		3	0.20	1.85		3.3	15	44		0.23
Lactate	0.156 ± 0.004	g	0.081	1.9	2.5	4.5	13	34	31 ± 4	0.92
		1	0.099	1.8		4.5	14	35		0.90
		2	0.23	1.1		3.8	15	41		0.77
		3	0.29	1.68		3.5	15	45		0.71
Ethanol	0.205 ± 0.002	g	0.00	5.7	5.2	11	8.9	19	31 ± 2	>1
		1	0.00	3.6		8.8	14	23		>1
		2	0.00	1.1		6.4	26	32		0.84

(*Continued*)

Table 4.1   (*Continued*)

| Carbon source | Growth rate[a] (1/h) | P/O ratio | Fluxes[b] | | | | Y_{ATP}^{max}[c] | theo Y_S^d | exp Y_S^e | Yield index[f] |
			PP pathway	oxidative catabolic	anabolic	total carbon				
Acetate	0.16 ± 0.01	g	0.00	6.3	4.2	10	8.7	16	17 ± 3	>1
		1	0.00	5.5		9.6	10	17		>1
		2	0.00	3.1		7.3	11	23		0.77
		3	0.00	1.6		5.8	13	28		0.62

[a]Experimental growth rate (μ) $\pm$ SEM used for calculation of the flux through the PP pathway, oxidative catabolism and the total (catabolic plus anabolic) flux of substrate. It is equivalent to ln 2/(doubling time).

[b]Expressed as mmol carbon substrate $h^{-1} 1 g^{-1}$ dw.

[c]Y_{ATP}^{max} is the theoretical yield calculated from the known requirements of ATP for synthesis, polymerization and transport (see *Materials and Methods*) expressed as g yeast dry weight per mol ATP.

[d]Molar growth yield defined as the amount of biomass (g yeast dry wt.) synthesized from 1 mol of carbon substrate.

[e]Experimentally determined molar growth yield for CH1211 strain of *S. cerevisiae* growing in minimal medium on different carbon substrates. It is expressed as g dw mol^{-1} carbon substrate $\pm$ SEM.

[f]Yield index defined as the ratio of the theoretical substrate flux required to sustain the experimental growth rate to the experimental flux of substrate consumption. It may also be obtained from the ratio exptl Y_S (column 10) over theor Y_S (column 9) since the latter are inversely proportional to the substrate consumption rates.

[g]Calculation performed without taking into account the oxidation of NADH produced in anabolism by the mitochondrial electron transport chain.

Source: Cortassa *et al.* (1995) ©copyright 1995 John Wiley & Sons, Inc. Reprinted by permission of Wiley-Liss, Inc., a subsidiary of John Wiley & Sons, Inc.

a negative figure, this value was omitted (e.g. P:O = 3 in glucose, P:O = 2 or 3 in glycerol) (see Cortassa *et al.*, 1995). An increase in the efficiency of OxPhos will result in a lower catabolic flux since the same amount of ATP will be generated by a comparatively lower amount of substrate. This decrease may become very significant in yeast cells growing on glycerol at P:O larger than 1, or glucose or ethanol for P:O = 3. If all the NADH produced in anabolic steps I and II were reoxidized to produce ATP, this would largely overwhelm the requirement of phosphorylation energy for biosynthesis. Under those conditions, no extra carbon would be required to be catabolized to fulfill anabolic demand of high-energy transfer phosphate bonds ($\sim$P) (Table 4.1). The catabolic utilization of the PP pathway in a cyclic-function mode (i.e. generation of NADPH with the concomitant oxidation of the substrate to CO_2) was altered depending on the P:O ratio considered (Table 4.1). The latter can be understood by considering that the complete oxidation of the carbon substrate involves TCA cycle function and NADPH generation because the isocitrate dehydrogenase was assumed to use NADP as cofactor (Table 4.1). The PP pathway had to be used in order to complete the demand for NADPH at P:O ratios of one or more, as for lactate or pyruvate (Table 4.1).

A General Formalism for MFA

In this section, a general formalism to perform MFA developed by Savinell and Palsson (1992) is presented. Some further developments of this formulation as well as extension to reconstruct metabolic networks will be provided in Chap. 5. The formalism for MFA is postulated in algebraic terms of matrices and vectors. The dynamic behavior of the intracellular metabolites is represented by a system of equations:

$$\frac{d\mathbf{X}}{dt} = \mathbf{S} \cdot v - \mathbf{b}, \tag{4.2}$$

where $\mathbf{X}$ being the vector of n metabolite concentrations, $\mathbf{S}$, the stoichiometry matrix of dimensions $n \times m$, $\mathbf{v}$ is the vector of m metabolic fluxes and $\mathbf{b}$ the vector of known biosynthetic fluxes. The stoichiometric matrix accounts for the way n metabolites participate in m metabolic steps considered in the network of processes under study. We will use it further for calculating the matrix of control coefficients in metabolic control analysis as well as in all stoichiometric modeling methods in Chap. 5.

Under conditions of balanced or steady-state growth, the vector of metabolite concentrations may be considered constant and thus time invariant; so Eq. (4.2) reduces to:

$$\mathbf{S} \cdot \mathbf{v} = \mathbf{b}. \tag{4.3}$$

Usually, the number of metabolic fluxes, m, is much larger than the number of metabolites, n. Among the fluxes, at least a number $(m - n)$ of them must be known (accessible to experimental determination) in order to be able calculate the remaining n fluxes. If the vector of fluxes, $\mathbf{v}$, and the stoichiometric matrix, $\mathbf{S}$, are partitioned between measured and computed fluxes:

$$\mathbf{v} = \left| \frac{\mathbf{v}_c}{\mathbf{v}_e} \right| \quad \mathbf{S} = |\mathbf{S}_c| \mathbf{S}_e|, \tag{4.4}$$

then Eq. (4.3) becomes:

$$\mathbf{S}_c \mathbf{v}_c + \mathbf{S}_e \mathbf{v}_e = \mathbf{b}, \tag{4.5}$$

and $\mathbf{v}_c$ can be calculated:

$$\mathbf{v}_c = \mathbf{S}_c^{-1}(\mathbf{b} - \mathbf{S}_e \mathbf{v}_e). \tag{4.6}$$

Therefore, at least $(m - n)$ fluxes have to be measured to be able to estimate the n remaining fluxes. Certainly, the level of accuracy in the estimated n fluxes will depend on the experimental values of $\mathbf{b}$ and $\mathbf{v}_e$ utilized for flux determinations (Savinell and Palsson, 1992). When the dimension of the vector $\mathbf{v}_e$ is lower than $(m - n)$, the system is said to be "underdetermined," and a set of vectors are solutions of the system. In such a case, linear optimization techniques allow restriction of the set of solution vectors (Savinell and Palsson, 1992).

MFA Applications

Most MFA methods are based on stoichiometric models of metabolism. They were developed in the 1990s and have been applied to growth and product formation of microbial and mammalian cells. Although their predictive capabilities are more limited than kinetic modeling techniques, they provide useful information about product yield and flux distribution of large metabolic networks.

MFA applied to prokaryotic organisms

The first stoichiometric models of metabolic networks were proposed by Holms and applied to the growth of *E. coli* (Holms, 1986). Later, Stephanopoulos and Vallino (1991) presented the concept of network "rigidity" to explain why the yields calculated from a stoichiometric model of *Corynebacterium* metabolism, to produce the amino acid lysine, were much higher than those attainable under experimental conditions.

Savinell and Palsson (1992) presented a general formalism for the calculation of internal metabolic fluxes based on algebraic operations using mass balance equations (see above). These authors named their approach, "metabolic flux balancing" (Varma and Palsson, 1994). Essentially, the method includes linear optimization, both for flux calculation and the sensitivity to experimental error (Savinell and Palsson, 1992). Linear optimization requires an objective function, e.g. the growth rate or the growth yield on a certain substrate, with respect to which the microbial system should find an "optimal" solution (Edwards *et al.*, 2001). To accomplish this task two variables were introduced, namely *shadow prices* and *reduced costs*, that in the context of linear optimization are defined as a function of the sensitivity of an objective function to a constraint (e.g. minimize ATP production or maximize biomass production). Such a method has been applied to bacteria, particularly to *E. coli*, and to hybridoma cell metabolism (Savinell and Palsson, 1992). Also, it has been applied to the analysis of *Arabidopsis thaliana* metabolic modes (Fiehn *et al.*, 2000) and in combination with kinetic modeling to the study of the phenotype of *Petunia hybrida* regarding the pathways involved in benzenoid compounds (Colon *et al.*, 2010).

The first *in silico* reconstruction of a metabolic network was performed by Edwards *et al.* (2001) based on a cellular inventory of *E. coli* metabolic gene products (Reed *et al.*, 2003). The optimization criterion for maximizing the growth rate was found to be consistent with experimental data from bacteria growing on acetate. Figure 4.1 shows the modeling results as well as the experimental points. The latter were found to correspond to the line

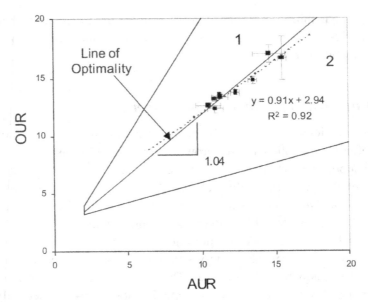

Figure 4.1.   *In silico* predictions of growth and substrate consumption rates and comparisons to experimental data. The acetate uptake rate (AUR) versus the oxygen uptake rate (OUR) (both rates in mmol g^{-1}dw h^{-1}) phase plane analysis (phenotype phase plane). All data points lie close to the line of optimality (separating regions 1 and 2) which defines the line of the highest growth yields on both substrates. The errors bars in each data point represent a single standard deviation. Regions 1 and 2 represent nonoptimal metabolic phenotypes.
Source: Reprinted with permission from Edwards *et al.* (2001).

of optimality defined by the phenotype phase plane, in this case namely the 3D plot of growth rate as function of both acetate- and oxygen-uptake rate (Edwards *et al.*, 2001). From a physiological point of view, the line of optimality corresponds to a maximal growth yield for both substrates, i.e. acetate and oxygen, under those particular culture conditions.

Metabolic reconstructions frequently calculate modes of functioning of the metabolic networks without considering the thermodynamic feasibility of a reaction or pathway participating in the scheme. Recent approaches utilizing genetic algorithms have shown that thermodynamic constraints are necessary for having a more realistic appraisal of the fluxes sustained by microorganisms (Boghigian *et al.*, 2010). Genetic algorithms allow to search for optimal solutions with respect to an objective, e.g. fluxes in

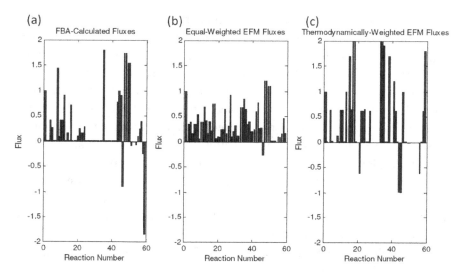

Figure 4.2. A bar plot of fluxes for the ethanol model calculated using (a) FBA linear optimization in which the objective function is the production of biomass, (b) equally weighted EMs, and (c) thermodynamically weighted EMs. In the last two cases, the network is optimized for the production of ethanol instead of biomass, either taking or not taking into account the thermodynamics of the processes involved in each case.

Source: Reproduced from Boghigian BA, Shi H, Lee K, Pfeifer BA. (2010) Utilizing elementary mode analysis, pathway thermodynamics, and a genetic algorithm for metabolic flux determination and optimal metabolic network design. *BMC Syst Biol* **4**, 49.

a network, by "evolutionary" methods (e.g. via mutation and crossover through multiple generations). A genetic algorithm was applied to *E. coli* during ethanol or lycopene production (Fig. 4.2) (Boghigian *et al.*, 2010). The method accounts for known topological properties of the metabolic pathways involved in *E. coli* growth, and synthesis and excretion of the products as well as thermodynamic constrains that indicate the energy balance in each elementary mode (EM).

Sauer *et al.* (1998) applied a metabolic network optimization approach to the production of purine nucleotide, riboflavin, or folic acid by *Bacillus subtilis*. The metabolites of interest were all derived from pentose phosphates as intermediary metabolites. Two criteria were used as optimization principles based on: (i) a minimal energy expense, and (ii) the so-called stoichiometric criteria that searches to improve the yield of biomass on carbon substrate, minimizing losses as CO_2 and allowing ATP to accumulate

(Sauer *et al.*, 1998). The first optimization criterion was dependent on the P:O ratio chosen to perform the calculations and the yield of ATP, Y_{ATP}. The Y_{ATP} parameter adopted in the work of Sauer *et al.* (1998) takes into account the energetic efficiency of the metabolic machinery and encompasses the biomass and product yields on substrate together with the maximal ATP formed per mole of substrate consumed. The parameter Y_{ATP} is useful for distinguishing between an energy-limited biological system and a limitation arisen from the stoichiometry of the pathways involved. As a matter of fact, certain anaplerotic pathways will be activated with final product yields (in moles of product per 6 carbon mole of substrate), varying around 0.6 for guanosine and 0.16 for folic acid, according to the metabolic routes utilized by the carbon substrate (irrespective of whether the uptake mechanisms involve a phosphotransferase system or not).

Extracellular metabolite production is often a consequence of intracellular redox imbalance: examples where this is so have been analyzed in many prokaryotes where emphasis was on NADH/NADPH interconversion (by transhydrogenases, malic enzymes, and isocitrate dehydrogenases) (Fuhrer and Sauer, 2009). A combined experimental — flux balance analysis was carried out with ^{13}C-detected intracellular fluxes and metabolite profiles. The conclusions questioned the validity of the assumption that the pentose phosphate pathway is exclusively functioning with NADP as cofactor, due to the lack of specificity for the cofactor in the pathway's enzymes in microorganisms (Fuhrer and Sauer, 2009).

MFA applied to lower eukaryotic organisms

A metabolic network method was validated through analysis of the growth of *S. cerevisiae* in mixed substrate media, with glucose and ethanol (Vanrolleghem *et al.*, 1996). MFA was performed at the steady state for 98 metabolites, the elemental balance, and 99 reaction rates, including both conversion and transport reactions. According to the amount of glucose in the feed, four functional regimes of the metabolic network were observed; the onset of each regime was predicted by the numerical results. At large glucose fractions, the AcCoA pool was mainly derived from glucose, via pyruvate dehydrogenase, and from ethanol, through the AcCoA synthetase. When the ethanol fraction increases, anaplerotic reactions are fed via a glyoxylate shunt. At even lower glucose fractions, ethanol-derived intermediates enter gluconeogenesis and PEPCK becomes operative. Finally, gluconeogenesis and the pentose phosphate pathway function with ethanol-derived metabolites, above a given ethanol fraction in the feed. The method was validated by a statistical treatment of experimental measurements of fluxes and *in vitro* enzyme activities (Vanrolleghem *et al.*, 1996). The main outcome of the model was the estimation of two important physiological parameters: the P:O ratio and the maintenance energy, i.e. that amount of energy dissipated not associated with growth. Operational P:O ratios ranging from 1.07 to 1.11 mol ATP/O and maintenance between 0.385 and 0.445 mol ATP/C mol biomass, were obtained. The latter value was interpreted by the authors to be independent of C-substrate supported growth.

Concerning the yeast *S. cerevisiae* growth on mixed substrates, dos Santos *et al.* (2003) have studied the flux patterns at different ratios of acetate and glucose in a chemostat culture system. Special attention was paid to the role of malic enzyme assessed through the labeling of pyruvate and PEP from ^{13}C-glucose in wild type and a malic enzyme deficient yeast strain. The MFA performed indicated that at low glucose to acetate ratio gluconeogenesis is employed simultaneously with glycolysis and at sufficiently large acetate fraction in the feed, glucose is used in the pentose phosphate pathway to generate NADPH for biosynthesis (dos Santos *et al.*, 2003).

Amphibolic pathways flux in different carbon sources

Carbon sources that share most enzymes required to transform the substrates into key intermediary metabolites under similar growth rates, bring about similar fluxes through the main amphibolic pathways. Fluxes through the central metabolic pathways of *S. cerevisiae* have been quantified by MFA (Aon and Cortassa, 1997; Cortassa *et al.*, 1995) under balanced growth conditions in the presence of fermentative (glucose) or gluconeogenic (acetate, pyruvate, lactate, ethanol, glycerol) substrates. Similar fluxes through the central pathways were exhibited by yeast in the presence of pyruvate plus lactate or ethanol plus acetate that share most metabolic pathways for their assimilation and also result in similar growth rates (Table 4.1).

As expected, widely different fluxes through specific enzymatic steps were determined when pathways for utilization of the carbon source differ.

This was the case of, e.g. acetate, lactate, or glucose. A large flux through the AcCoA synthase-catalyzed step was required by acetate as well as an extensive use of the glyoxylate cycle and of the NADP-dependent malic enzyme to supply key intermediary metabolites. On the other hand, lactate sustained large fluxes through the pyruvate carboxylase step, which was also used to a lesser extent by glucose or glycerol to replenish OAA and αKG, although not required at all for growth on acetate (Cortassa *et al.*, 1995).

Interaction between carbon and nitrogen regulatory pathways in S.cerevisiae

Gluconeogenesis from glutamate as N-source may simultaneously operate with a high glycolytic flux in glucose-limited chemostat cultures (Aon and Cortassa, 2001). During growth on fermentable carbon sources such as glucose, most gluconeogenic enzymes, e.g. phospho-enolpyruvate carboxykinase (PEPCK), fructose 1,6 bisphosphatase and malic enzyme, are repressed at the transcriptional level. Nevertheless, the latter has always been assayed in media with ammonia as N-source, or in rich nutrient agar with glutamine. In the presence of glutamate, we found that *S. cerevisiae* displayed an almost complete fermentation of the glucose consumed with biomass being synthesized from glutamate under C-limiting conditions in the respiro-fermentative growth mode of glucose-limited chemostat cultures (Figs. 4.3 and 3.9). The latter implies that the enzymes of the gluconeogenic pathway, namely PEPCK and the malic enzyme are not repressed (activity levels of 22 and 49 nmol min^{-1} mg^{-1} protein, respectively, were measured). This situation contradicts the traditional concept of catabolite repression, further suggesting the interplay between carbon and nitrogen regulatory pathways. The derepression of gluconeogenic enzymes could be perhaps elicited by the presence of N-source such as glutamate. This amino acid also provoked a much higher qO_2 under N-limitation than that displayed in ammonia-fed cultures (Aon and Cortassa, 2001).

At the molecular level, a dual regulation of NADP-GDH by C and N sources has been reported, suggesting an interaction between both routes of metabolism, although the physiological consequences of this dual control were not addressed (Coschigano *et al.*, 1991). The simultaneous operation of glycolysis and gluconeogenesis suggests that the dual regulation

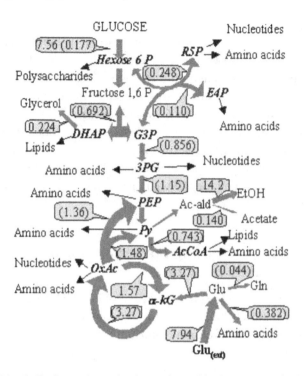

Figure 4.3. Metabolic fluxes determined or estimated in *S. cerevisiae* in the presence of glutamate as N source, during the respiro-fermentative regime of C-limited aerobic chemostat cultures. The flux values at $D = 0.36\,h^{-1}$ are shown. The fluxes experimentally determined are indicated without parenthesis in the oval, whereas estimated ones are depicted between brackets. The ovals point to the metabolic step for which the flux has been estimated or measured. The thickness of the arrows attempt to give a qualitative indication of the magnitude of the overall flux (anabolic plus catabolic) through a given reaction step, or lumped reaction steps within a metabolic pathway. Monomer precursors of macromolecules such as amino acids and nucleotides, polysaccharides, or lipids, as final products of the anabolism are emphasized on a dashed background.

Source: Reproduced from *Metabolic Engineering*, Aon, J.C. and Cortassa, ©copyright 2001 Academic Press, with permission.

of a nitrogen repressible gene such as *GDH2* may be extended to gene products expressed differentially according to carbon catabolite repression. A dual regulation at biochemical and genetic levels was proposed for glutamine synthesis that would act as a regulatory signal for glucose breakdown through the ATP/ADP ratio (Flores-Samaniego *et al.*, 1993). The box-binding HAP transcription factor has also been proposed as a new regulatory crossroad between nitrogen and carbon metabolism (Dang *et al.*, 1996). From the results of Aon and Cortassa (2001), we hypothesize that both nitrogen and carbon catabolite repression are interlinked phenomena and that many genes under the control of each of these general mechanisms are subjected to a dual control by C and N sources.

S. cerevisiae, E. coli, and *C. glutamicum* are the model organisms most used as a test bench of the numerous methodological MFA developments, that would be too long to even mention at this point. A brief account of additional tools developed for MFA and metabolic networks reconstructions will be discussed in Chap. 5.

MFA as applied to studying the performance of mammalian cells in culture

Zupke and Stephanopoulos (1995) have applied MFA to a hybridoma cell line based on material balance of the biochemical network of reactions leading to the synthesis of biomass and antibodies. The method is based on an algebraic expression of fluxes at a pseudo-steady state. The calculation was validated by ^{13}C NMR measurement of lactate production. The validation was performed through labeling experiments with Glucose-1-^{13}C and analysis of the label recovery in lactate, focusing on the label distribution among its three carbons. They found reasonable good quantitative agreement between predicted and determined values of intracellular fluxes in cells consuming glucose and glutamine, confirming the assumed biochemistry of the metabolic network.

Maximal fluxes through the main catabolic pathways in a myeloma cell line grown in chemostat cultures under several experimental conditions: (i) glutamine-limited, which is equivalent to an energy limitation (i.e. mammalian cells in culture-driven amino acids into the TCA cycle for complete combustion), (ii) glucose-limited, and (iii) oxygen-limited. The measured fluxes were confirmed by measurements of enzyme activities involved in the central metabolic pathways of carbon and nitrogen metabolism (Vriezen and van Dijken, 1998).

An alternative mass balance method based on a stoichiometric matrix was applied to the growth and antibody production of hybridoma cells in different culture media (Bonarius *et al.*, 1996). These authors introduced a heuristic algebraic method to overcome the indetermination in the set of fluxes considered. Such indetermination arises because the number of mass and energy balances is smaller than the number of fluxes.

Cybernetic models, introduced by Ramkrishna and coworkers (Varner and Ramkrishna, 1999), are based on the hypothesis that metabolic systems have evolved optimal goal-oriented strategies. This modeling approach has been applied to the growth of *S. cerevisiae* (Giuseppin and van Riel, 2000). It uses linear optimization toward several objectives through cost functions, among which the maintenance of homeostasis is of utmost importance. The main difference with flux balancing analysis is that the assumption of steady state or balanced growth condition is eliminated allowing the calculation of time-dependent phenomena. At each integration step the optimal rates have to be calculated using a constrained optimization algorithm (Giuseppin and van Riel, 2000). The behavior of a yeast culture subjected to glucose pulses has been studied and the objective of maximizing glucose uptake was found to dominate over the other objective functions during the transient (Giuseppin and van Riel, 2000).

This survey does not pretend to be exhaustive and some outstanding papers may have not been included, but its purpose is to illustrate how general are the uses of metabolic stoichiometric algebraic methods, along with their potentiality for diagnosis of limitations in an optimization program.

Bioenergetic and Physiological Studies in Batch and Continuous Cultures: Genetic or Epigenetic Redirection of Metabolic Fluxes

Introduction of heterologous metabolic pathways

The development of arabinose- and xylose-fermenting *Z. mobilis* strains are a conspicuously successful example of metabolic engineering by introduction of heterologous metabolic pathways (Deanda *et al.*, 1996; Zhang *et al.*, 1995). *Z. mobilis* has become an important fuel ethanol-producing microorganism because of its 5% to 10% higher yield and up to five-fold higher specific productivity compared with traditional yeast fermentations. As its substrate range is restricted to glucose, sucrose, and fructose fermentation, the introduction of the ability to ferment pentose sugars (derived from lignocellulose feedstocks) through ME is a biotechnological achievement of great impact.

Additional advantages of Z. *mobilis* as an ethanol-producing microorganism are its high ethanol yield and tolerance (up to 97% of ethanol theoretical yield and ethanol concentrations of up to 12% w/v from glucose) along with considerable resistance to the inhibitors found in lignocellulose hydrolysates. The "generally regarded as safe" (GRAS) character (shared with yeast) allows its use as an animal feed.

Rapid and efficient ethanol production by Z. *mobilis* has been attributed to glucose fermentation through the Entner–Doudoroff pathway which produces only 1 mol of ATP per mole of glucose and therefore reduced biomass formation. A facilitated diffusion sugar transport system coupled with its highly expressed pyruvate decarboxylase and alcohol dehydrogenase genes have also been implied (Zhang *et al.*, 1995).

Xylose-fermenting strains of Z. *mobilis* were developed by introducing genes that encode the xylose isomerase, xylulokinase, transaldolase, and transketolase activities (Zhang *et al.*, 1995). Two operons comprising the four xylose assimilation and pentose phosphate pathway genes were simultaneously transferred into Z. *mobilis* CP4 on a chimeric shuttle vector. Enzymatic analysis of Z. *mobilis* CP4 grown in a glucose-based medium demonstrated the presence of the four enzymes, whereas these were largely undetectable in the control strain that contained the shuttle vector alone.

The expression of these genes allowed for the completion of a functional metabolic pathway that converts xylose to central intermediates of the Entner–Doudoroff pathway, enabling *Zymomonas* to ferment xylose to ethanol (Fig. 4.4). Xylose is presumably converted to xylulose-5-phosphate and then further metabolized to glyceraldehyde-3-phosphate and fructose-6-phosphate, which effectively couples pentose metabolism to the glycolytic Entner–Doudoroff pathway. The recombinant strain obtained was capable of growth on xylose as the sole carbon source and efficient ethanol production (0.44 g per gram of xylose consumed: 86% of theoretical yield). Furthermore, in the presence of a mixture of glucose and xylose, the recombinant strain fermented both sugars to ethanol at 95% of theoretical yield within 30 h (Zhang *et al.*, 1995).

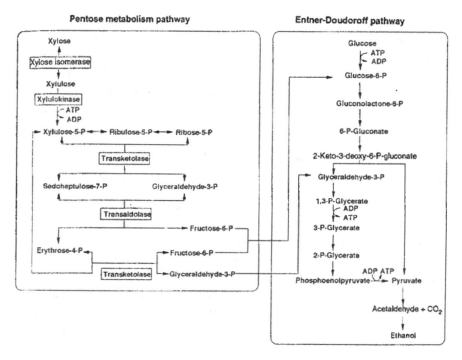

Figure 4.4.   Proposed pentose metabolism and Entner–Doudoroff pathways in engineered *Z. mobilis*. The xylose-fermenting strains of *Z. mobilis* were developed by introducing genes that encode the xylose isomerase, xylulokinase, transaldolase, and transketolase activities. The expression of these genes allowed for the completion of a functional metabolic pathway that converts xylose to central intermediates of the Entner–Doudoroff pathway enabling *Zymomonas* to ferment xylose to ethanol.

Source: Reprinted with permission from Zhang *et al.* (1995) ©copyright 1995 American Association for the Advancement of Science.

A similar strategy was followed by the same team of researchers to expand the substrate fermentation range of *Z. mobilis* to include the pentose sugar, L-arabinose (Deanda *et al.*, 1996). Five genes, encoding L-arabinose isomerase, L-ribulokinase, L-ribulose-5-phosphate-4-epimerase, transaldolase and transketolase, were isolated from *E. coli* and introduced into *Z. mobilis* under the control of constitutive promoters that permitted their expression even in the presence of glucose. The engineered strain grew on and produced ethanol from L-arabinose as a sole carbon source at 98% of the maximum theoretical ethanol yield, based on the amount of sugar consumed.

Since the plasmid containing the heterologous genes was episomal, the recombinant strain with the ability to ferment L-arabinose exhibited plasmid instability. Only 40% of the cells retained the ability to ferment arabinose after 20 generations of growth in complex media without selection pressure at 30°C, and they completely lost the ability to ferment arabinose within 7 generations at 37°C (Deanda *et al.*, 1996).

Metabolic engineering of lactic acid bacteria for optimizing essential flavor compounds production

The work of Platteeuw *et al.* (1995) shows how, through a combination of molecular biology studies and the establishment of appropriate environmental conditions, the metabolic path leading to diacetyl production from pyruvate can be optimized in *Lactococcus lactis*. Diacetyl is an essential flavor compound in dairy products such as butter, buttermilk, and cheese, and in many nondairy products, where a butter-like taste is desired. The pyruvate pathway leading to diacetyl production in *L. lactis* is shown in Fig. 4.5. From two molecules of pyruvate, diacetyl is formed by an oxidative decarboxylation of the intermediate α-acetolactate by the activity of α-acetolactate synthase. α-acetolactate is further decarboxylated to acetoin by the action of α-acetolactate decarboxylase.

Four enzyme activities are known to metabolize pyruvate under different physiological conditions (Fig. 4.5): (i) α-acetolactate synthase, which is active at high pyruvate concentrations and low pH; (ii) L-lactate dehydrogenase, with maximal activities at high sugar concentrations and high intracellular NADH levels; (iii) pyruvate-formate lyase, which is active at a relatively high pH (above 6) and under anaerobic conditions; and (iv) pyruvate dehydrogenase, which is active under aerobic conditions and low pH.

Taking into account the above metabolic features, strains of *L. lactis* that overproduce α-acetolactate synthase in a wild type MG5276, or in a strain deficient in lactate dehydrogenase, were analyzed for their fermentation pattern of metabolites at different pH and aeration conditions.

Tables 4.2 and 4.3 show the product formation and lactose consumption by both types of recombinant strain of *L. lactis*. The *L. lactis* strain without α-acetolactate synthase (als) overproduction shows a basal activity of the enzyme. When the plasmid pNZ2500 harboring the lactococcal *als* gene

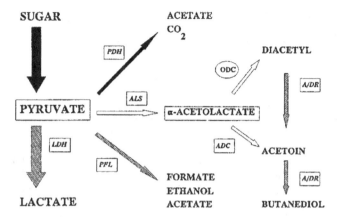

Figure 4.5. Pyruvate pathway in *L. lactis*. Black and shaded arrows indicate conversions that generate and consume NADH, respectively. Enzymatic and chemical conversions are indicated by boxes and circles, respectively. Abbreviations: ALS, α-acetolactate synthase; LDH, lactate dehydrogenase; PFL, pyruvate formate lyase; PDH, pyruvate dehydrogenase; ADC, α-acetolactate decarboxylase; A/DR, acetoin and diacetyl reductase; ODC, oxidative decarboxylation.
Source: Reprinted with permission from Platteeuw *et al.* (1995) ©copyright 1995 American Society for Microbiology.

Table 4.2 Product formation and lactose consumption by *L. lactis* MG5267 harboring pNZ2500

Fermentation condition		Product formed (mM)						Lactose utilized (mM)
Initial pH	Aeration	Lactate	Formate	Acetate	Ethanol	Acetoin	Butanediol	
6.8	−	54.6 (100)[b]	N D[c]	ND	N D	N D	N D	18.1
6.0	−	41.2 (99.7)	N D	0.1 (0.2)	N D	N D	N D	9.7
6.8	+	52.0 (70.8)	N D	2.3 (3.1)	N D	9.6 (26.1)	N D	19.2
6.0	+	42.5 (53.4)	N D	2.0 (2.6)	N D	16.1 (42.0)	N D	18.7

[a]Data represent the average of at least two experiments.
[b]Values in parentheses are percentages of pyruvate converted into the product.
[c]ND; not detected.
Source: Reprinted with permission from Platteeuw *et al.* (1995) ©copyright 1995 American Society for Microbiology.

Table 4.3   Product formation by lactate dehydrogenase-deficient strain *L. lactis* NZ2007 harboring pNZ2500

Fermentation condition		Product formed (mM)							Lactose utilized (mM)
Initial pH	Aeration	Lactate	Formate	Acetate	Ethanol	Acetoin	α-Al[b]	Butanediol	
6.8	−	3.5	29.1	0.8	24.1	5.5	0.1	22.3	16.6
		(3.1)[c]	(25.7)	(0.7)	(21.3)	(9.7)	(0.2)	(39.3)	
6	−	2.3	21.9	0.1	17.8	5.5	0.3	14.4	7
		(2.8)	(26.6)	(0.1)	(21.6)	(13.3)	(0.7)	(34.9)	
6.8	+	1.2	17.6	11.8	10.5	35.0	1.2	ND[d]	20.2
		(1.1)	(15.5)	(10.4)	(9.4)	(61.7)	(2.1)		
6	+	1.1	ND	5.9	8.7	36.6	0.6	10.5	20.2
		(1.1)		(5.3)	(7.8)	(65.7)	(1.1)	(18.9)	

[a]Data represent the average of at least two experiments.
[b]α-AL, α-acetolactate.
[c]Values in parentheses are percentages of pyruvate converted into the product.
[d]ND; not detected.
Source: Reprinted with permission from Platteeuw *et al.* (1995) ©copyright 1995 American Society for Microbiology.

(isolated and cloned under the control of the strong lactose-inducible lacA promoter) was introduced into the wild type strain, an increase of 124-fold in activity over the basal level was observed under inducing conditions in lactose-containing media (Platteeuw *et al.*, 1995).

When cultivated aerobically, *L. lactis* MG5276 harboring pNZ2500 was found to produce acetoin in addition to lactate. At an initial pH of 6.8, 26% of the pyruvate appeared to be converted to acetoin, while at an initial pH of 6.0, almost twice the acetoin production was observed (Table 4.2).

When a lactate dehydrogenase deficient strain (NZ2007) constructed from the lactose-fermenting strain MG5267 was modified with the plasmid pNZ2500 for the *als* overproduction, a range of metabolites were produced under anaerobic conditions (Table 4.3). Approximately half of the pyruvate was converted into formate and ethanol. The major part of the acetoin synthesized was further reduced to butanediol. Under aerobic conditions and an initial pH of 6.8, approximately 64% of the pyruvate was converted mainly into formate (15% of the pyruvate converted), acetate (10%), and acetoin

(62%). At a lower initial pH of 6.0, the vast majority (85%) of the pyruvate was converted into products from the α-acetolactate synthase pathway (Table 4.3). These results clearly illustrate that the availability of an electron acceptor, and the intracellular redox status are main determinants of the fermentation pattern. In agreement with the latter, the transformation of *S. marcescens* with a Vitreoscilla (bacterial) hemoglobin gene shifted the fermentation pathway toward the production of 2,3-butanediol and acetoin (Table 1.1) (Wei *et al.*, 1998). Moreover, the introduction of NADH oxidase into *L. lactis* altered the fermentation pattern from homolactic to mixed acid (Table 1.1) (Lopez de Felipe *et al.*, 1998).

The work of Platteeuw *et al.* (1995) shows that under optimal fermentation conditions and with appropriately constructed strains of *L. lactis*, up to 85% of the pyruvate (and thereby also the carbon source lactose) was converted via the α-acetolactate synthase pathway to acetoin and butanediol. It also became clear that to produce high yields of diacetyl, the conversion of α-acetolactate into acetoin by the activity of α-acetolactate decarboxylase should be prevented.

High throughput Bioenergetic and Physiological Studies of H_2 Production by Algae

A metabolic switch elicited by the limitation of sulfur

In green algae, it has been revealed that the level of sulfur, S, from the growth medium acts as a metabolic switch one that selectively and reversibly turns on/off photosynthetic O_2 production. In the presence of S, green algae do normal photosynthesis (water oxidation, O_2 evolution, and biomass accumulation) but in the absence of S and O_2, photosynthesis in *Chlamydomonas reinhardtii* slips into the H_2 production mode (Fig. 4.6).

In the alga *C. reinhardtii*, H_2 can be produced from water as a source of electrons in photosystem II (PSII) with 2:1 stoichiometric amounts of $H_2:O_2$. Initially high rates of H_2 production (2:1 stoichiometry $H_2:CO_2$) can be detected for a short time (seconds to a few minutes) upon illumination after anaerobic incubation in darkness of a *C. reinhardtii* culture in the presence of a PSII inhibitor (Melis and Happe, 2001). Lack of S from the growth medium of *C. reinhardtii* causes a specific but reversible decline in

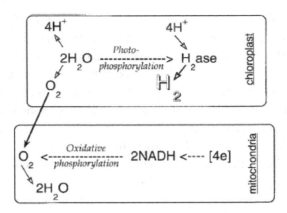

Figure 4.6. Coordinated photosynthetic and respiratory electron transport and coupled phosphorylation during H_2 production. Photosynthetic electron transport delivers electrons upon photo-oxidation of water to the hydrogenase (H_2ase), leading to photophosphorylation and H_2 production. The O_2 generated by this process serves to drive the coordinate oxidative phosphorylation during mitochondrial respiration. Electrons for respiration are derived upon endogenous substrate catabolism, which yields reductant and CO_2. Release of H_2 by the chloroplast enables the sustained operation of this coordinated photosynthesis–respiration function in green algae and permits the continuous generation of ATP by both organelles in the cell.

Source: Reproduced from Melis A, Happe T. Hydrogen production. Green algae as a source of energy. *Plant Physiol* **127**, 740–748. © 2001 The American Society for Plant Biologists.

the rate of oxygenic photosynthesis (Wykoff *et al.*, 1998) without affecting the rate of mitochondrial respiration (Melis *et al.*, 2000).

Imbalance in the photosynthesis–respiration relationship by S deprivation in sealed cultures resulted in net consumption of O_2 by the cells causing anaerobiosis in the growth medium, a condition that automatically elicited H_2 production by the algae (Melis *et al.*, 2000) (see Figs. 4.6 and 4.8). Since O_2 is a powerful inhibitor of the hydrogenase (Ghirardi *et al.*, 2000), S deprivation appears to temporally separate the reactions of O_2 and H_2 photoproduction thereby circumventing the O_2 sensitivity of the hydrogenase. This process has been called "two-stage photosynthesis and H_2 production" (Melis *et al.*, 2000). The switch elicited by S deprivation is reversible so that repeated application of presence/absence of S permits the algae to alternate between O_2 production and H_2 production (Ghirardi *et al.*, 2000), thus bypassing the incompatibility and mutually exclusive nature of the O_2- and H_2-producing reactions (Melis and Happe, 2001). Under S deprivation conditions, sealed (anaerobic) cultures of *C. reinhardtii* produce H_2 gas in the light with a sustained rate of 2.0 to 2.5 ml H_2 production L^{-1} culture h^{-1} for a 24 h to 70 h period, gradually declining thereafter (Melis and Happe, 2001).

MFA under different growth conditions and high throughput metabolomics in plant cells

Together with light and carbon, nitrogen and phosphorous are the most important inputs for biofuel production by microalgae. Under photoautotrophic conditions, the photosynthetic machinery requires solar energy to fix CO_2 into organic molecules. However, biomass production can be greatly improved if reduced forms of carbon can be taken up from the medium by the microalgae and used for biomass production during photoheterotrophic or mixotrophic growth, constituting an indirect form of solar energy harvesting.

C. reinhardtii is considered a representative algal species used as a model organism to study numerous cellular functions. Most recently *C. reinhardtii* has been utilized for genome annotation using data obtained from high throughput studies of metabolomics and proteomics. Importantly, *C. reinhardtii* can grow autotrophically in the presence of light or heterotrophically on acetate under dark conditions (Boyle and Morgan, 2009). More interesting, in the presence of light *C. reinhardtii* can exhibit mixotrophy utilizing pentoses, hexoses, and acetate as carbon sources for heterotrophy and, concomitantly, CO_2 for supporting autotrophic growth.

A metabolic network of *C. reinhardtii* was modeled comprising 458 metabolites and 484 metabolic reactions distributed in three compartments (cytoplasm, chloroplast, and mitochondria) were reconstructed. Roughly half of the metabolites included are present in the chloroplast (212 out of 458), one-third localizes in the cytoplasm with the remaining ones assigned to mitochondria (Boyle and Morgan, 2009). The reactions (enzymes and transporters) were distributed as follows: chloroplast (44%), cytoplasm (20%), mitochondria (10%), and transport (26%).

MFA was conducted under three growth conditions (auto-, hetero-, and mixo-trophic). During autotrophic growth, the cell fixes CO_2 by converting light into cellular energy (reducing equivalents and ATP). Heterotrophic growth is defined as aerobic growth on acetate in the dark; the cell uses acetate as both carbon and energy sources. Mixotrophic growth is intermediate between both auto- and heterotrophic conditions with three inputs: light, acetate, and CO_2.

In the autotrophic case, the cell produces most of its energy from the conversion of light energy, which occurs in the chloroplast. The majority of the carbon flux is conducted through the Calvin Cycle (Fig. 4.7). The energy required for the regeneration of GAP from 3PG to run the Calvin Cycle is supplied by photophosphorylation. Outside the chloroplast, ATP and NADH are generated from GAP which is transported from chloroplast into mitochondria where it is subsequently oxidized into 3PG releasing ATP and NADH (Fig. 4.7).

Under heterotrophic growth conditions on acetate, *C. reinhardtii* directs most of the carbon flux through the TCA cycle (Fig. 4.7). Since the cell is not capable of metabolizing external sugars in the dark, almost all the energy is produced via respiration of acetate in the mitochondria. The Pentose Phosphate and Glyoxylate pathways are also active under heterotrophic conditions. Synthesis of G6P occurs via gluconeogenesis; due to the absence of ATP and NADH in the cytosol, the regeneration of GAP from 3PG takes place in the mitochondria and GAP is exported into the cytosol to be transformed into G6P. The glyoxylate shunt is also active, which is needed to metabolize acetate in several organisms.

C. reinhardtii is also capable of mixotrophic growth, utilizing acetate, light, and carbon dioxide for growth. Depending on the light levels two distinct regions of growth could be distinguished; at very low light levels, a regime resembling heterotrophic growth happens, i.e. carbon fixation does not occur and the cell produces CO_2. When light intensity increases enough to send flux through Ribulose Bisphosphate Carboxylase (Rubisco), the cell is capable of producing enough NADPH through the noncyclic ETC to supply metabolism with NADH via transhydrogenases. At light intensities adequate to initiate metabolic flux via autotrophic growth, the glyoxylate shunt flux steadily decreases while the Rubisco flux increases rapidly with increasing light (Boyle and Morgan, 2009).

Comparing biomass yields between autotrophic and heterotrophic modes of growth, 28.9 versus 15 g biomass per mole of carbon, respectively, were determined. Based on the elemental analysis of *C. reinhardtii*, 100% of the carbon was fixed into cell biomass under autotrophic conditions, due

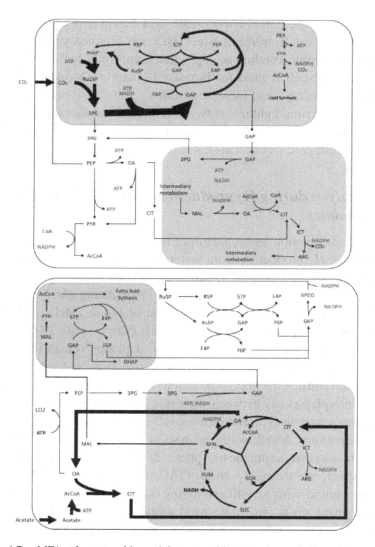

Figure 4.7. MFA of autotrophic and heterotrophic central metabolism flux maps in *C. reinhardtii*. Top panel, in the autotrophic case the majority of the carbon flux is conducted through the Calvin Cycle in the chloroplast. Bottom panel, under heterotrophic growth conditions on acetate, *C. reinhardtii* directs most of the carbon flux through the TCA cycle in the mitochondrion. The thickness of the arrows has been normalized to the total carbon dioxide uptake of 100 mol. The gray compartments represent the chloroplast (top) and the mitochondrion (bottom).
Source: Reproduced from Boyle NR, Morgan JA. (2009) Flux balance analysis of primary metabolism in *Chlamydomonas reinhardtii. BMC Syst Biol* **3**, 4.

mainly to the production of energy from light during photosynthesis, without net carbon loss from respiration. The lower biomass yields registered under heterotrophic conditions imply that about only half of the carbon taken in by the cell is used for biomass formation. During mixotrophic growth, the biomass yield increased from 13.5 to 22.9 g per mole carbon since with increasing light intensity, the cell can direct more carbon toward biomass and less toward energy production. The maximum yield is lower than the autotrophic yield because it is energy-limited.

Flux analysis during H_2 production using data from plant metabolomics

The cyclical depletion and repletion of liquid *C. reinhardtii* cultures with S, facilitates H_2 production from water (Melis *et al.*, 2000). Essentially, sulphur deprivation inactivates PSII; thus, the water-splitting reaction and O_2 evolution are inhibited (Wykoff *et al.*, 1998; Zhang and Melis, 2002). This in turn releases the oxygen inhibition of H_2ase (Melis and Happe, 2001) thereby starting to actively produce H_2 (Kruse *et al.*, 2005).

C. *reinhardtii* growth on acetate after S depletion occurs in three distinct phases: aerobic, anaerobic with H_2 production, and a phase of pH stabilization and decline in H_2 production (Fig. 4.8). During H_2 production, a close interplay exists between oxygenic photosynthesis, mitochondrial respiration, catabolism of endogenous substrate and electron transport via the H_2ase pathway. Metabolomic analysis of *C. reinhardtii* under these conditions showed that during the first phase (24 h after S depletion) cells accumulate starch and triacylglycerides (TAG) (Matthew *et al.*, 2009). Starch and some amino acids are utilized during the following anaerobic, H_2 production, phase while initiating mixed fermentative metabolism (formate, ethanol). A participation of starch as an important factor in H_2 production has been inferred from a strong correlation between starch consumption and H_2 production, although the mechanisms remain unknown (Matthew *et al.*, 2009). As expected in anaerobiosis, an overall reduction of the flux through the TCA cycle was also registered, with an increase of metabolites upstream pyruvate (sugars, sugar alcohols, and some amino acids). Apparently, the redox hurdle posed to the cells by the anaerobic condition, i.e. NAD(P)H oxidation, is solved by activating the fermentative pathways and amino acids, succinate, and sugar alcohols synthesis.

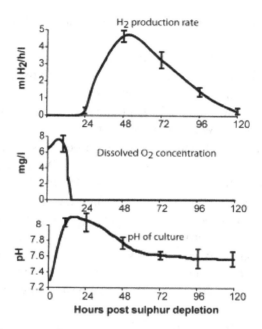

Figure 4.8.   H₂ production, change in dissolved O₂ concentration, and pH of *C. reinhardtii* cultures following sulphur depletion.
Source: Reproduced from *J Biol Chem* 284, Matthew T, Zhou W, Rupprecht J, Lim L, Thomas-Hall SR, Doebbe A, Kruse O, Hankamer B, Marx UC, Smith SM, Schenk PM. The metabolome of *Chlamydomonas reinhardtii* following induction of anaerobic H₂ production by sulfur depletion, 23415–23425. © 2009 The American Society for Biochemistry and Molecular Biology.

Probably, mixotrophic growth is optimal for H_2 production because the latter can only be observed in trace amounts without acetate or light (Doebbe *et al.*, 2010; Matthew *et al.*, 2009). Channeling of acetate to TAG production could be also possible during the aerobic initial period, in which a large increase in TAG has been noticed (Matthew *et al.*, 2009). This pathway represents an important potential energy reserve for H_2 and oil production in algal systems.

The metabolite profiles of *C. reinhardtii* WT (*cc406*) and the high H_2-producing mutant strain (*Stm6Glc4*) growing heterotrophically in the presence of acetate were characterized during defined stages of the H_2 production process. When cells are S-deprived, acetate from the medium is consumed, and starch accumulates along with a sharp decline of Rubisco protein levels during the first 24 h, which results in an inhibition of CO_2

fixation. The aerobic consumption of acetate during this period leads to starch, fatty acid, and neutral lipids synthesis.

This mitochondria-driven utilization of acetate, as can be judged from the increase of the intermediates of the citric acid/glyoxylate cycles, leads to anaerobiosis and the start of H_2 production. At this stage, starch appears to serve as a substrate source for H_2 production, decreasing to 70% of the maximum levels in the WT and to 42% of the maximum level in *Stm6Glc4*. Glycolytic intermediates and fermentative products such as formate and ethanol increase during the H_2 production phase. During glycolytic starch breakdown, $NAD(P)^+$ is reduced to $NAD(P)H$, which can then serve as the electron source for H_2 production, because it can be used to reduce the PQ-pool, catalyzed by the $NAD(P)H$ dehydrogenase (Doebbe *et al.*, 2010).

According to the interpretation of Doebbe *et al.* (2010) competition for pyruvate among fatty acids, lipids, and starch accumulation during the initial 24 h of S-starvation is a key factor for H_2 production. Under aerobic growth, pyruvate is diverted to many biosynthesis pathways, but when the culture is subjected to S limitation but still aerobic, starch, fatty acid, and lipid synthesis are strongly induced. In contrast, under anaerobic growth, it appears that pyruvate metabolism through amino acid synthesis (especially alanine) and fermentative pathways modulates the cellular redox balance (Doebbe *et al.*, 2010).

Metabolic engineering of microalgae

A major focus of the microalgal biofuels industry is to develop photoautotrophic production systems. However, there are configurations (e.g. use of wastewater resources or industrial carbon sources) in which photoheterotrophic growth can offer economic benefits and in which organic carbon sources can be recycled.

ME is increasingly being applied to optimize biofuel production in algal systems. DNA recombinant-based technologies together with optimized culture conditions are helping to achieve a more efficient conversion of sunlight. However, still a comprehensive understanding of the biosynthesis and degradation of precursors, intermediates, and metabolic end products, and the identification of the regulatory networks that control metabolic flux is lacking. This understanding is central to establish informed engineering strategies for optimizing biofuel production in microalgae.

The discovery of sustainable H_2 production that bypasses the sensitivity of the reversible hydrogenase to O_2 is a significant development in the biofuel field. It may lead to exploitation of green algae for the production of H_2 gas as a clean and renewable fuel. It has been suggested that the relatively slow rate of H_2 production and low yield (15% to 20% of the photosynthetic capacity of the cells at best) could signify that there is room for significant improvement, by as much as one order of magnitude. Long-term operation requires reversion to normal conditions for photosynthesis and restoration of endogenous stored substrate after about 100 h of S starvation (Ghirardi *et al.*, 2000).

In *C. reinhardtii*, the light activation of photosystems I (PSI) and II (PSII) is regulated by the state transition process which involves the adjustment of the respective light harvesting complex (LHC) antenna sizes (Bonaventura and Myers, 1969; Murata, 1969). In state 1, the antenna of PSII is larger than that of PSI. In state 2, the PSII antenna size is reduced, with mobile LHC-II proteins migrating to PSI. By shuttling LHC-II proteins between PSI and PSII, antenna sizes are adjusted (Beckmann *et al.*, 2009; Kruse *et al.*, 2005). Normally, in *C. reinhardtii*, state transitions have the valuable property of determining the functioning of cyclic e^- transport: inhibited in state 1 or accounting for most of the overall photosynthetic e^- flow in state 2. This is of crucial importance in terms of photobiological H_2 production, because under anaerobiosis, after S deprivation (when H_2 production starts), state 2 and cyclic e^- transport are activated representing an additional e^- sink with which H_2ase must compete. This is likely to restrict the H_2 production capacity of wild type *C. reinhardtii* in S-depleted anaerobic cultures.

In an attempt to increase the rate of e^- supply to the H_2ase and thus of H_2 production, several investigators have performed a systematic screening for mutants locked in state 1 under anaerobiosis. State transition mutants (*Stm*) of *C. reinhardtii* have been identified that produce high levels of H_2.

The high H_2 producing phenotype of *Stm6* was based upon an increased rate of O_2 consumption (altered mitochondrial metabolism) and at the same time a decreased rate of O_2 evolution due to reduced number of active PSII complexes under illumination with moderate light intensity (Fig. 4.9). The combined effect of these two modifications was that the dissolved O_2 concentration in illuminated *Stm6* cultures was only 30%–40% of WT levels. *Stm6* also deposited large amounts of starch in the chloroplast, which

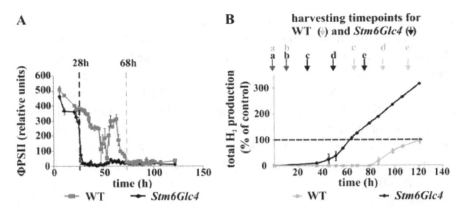

Figure 4.9.   Correlation between photosynthetic quantum yield and H_2 production rates of *C. reinhardtii* WT and *Stm6Glc4* during the shift from aerobiosis to anaerobiosis (induced by S-deprivation). (A) quantum yield of PSII of sealed, sulphur-depleted *C. reinhardtii* cultures. (B) total H_2 production during S-deprivation. Harvesting time points (a–e) are labeled. The amount of H_2 produced in the WT was set to 100%.
Source: Reproduced from *J Biol Chem* 285, Doebbe A, Keck M, La Russa M, Mussgnug JH, Hankamer B, Tekce E, Niehaus K, Kruse O. The interplay of proton, electron, and metabolite supply for photosynthetic H_2 production in *Chlamydomonas reinhardtii*, 30247–30260. © 2010 The American Society for Biochemistry and Molecular Biology.

appears to be linked to the inhibition of energy consumption by mito-chondrial respiration under anaerobic conditions during illumination (Kruse *et al.*, 2005). The increased levels of stored starch are likely to be respon-sible for the enhanced duration of H_2 production observed in long term experiments.

The engineered metabolome of C. reinhardtii

Starch synthesis and catabolism in *C. reinhardtii* has been thoroughly reviewed (Ball, 2002); however, further research is required to better under-stand the partitioning of carbon. Metabolomic analysis of *C. reinhardtii* rep-resents such an approach directed toward the understanding of how carbon partitions into increased production of starch for subsequent fermentation into H_2 or ethanol, or the redirection of photosynthate from starch into lipids for conversion to diesel fuels.

The metabolome of *C. reinhardtii* has been analyzed under condi-tions of heterotrophic growth on acetate after S depletion (Doebbe *et al.*, 2010; Matthew *et al.*, 2009). The metabolite profiles of wild type (WT, *cc406*) *C. reinhardtii*, and the high H_2-producing mutant strain (*Stm6Glc4*) growing heterotrophically in the presence of acetate were characterized during defined stages of the H_2 production process (Fig. 4.9) (Doebbe *et al.*, 2010).

In the *Stm6*, starch over accumulates, rates of cellular respiration are increased, and cyclic electron transfer around PSI is inhibited leading to increased H_2-production rates (Kruse *et al.*, 2005). Hydrogen production has been further enhanced in the *Stm6* mutant through heterologous expression of the hexose/H^+ symporter (HUP1). This symporter further enhanced H_2 production by enabling *C. reinhardtii*, which lacks extracellular glucose uptake transporters, to couple glucose oxidation to H_2ase activity (Doebbe *et al.*, 2007). Because the production of fatty acids, lipids, and fermentative products are possible competitors for H_2 production, it is a goal for the future to diminish these competitors by ME.

The quantum yield of PSII in *Stm6Glc4* rapidly declined to nearly zero within 28 h after transfer into sulphur-depleted medium. With the drop of photosynthetic activity, anaerobiosis was gradually established in the culture (Doebbe *et al.*, 2010). Interestingly, the WT showed a much slower and delayed decline phase lasting over a period of around 68 h (Fig. 4.9(A)). This phenotype may be caused by an impaired PSII repair mechanism in the mutant, resulting in an earlier establishment of anaerobiosis. As a consequence, H_2 production by the oxygen-sensitive hydrogenase(s) started 51 h earlier in *Stm6Glc4* compared with the WT strain (after 35 h for *Stm6Glc4* *versus* 86 h for WT, Fig. 4.9(B)). In addition, the *Stm6Glc4* H_2 production phase was maintained for much longer (215 h *versus* 79 h in WT) with higher peak H_2 production rates (4.9 ml/h *versus* 2.8 ml/h in WT). The experiment was stopped after 120 h during the peak H_2 production phase, before the H_2 production cycle was completed. Until then, the WT had produced a total amount of 86 ml H_2/liter cell culture (set to 100% in Fig. 4.9(B)), whereas *Stm6Glc4* had produced 284 ml H_2/liter cell culture.

Rational Design of Microorganisms: Two Case Studies

The TDA approach (Fig. 1.2) provides the basis for the rational design of microorganisms or cells. In the following two sections we present two examples: ethanol production in *S. cerevisiae*, and L-threonine production in *E. coli*.

Increase of Ethanol Production in Yeast

Phase I: Physiological, metabolic, and bioenergetic studies of different strains of S. cerevisiae

According to the flow diagram of Fig. 1.2, the first step in the TDA approach is to investigate the conditions under which yeast produces the maximal output of ethanol. A comparison between two strains, a wild type and a *mig1* mutant, is shown. A broader screening was performed between several isogenic strains carrying disruption genes involved in the expression of glucose-repressible genes namely *SNF1*, *SNF4*, and *MIG1*, for comparisons with the wild-type strain CEN.PK122. *SNF1* (*CCR1, CAT1*) and *SNF4* (*CAT3*) code for proteins that exert a positive regulatory function for the derepression of gluconeogenic, glyoxylate cycle, or alternative sugar-utilizing enzymes as shown in the scheme of Fig. 4.10. Recently,

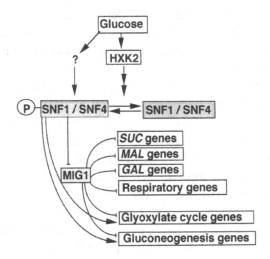

Figure 4.10. Carbon source-dependent gene regulation of glucose repressible genes. In *S. cerevisiae* the *SNF1* (sucrose nonfermenting) gene encodes a serine/threonine protein kinase (72 kDa) (Celenza and Carlson, 1986). SNF1 is physically associated with a 36 kDa polypeptide termed SNF4 which is also required for the expression of many glucose repressible genes and is thought to function as an activator of SNF1. The SNF1/SNF4 protein kinase complex in its active, probably phosphorylated, form is essential for the derepression of possibly all glucose repressible genes (Celenza and Carlson, 1986; Entian and Barnett, 1992; Gancedo, 1992; Ronne, 1995). Under conditions of glucose repression, the SNF1/SNF4 protein kinase is inactive.

the interactions between the known genes/gene products participating in the signaling, according to the scheme shown in Fig. 4.10, has received theoretical and experimental support. The network of signaling processes involved in catabolite repression in *S. cerevisiae* was reconstructed and modeled according to a Boolean approach (Christensen *et al.*, 2009). The model considered *SNF1/SNF4*, *SNF3*, *MIG1*, *MIG2*, *GRR1*, *RGT1*, and *RGT2* interactions with the targets of repression, e.g. transcription of *MAL* genes, *SUC* genes, and *GAL* genes. The method of Boolean modeling was tested for detection of inconsistencies and prediction of the behavior of mutants regarding substrate selectivity (Christensen *et al.*, 2009).

The onset of fermentative metabolism depends upon the glucose repression features of the strain under study (Cortassa and Aon, 1998). Isogenic yeast strains disrupted for *SNF1*, *SNF4*, and *MIG1* genes were grown in aerobic, glucose-limited, chemostat cultures, and their physiological and metabolic behavior analyzed for comparisons with the wild type strain (Aon and Cortassa, 1998; Cortassa and Aon, 1998) (see Chap. 3 section: *Catabolite repression and cell cycle regulation in yeast*). Under these conditions, the wild-type strain displayed a critical dilution rate (Dc, beyond which ethanolic fermentation is triggered) of $0.2\,h^{-1}$, whereas the *MIG1* mutant exhibited a Dc of $0.17\,h^{-1}$. Figure 4.11 describes main physiological variables measured, relevant to the present analysis, e.g. the specific glucose consumption and ethanol production fluxes, and the biomass, as a function of the dilution rate, D. Comparison of Wild-type and *MIG1* indicates that they satisfy the aim of the project, i.e. maximal ethanol output. This aim was achieved at high $D(=0.3\,h^{-1})$ (Fig. 4.11), and was higher for the wild-type. In fact, 3.4 and 3 g/l of ethanol were attained at $0.3\,h^{-1}$ for the wild type and mutant, respectively (see also Table 5.4). Thus, the expected outcome of phase I of TDA has been attained.

Phase II: Metabolic control analysis and metabolic flux analysis of the strain under the conditions defined in phase I

Yields and flux analysis

MFA may help determine the theoretical as well as the actual yields of ethanol, in order to determine the "ceiling" of the improvement. Otherwise

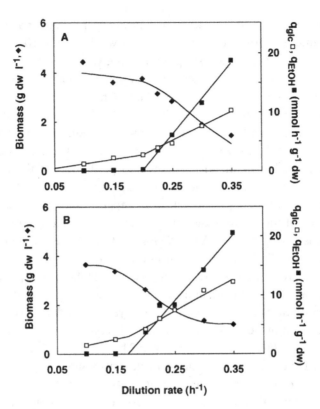

Figure 4.11.   Physiological behavior of the Wild type strain CEN.PK122 (A) and of the *MIG1* (B) mutant of *S. cerevisiae* in aerobic, glucose-limited chemostat cultures. Biomass yield, glucose consumption, and ethanol production rates are plotted as a function of the dilution rate, D, after attaining the steady state judged through the constancy in biomass concentration in the reactor vessel. The values of all physiological variables plotted were obtained from triplicate determinations and the standard deviation was not higher than 5% of the values shown in the figure.
Source: Reprinted from Cortassa and Aon (1998) ©copyright 1998, with permission from Elsevier Science.

stated, we seek to know the upper limit of increased ethanol production that we may expect by performing the ME of the yeast strain chosen, under the conditions already described in chemostat cultures (Fig. 4.11).

For the particular case of ethanol production in yeast, the calculation of the actual as well as potential ethanol yields at high growth rates is quite simple. Nevertheless, it must be stated that these calculations could be made more involved especially where several alternative competing pathways of synthesis of the product desired (e.g. metabolite, macromolecule) are possible. This is the case in *E. coli* for carbon flux redirection toward aromatic amino acids (Gosset *et al.*, 1996; Liao *et al.*, 1996), lysine (Koffas *et al.*, 2002), or L-threonine (Dong *et al.*, 2011) synthesis. Several engineering strategies are possible in the latter cases (see below and Chap. 3).

According to the biomass composition of *S. cerevisiae* under conditions suitable for the purposes of the engineering (phase I), 7.381 mmol glucose are required for the synthesis of 1 g biomass (see Table 4.1). Then, the glucose flux directed to anabolism, i.e. that devoted to replenish key intermediary metabolite precursors of biomass macromolecules, is equal to 7.381 times the growth rate; for a growth rate of $0.3\,h^{-1}$, this gives 2.214 mmol glucose $h^{-1}\,g^{-1}$ dw. The rate of glucose consumption was determined to be 7.51 mmol glucose $h^{-1}\,g^{-1}$ dw (Cortassa and Aon, 1997). The glucose flux not directed to anabolism was then, $7.51 - 2.214 = 5.296$ mmol glucose $h^{-1}\,g^{-1}$ dw. The yield of ethanol could be therefore calculated from the flux of ethanol production divided by the nonbiomass-directed glucose flux ($5.793/5.296 = 1.09$) with a maximum stoichiometric limit of 2.0 mol ethanol/mol of glucose. Therefore, according to the present analysis, there is still scope for greatly increasing ethanol production, thereby making ME worthwhile.

Metabolic control analysis

Once the growth and metabolic characteristics, and the attainable improvement in product yield have been defined, the task is now to search for the rate-controlling steps in ethanol production (see a detailed account of the analysis in Chap. 5). The Westerhoff–Sauro method of MCA (Sauro *et al.*, 1987; Westerhoff and Kell, 1987) was applied to elucidate the control of glycolysis in *S. cerevisiae* (Cortassa and Aon, 1994a; Cortassa and Aon, 1997).

The sugar transport and glycolytic models used to interpret the experimental data are depicted in Fig. 4.12. Important parameters of the model are the gradient of glucose across the plasma membrane, i.e. the difference in concentration of intra- and extra-cellular glucose, and the inhibitory constant, Kin_e, for G6P. Sugar transport in yeast appears to be a significant

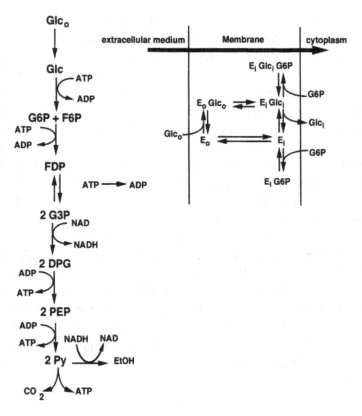

Figure 4.12.   Scheme of the glycolytic model and the branch toward the TCA cycle and ethanolic fermentation. The kinetic structure of the model is shown on the left, whereas a scheme of the sugar transport step and its inhibition by G6P is depicted on the right. The transporter molecules bound to G6P, EiG6P, and EiGlciG6P are no longer available for sugar transport.

Source: Cortassa and Aon (1997) ©copyright 1997, with permission from Elsevier Science.

rate-controlling step of the glycolytic flux, and important understanding of the molecular biology of the sugar transporters and glucose sensors has been obtained in the past (Ko *et al.*, 1993; Kruckeberg, 1996; Palma *et al.*, 2009; Reifenberger *et al.*, 1995; van Dam, 1996). However, it is not clear yet how these different transporters are regulated *in vivo* (Otterstedt *et al.*, 2004; Walsh *et al.*, 1996). The equations of the model are described in Cortassa and Aon (1994a) and Cortassa and Aon (1997); model parameter optimization and simulation conditions are described in the Appendix.

In order to apply the Westerhoff–Sauro method of MCA to determine rate-controlling steps of ethanol production flux (i.e. the steps that exert a significant quantitative effect on the overall flux of ethanol generation), we need to quantify the intracellular metabolite concentrations required by the model (Fig. 4.12). The latter are introduced into the derivatives of the rate equations, in order to calculate the elasticity coefficients for constructing the corresponding matrix **E**. The inversion of matrix **E** results in the matrix of control coefficients **C**, the first column of which contains the flux control coefficients (see Chap. 5).

According to the results obtained by using MCA, in the wild-type at high growth rates, most of the flux control toward ethanol production was exerted by the PFK step (flux control coefficient = 0.93) (Fig. 4.13). A gradual shift of the control from PFK toward HK and the uptake step

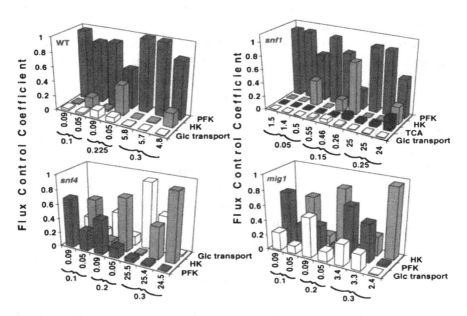

Figure 4.13.   Flux control coefficients of the yeast wild type strain CEN.PK122 (WT) and the catabolite-(de)repression mutants *SNF1*, *SNF4*, and *MIG1* (B) of yeast simulated under the conditions described in Fig. 4.10. In all cases, the value of the extent of inhibition constant of the glucose transport by G6P, Kin_e, was fixed at 12 mM while that of the intracellular glucose was allowed to vary, thereby resulting in the glucose gradients indicated (in mM) next to the axis.

Source: Cortassa and Aon (1997) ©copyright 1997, with permission from Elsevier Science.

occurs when the gradient of glucose through the plasma membrane is decreased (Fig. 4.13) (see Aon and Cortassa, 1998; Cortassa and Aon, 1997). Thus, the rate-controlling step, the target for genetic engineering, has been identified.

Phases III and IV: To obtain a recombinant yeast strain with an increased dose of PFK, and to assay the engineered strain in chemostat cultures under the conditions specified in phase I

These steps have not yet been attempted for the strain under analysis. Thus we simulated the results with the same model used to interpret the experimental data. We ask whether an effective increase of ethanol production is achieved as can be predicted from MCA. As a first step, we reproduced the experimental steady-state values of the glycolytic flux (q_{glc}) and intermediates at $D = 0.3 \, h^{-1}$. Taking this state as the reference one, PFK was increased 10% or 100% (i.e. V_{max}) and the flux toward ethanol augmented 3.8% or 7.2%, respectively, at the new steady state achieved. For each case above, we tested again whether the control had shifted toward another step. In fact, the control became based on the HK-catalyzed and the uptake steps (0.33 and 0.05 for the 10% increase, or 0.95 and 0.076 for the 100%, respectively).

Since the increments in ethanol production were relatively modest, we sought further improvement. Thus, we took that steady state characterized by a 7.2% increase of the flux to ethanol as the reference state. As HK is the main rate-controlling step in that steady state, we increased the flux through this step by 10% or 100%. With a 10% increase, an additional 4.8% increase in the ethanolic flux could be obtained. Under these conditions, an additional (modest) increase of 20% in the ATPase rate was necessary because of ADP depletion.

Finally we found that a 100% increase in ethanolic fermentation could be theoretically achieved when the fluxes through the glycolytic pathway were increased as follows: through the HK (100%), GAPDH (66%), PGK (73%), ADH (100%), together with the fluxes through the TCA cycle (33%) and the ATPase (100%). It must be stressed that the gains in each of these steps were performed under steady-state conditions.

Our results show the feasibility of a substantial increase in the ethanol flux with the simultaneous and coordinated overexpression of most of the enzymes in the glycolytic pathway. Similar results were obtained in the tryptophan synthesis pathway in yeast, in which the simultaneous overexpression of five enzymes led to more than additive elevation in the flux to the amino acid (Niederberger *et al.*, 1992). These results indicate coordinated changes in the level of activity of pathway enzymes in order to change flux levels (Fell, 1998). Nowadays, the task of simultaneously increasing and decreasing several enzymatic and transport steps represents a realistic goal, as shown in our next example.

Increase of L-threonine Production in *E. coli*

The metabolic pathway leading to L-threonine (Thr) production in *E. coli* is a challenging one for ME since it presents several steps in which the flux could be potentially diverted, and it exhibits feedback inhibition by product. The pathway depicted in Fig. 4.14 consists of five enzymatic steps, of which the first, third, and fourth reactions catalyzed by aspartate kinase, homoserine dehydrogenase, and homoserine kinase, respectively, are subjected to feedback inhibition by Thr, the end product of the pathway. As can be seen in the figure, four competing pathways (i.e. L-lysine, L-methionine, L-isoleucine, and glycine) affect the synthesis and accumulation of Thr (Dong *et al.*, 2011).

In order to develop a genetically defined Thr overproducer in *E. coli*, Lee *et al.* (2007) adopted a systems ME strategy. Regulatory and metabolic circuits were genetically modified with the aim to maximize the flux toward Thr, minimizing feedback inhibition by the amino acid and favoring its subsequent export from the cell. Concomitantly, other fluxes involving diversion of intermediates from the pathway depicted in Fig. 4.14 to other competing pathways were lessened. More specifically, Lee *et al.* (2007) released the feedback inhibitions of aspartate kinase I and III through site-directed mutagenesis of their coding genes, and the feedback repression to the chromosomal *Thr* operon; the deregulated *Thr* operon was overexpressed on an episomal vector. The biosyntheses of L-lysine and L-methionine were blocked, and the Thr consumption reduced. With this "rewiring" in place, the engineered strain was cultured in batch in a medium containing 50 g/l glucose for 48 h. Under these conditions, a 10.1 g/l of Thr

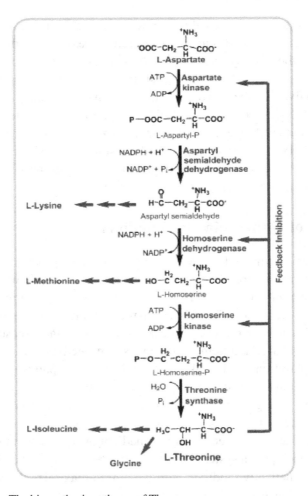

Figure 4.14.   The biosynthesis pathway of Thr.
Source: Reprinted from *Biotechnol. Adv.* 29, Dong X, Quinn PJ, Wang X. Metabolic engineering of *Escherichia coli* and *Corynebacterium glutamicum* for the production of L-threonine, 11–23. © (2011) with permission from Elsevier.

concentration was achieved (equivalent to a yield of 0.202 g Thr g^{-1} glucose) (Lee *et al.*, 2007).

In order to further improve the yield in Thr, the authors combined transcriptome analysis and MFA (see Chap. 5). The *in silico* flux response was performed using a genome scale metabolic model for *E. coli* (Lee *et al.*, 2005; Reed *et al.*, 2003).

The transcriptome analysis showed down-regulation of the enzyme phosphoenol pyruvate carboxylase (PPC) indicating restricted supply of oxalacetate (OAA), the intermediary precursor of aspartate. Transcript analysis also revealed an up-regulation of enzymes from the glyoxylate shunt suggesting the increasing demand for OAA. With this information at hand, the authors further investigated *in silico* the effect of an increase in flux through PPC or the glyoxylate shunt on Thr production. Steady-state flux evaluation, optimization according to an objective function, and constraint-based flux analysis on the engineered *E. coli* strain were utilized. The authors investigated the most effective flux change of maximizing Thr while reducing acetic acid production. The simulation results suggested that the expression of the *PPC* gene should be increased up to an optimum after which a gene-dosage increase will be counter-productive. Also, the flux through the glyoxylate shunt was enhanced by gene knock-out of malate synthase and the repressor of isocitrate lyase. Furthermore, the Thr uptake-carrier was deleted to cut down Thr uptake, and concomitantly genes related with Thr secretion were overexpressed.

The constructed strain TH28C (pBRThrABCR3) could produce 82.4 g/L Thr in 50 h fed-batch fermentation. The Thr/glucose conversion ratio was 39.3%. No lactate was formed and the accumulation of acetate was 2.35 g/L (Lee *et al.*, 2007). The Thr yield was close to the maximum achievable theoretical limit of 0.81 g g^{-1} glucose.

The systems level ME adopted in the work just described shares with the TDA approach the multidisciplinarity, as well as a clear engineering objective, i.e. increasing Thr production by *E. coli*. Further enhancement of the amino acid yield was achieved by utilizing a combined theoretical (MFA) and experimental (high throughput transcriptomics) to guide the genetic modification of the organism (Lee *et al.*, 2007).

Appendix

Conditions for parameter optimization and simulation of a mathematical model of glycolysis

It is a fair criticism that kinetic parameters of glycolytic enzymes from *in vitro* data could not be representative of their values *in vivo*. In order to address this problem, the steady states achieved by *S. cerevisiae* in chemostat cultures were simulated in the model depicted in Fig. 4.12, as described by 10 ordinary differential equations (ODEs) with the aim of optimizing the kinetic parameters of the individual enzymatic rate laws. The set of ODEs describes the temporal evolution (dynamics) of 10 metabolites of glycolysis as a function of the individual rate laws determined from the enzyme kinetics *in vitro* (Cortassa and Aon, 1994a; Cortassa and Aon, 1997). The V_{max} of each rate equation was considered an adjustable parameter. Three criteria were chosen to perform a parameter optimization: (1) the system of ODEs should attain a steady state; (2) the flux through glycolysis should be equal to the experimentally-measured flux of glucose consumption; and (3) the metabolite levels should match the experimentally determined steady-state concentrations. This optimization procedure along with the criteria chosen are very important in order to validate the results of the control of glycolysis *in vivo*. In our hands, it was seen that V_{max} values were the most sensitive to the optimization procedure, rather than Km, suggesting that the former would be more affected by differences between *in vitro* and *in vivo* data.

Chapter 5

Modeling Networks: Concepts and Tools

Cells as Networks of Processes

The organization of processes of different natures, e.g. biological, geophysical, social, and economic, may be analyzed through the concept of networks. Biological systems (cells, tissues, organs, and organisms) are considered as multilayered networks of processes organized on different spatiotemporal scales (Aon *et al.*, 2006) (Fig. 5.1).

As already pointed out (Chap. 1), the study of networks can be approached from four different perspectives (Aon, 2010; Aon *et al.*, 2007): (i) architectural (structural morphology), (ii) topological (connectivity properties), (iii) dynamical, and (iv) molecular (Xia *et al.*, 2004). In the case of metabolic networks, the architectural view would correspond to the localization and distribution of metabolic pathways among different cellular compartments; the topological approach will emphasize the organization of participant reactions in a pathway and their interactions with other pathways, whereas a dynamic approach will focus on systemic flux patterns sustained by metabolic and transport pathways. The molecular approach more specifically analyzes biomolecular interactive networks involving different components which result in varied functional outputs: protein–protein, genetic expression (multiarrays), regulatory (protein–DNA interactions, combinatorial transcription factors), metabolic (nodes represent metabolic substrates and products, and links represent metabolic reactions), and signaling (signal transduction pathways through protein–protein and protein–small molecule interactions).

The organization of complex systems is heterarchical (as opposed to hierarchical) (Yates, 1993) comprising nested networks, i.e. networks

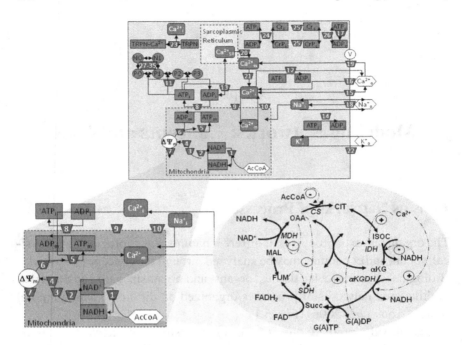

Figure 5.1. Heterarchical control and regulation in the network of energetic and electromechanical processes of the heart cell. Three degrees of detail in the description of the Excitation–Contraction coupling and Mitochondrial Energetic model (ECME) can be applied by different levels of "magnification" on the network of integrated cardiomyocyte function (Cortassa *et al.*, 2006; Cortassa *et al.*, 2009a; Cortassa *et al.*, 2009b). From the top panel, depicting the overall network of the ECME, it is possible to zoom in the mitochondria with its own reaction network (lower left panel) or yet deeper into the tricarboxylic acid (TCA) cycle (lower right panel) with its specific biochemical circuitry. In the top panel, rectangular (ion or metabolites) or circular (myofibril conformations) boxes indicate state variables network. Hexagonal boxes denote inputs (ions or carbon substrate) that correspond to parameters. Arrowheads point to the products of the numbered processes, whereas lines without arrowheads indicate inputs to those processes. In the mitochondrial energetics scheme, the TCA cycle was considered as a single step in the stoichiometric matrix. In the scheme of the TCA cycle (lower right panel) solid lines represent mass–energy transformation reactions, whereas regulatory interactions (with negative signs indicating inhibition, and positive signs activation) are denoted by dashed lines.

Source: Reproduced from Aon *et al.* (2011).

within networks (Fig. 5.1) (Aon and Cortassa, 2011; Aon *et al.*, 2011). Heterarchical control and regulation in complex networks of reactions is bidirectional, i.e. bottom-up as well as top-down. In a heterarchy, as applied to metabolic and transport networks, every reaction, metabolite, ion, and process contributes, although to a different extent, to the overall control and regulation of the network, undermining existing concepts of "central controllers" and "rate-limiting" processes as well as cause–effect relationships (Aon and Cortassa, 2011; Cortassa *et al.*, 2009a; Cortassa *et al.*, 2009b).

High throughput technologies such as proteomics, genomics, transcriptomics, and metabolomics, render massive amounts of information regarding protein, gene expression, mRNAs, and metabolite profiles, respectively, under specific (patho)physiological conditions. The large volume of information available, now stored in databases and repositories, has driven the development of algorithms allowing its interpretation and organization ("data mining"), and promoting understanding (reviewed in Kell (2006) and Winslow *et al.* (2005)).

In this chapter, we will discuss the types of algorithms that have been so far developed, their use and limitations as applied to MCE. Discussing high throughput methods is outside the scope of this book, and will not be addressed. Our main focus will be directed to analyze how the "-omics" information in combination with mathematical modeling can be used for the specific and purposeful understanding, prediction, and modification of cells with a special emphasis on the topological and dynamic views of networks.

From "-Omics" to Functional Impact in Integrated Metabolic Networks

For the sake of clarity and precision, let us analyze the challenge raised by the "omics" technologies with an example. The scheme in Fig. 5.2 represents a network of biochemical reactions inside a mitochondrion. Topologically, the network consists of nodes (usually metabolites) and edges (enzymes). If we want to characterize the mitochondrial proteome under a certain (patho)physiological situation, we will follow a meticulous high throughput procedure of protein electrophoretic separation, isolation, identification, and analysis, called Proteomics. The outcome of Proteomics is a list of proteins, some of them with post-translational modifications (e.g. phosphorylation, glycation, S-glutathionylation). The protein amounts and status of post-translational modifications may be analyzed with respect to a reference state.

Missing from the proteins list will be the network topology, functional relationships, and regulatory interactions (feedback and feedforward). Also absent from the list is the node "character" in the network, since some

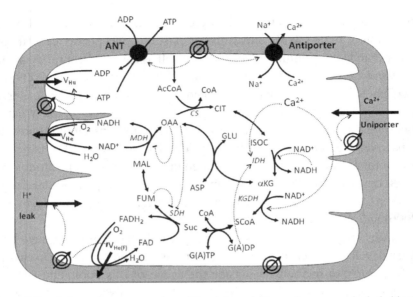

Figure 5.2. Schematic representation of the state variables and processes included in the "mitochondrial energetics" computational model. Physiological and metabolic processes in heart mitochondria and their interactions are taken into account including oxidative phosphorylation and matrix-based processes, namely the TCA cycle. Acetyl CoA (AcCoA) is the point of convergence for the oxidation of fatty acids and glucose, the two main substrates of the heart. The TCA cycle completes the oxidation of AcCoA to CO2 and produces NADH and FADH2, which provide the driving force for oxidative phosphorylation. NADH and FADH2 are oxidized by the respiratory chain and the concomitant pumping of protons across the mitochondrial inner membrane establishes an electrochemical gradient, or proton motive force, composed of an electrical gradient ($\Delta \Psi_m$) and a proton gradient (ΔpH). This proton motive force drives the phosphorylation of matrix ADP to ATP by the F1F0-ATPase (ATP synthase). The large $\Delta \Psi_m$ of the inner membrane also governs the electrogenic transport of ions, including the counter-transport of ATP and ADP by the adenine nucleotide translocator, Ca^{2+} influx via the Ca^{2+} uniporter and Ca^{2+} efflux via the Na^+/Ca^{2+} antiporter (Magnus and Keizer, 1997). Key to symbols: The concentric circles with an arrow across and located at the inner mitochondrial membrane, represent the $\Delta \Psi_m$. Dotted arrows indicate regulatory interactions either positive (arrowhead) or negative ($-$).

Source: Reproduced from Cortassa S, Aon MA, Marban E, Winslow RL, O'Rourke B. (2003) An integrated model of cardiac mitochondrial energy metabolism and calcium dynamics. *Biophys J* **84**, 2734–2755, by copyright permission of the Cell Press.

of them are "hubs" and exhibit multiple rather than single inputs and outputs, as the great majority of them. Importantly, also missing are ions, metabolites, that as we will see have important effects on the network function from a quantitative standpoint.

One of the main objectives of Systems Biology is to combine the output of "-omic" technologies with computational models, and analyses directed to the processing and interpretation, of massive amounts of data. Keeping this objective in mind, how do we go from the mitochondrial proteome list to analyze the functional impact of the multiple changes in protein amount and post-translational modifications under a certain condition (normal or disease)? The ultimate goal is to understand in an integrated way normal physiology or the mechanism(s) of a disease. In order to achieve this goal, we need quantitative tools such as computational models (Fig. 5.2). Mathematically, these models can be formulated in detailed kinetic equations, accounting for topological, functional and regulatory interactions involving ions, metabolites, and second messengers. Also mathematical models involving simple stoichiometric relationships can be formulated, in those cases where kinetic information is not available.

In this chapter, we present the basic concepts that can be applied to the analysis of experimental and computational data using Metabolic Control Analysis (MCA) or Metabolic Flux Analysis (MFA) (alternatively referred to as flux balance analysis or constrain-based flux analysis (Orth *et al.*, 2010; Otero and Nielsen, 2010; Ruppin *et al.*, 2010). We will use these quantitative tools, but specifically with the aim of analyzing properties that depend on network function, so as to simulate the physiological behavior of different cell types in response to genetic modification and culture conditions through integration of high throughput information combined with computational modeling.

Networks Analyzed with Stoichiometric Models

The analysis of fluxes displayed by an organism under specific conditions provides a true dynamic picture of the phenotype. It captures the metabolome as the overall result of the functional interaction between the metabolic/transport network with the genome and the environment (Aon and Cortassa, 1997; Cascante and Marin, 2008). The total set of fluxes through the metabolic and physiologic network of the cell ("the fluxome") depends both on the proteome (since protein enzymatic levels will determine the maximal rates of biochemical reactions) and the metabolome (or the set of all low-molecular-weight compounds present in the cell under specific physiological conditions). However, because of the nonlinear nature of the functional relationships of most biological processes, approaches based solely on topological and stoichiometric information can provide only an approximate picture of flux variations. The pattern of fluxes so determined may encompass desired and undesired ranges of values in a single solution space. The utilization of stable isotope tracer experiments allows to overcoming some of the limitations of MFA. This approach is based on ^{13}C-labeled substrates fed to a cell population in steady state, until the isotope label distributes throughout the network. The quantification of intracellular fluxes is performed according to specific labeling patterns of metabolic intermediates and reconstitution of the flux distribution from measurements (Cascante and Marin, 2008; Sauer, 2006). Fluxes can be obtained from labeled metabolites determined with nuclear magnetic resonance (Sauer et al., 1997) or mass spectrometry data (Klapa et al., 2003a), also called tracer-based metabolomics (Lee and Go, 2005). This is a useful and powerful approach for determining fluxes in vivo and is still evolving (Cascante and Marin, 2008; Cascante et al., 2010; Sauer, 2006).

MFA or flux balance analysis relies on optimization of metabolic networks with respect to certain objectives, e.g. maximization of growth rate or product formation from a specific substrate, or minimization of the use of a pathway leading to an undesired product. Linear programming is usually applied to find the solution space that defines a "volume" of possible solutions (e.g. fluxes through paths, or individual rates), thereby complying with the constraints used to optimize the network under study. Several algorithms exist for computing optimal fluxes in a network according to a predetermined objective function (Klamt et al., 2005; Otero and Nielsen, 2010; Price et al., 2004; Rocha et al., 2010).

Most packages for the analysis of metabolic networks rely on the concept of elementary flux modes developed by Schuster et al. (1999). Each elementary flux mode is characterized by a specific ratio of byproducts or yield of end-products with respect to the substrate feeding the pathway. However, in most cases biological systems operate in multiple elementary flux mode combinations giving rise to suboptimal yields. The definition of a metabolic elementary flux mode is given by the minimal set of enzymes that could operate at steady state. The enzymatic set is considered minimal because if only the enzymes belonging to this set were operating, complete inhibition of one of them would lead to cessation of any steady-state flux in the system (Schuster *et al.*, 2000). This concept has provided strict criteria for the definition of a metabolic pathway by introducing the "nondecomposability" property in opposition to the vague definition of metabolic pathways as a "simple set of biochemical reactions leading to a specific product" (Schuster *et al.*, 1999).

Besides conceptual precision, the use of elementary flux modes becomes evident in some applications aiming to (i) target a pathological organism with a drug that should cease all flux through a metabolic pathway vital for the pathogen's survival; (ii) compute maximal product yields; and (iii) genetically modify an organism to improve yield by introducing gene coding for enzymes favoring an elementary flux mode (Trinh and Srienc, 2009).

As an example, we present the stoichiometric analysis as applied to mitochondrial metabolism. Placed at the convergence of most anabolic and catabolic biochemical pathways, mitochondria appear as "hubs" in the intracellular metabolic network (Aon *et al.*, 2007a). Through the TCA cycle, mitochondria exhibit multiple links to other cellular pathways either as an input (source) or as an output (sink) (Mathews *et al.*, 2000; Nelson and Cox, 2005). Several of the TCA cycle intermediates such as oxaloacetate or alpha-ketoglutarate (2-αKG) feed several amino acid production pathways, or others such as succinyl CoA or fumarate (Fum) are fed by the phenylalanine, valine, leucine, and isoleucine or the tyrosine degradation pathways. Two of the main dehydrogenases catalyzing the conversion of isocitrate and 2-αKG, respectively, and responsible for producing the electron donor NADH are activated by Ca^{2+}, along with pyruvate dehydrogenase. Metabolic networks show other conspicuous examples of hubs such as the nodes represented by the intermediary key metabolite precursors of the biomass (see Chap. 2) (Cortassa *et al.*, 1995). Understanding the extent and connectivity of the cytoplasmic and mitochondrial components of the biochemical network has been the subject of recent developments in network theory (Barabasi and Oltvai, 2004) and is crucial to comprehending the web of linked reactions underlying the global metabolic status of an organism (see below).

Figure 5.3 shows the same scheme of the computational model of mitochondrial energetics introduced in the Cell Net Analyzer (CNA), one of the

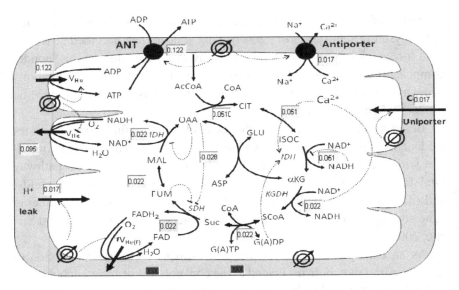

Figure 5.3.   Graphical network representation for the analysis of mitochondrial energetics fluxes in the Cell Net Analyzer (CNA) environment. Redrawn from the scheme of the computational model (Fig. 5.2) with the addition of gray square boxes where the fluxes calculated to operate under steady-state conditions are displayed. The boxes in the bottom of the figure (inside the gray area) stand for the state variables with respect to which the fluxes are being optimized, mitochondrial membrane potential, $\Delta\Psi_m$, and mitochondrial ATP, ATP_m.

most popular packages used to perform flux analysis of metabolic networks. This package (the authors call it "a structural analysis of cellular networks") operates in a MATLAB® environment and is able to analyze/create networks to account for either mass (metabolism) or information (signaling) flows. Network elements are presented in special tool boxes, and by importing a graphical representation from an external program or from databases such as KEGG or other sources, an interactive network map can be created. The toolboxes enable the analysis of stoichiometric networks, including steady-state methods such as flux and pathway analyses, flux optimization, and signal-flow networks among others. As in the case of the mitochondrial network there is no "biomass" as the main output, the optimization criterion used by the algorithm was not to maximize biomass or macromolecule production. Instead, we declared $\Delta\Psi_m$ and ATP_m to be maximized as objectives that are appearing in the light gray boxes at the bottom of Fig. 5.3.

The stoichiometric matrix is calculated from reactions specified in the "Network composer" box and is shown, gray level coded, in Fig. 5.4 top. The actual number of a reaction producing, consuming or participating in a reversible scheme for each metabolite, is shown in parentheses in the column

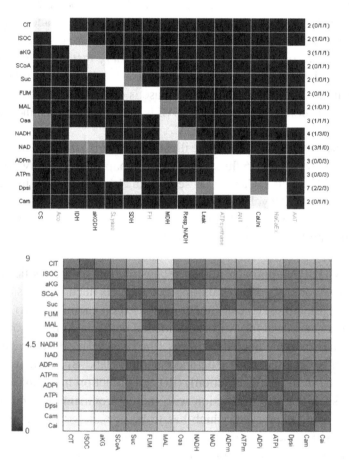

Figure 5.4. Stoichiometric (top) and distance (bottom) matrices of the mitochondrial model (Fig. 5.3) introduced in the CNA. In the matrix of stoichiometric coefficients, light gray indicates a positive coefficient; dark gray, a negative coefficient; black, a zero coefficient, and white a coefficient that may change sign according to direction of a reversible reaction. In the distance matrix are computed the steps that separate two nodes of the network: the lighter the shade of gray, the larger the distance between the two nodes. Since direction exists in networks (due to the irreversibility of some reactions) the distance from, e.g. CIT to MAL is different from the one from MAL to CIT (see text).

at the right. The gray-level coding distinguishes whether a metabolite or ion (rows) is produced (positive coefficient) or consumed (negative coefficient) by a reaction (columns) which in turn may be reversible or irreversible. The algorithm is able to compute the distance between nodes with the "graph theoretical path length." The result is a gray-level encoded matrix depicted in the panel at the bottom of Fig. 5.4; the darker the shade of gray, the closer two elements in the network structure are (columns and rows in the "distance" matrix). For example, the intermediary metabolite citrate (CIT) is shown to be close to ISOC and NAD (distance 1 or 2) and relatively far from malate (MAL) since citrate synthase is considered to be irreversible. However, the converse is not true, MAL is closer to CIT (distance) because of the direction of the flow in the TCA cycle.

The elementary flux modes are illustrated in Fig. 5.5, where the different panels portray 3 of the 15 flux modes that can happen in the network. Due to the approach adopted, Ca^{2+} is not seen as an element that will influence the activity of the TCA cycle (it acts as an activator of IDH and αKGDH). Then, several of the elementary flux modes do not even consider the operation of Ca^{2+} transport (Ca^{2+} uniporter and Na^+, Ca^{2+} exchanger) (top modes in Fig. 5.5).

The COBRA approach allows reconstruction of metabolic networks based on "-omics" data (Palsson, 2009). This technique assumes the following principles: (i) cell function is based on chemistry; (ii) annotated genome sequences along with experimental data enable the reconstruction of genome-scale metabolic networks (the authors have developed a collection of Biochemical, Genetic and Genomic data (BiGG) that should be represented in the reconstruction); (iii) cells function in a context-specific manner such that the interaction with the environment constrains the set of genes being expressed, which has to be considered in the reconstruction; (iv) cells operate under a series of constraints. Factors constraining cell function fall into four principal categories: physico-chemical (mass–energy conservation relations), topological (molecular crowding effects and steric hindrance), environmental, and regulatory (basically self-imposed constraints, or restraints); (v) mass and energy are conserved; (vi) organisms evolve under selection pressure in a given environment.

The application of each of the principles above gives rise to genome-scale models (GEM) which can be used for ME (including the improvement

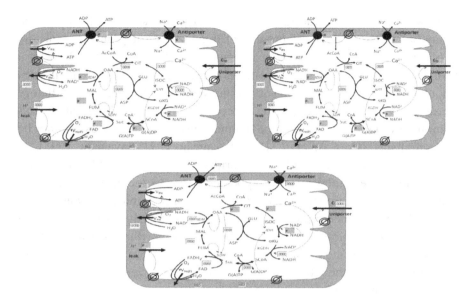

Figure 5.5. Elementary modes of the mitochondrial model as calculated by the CNA. Each panel shows the fluxes through each of the processes considered in the network either with a # sign or a number. The reactions or processes that have a number in each of the panels are able to operate together, leading to a steady state, according to the definition of an elementary mode (see text). For example, at the top left panel, the upper portion of the TCA cycle is operating (including citrate synthase, aconitase, isocitrate dehydrogenase, and aspartate amino transferase), providing NADH that will be consumed in respiration, which in turn will generate $\Delta\Psi_m$ that will be consumed by the leak of protons. The calculation of elementary fluxes is constrained so that the system should be able to evolve to a stable steady state. This elementary mode does not change the level of adenine nucleotides, since the enzymes catalyzing reactions in which ADP_m or ATP_m are involved, are not operational.

of a process or designing systems for a specific purpose), and for evolutionary studies by analysis of multispecies genomes (Palsson, 2009).

On a different note, Jamshidi and Palsson (2008), attempting to breach the evident limitations of COBRA and other stoichiometric approaches, proposed a method based on linear algebra to incorporate kinetic information in order to reconstruct networks at genome-scale. In this formalism, the flux matrix, represented by the Jacobian matrix of the network of interest, is decomposed into the stoichiometric, rate constants, and thermodynamic, matrices containing all kinetic information and thermodynamic constraints of the system. This approach would be useful if all the processes in the cell were linear or operating at or close to a steady state. In this case, one may be able to express the product of the enzyme rate constant times its concentration and its dependence on substrates and effectors following a Michaelis–Menten–Henri or an allosteric mechanism. This formalism does not apply to transients or nonasymptotic steady conditions, e.g. oscillatory states, since the reduction of the process to a linear combination of stoichiometric, kinetic, and thermodynamic aspects of the process cannot be accomplished (Cortassa *et al.*, 1991; Savageau, 1991; Savageau, 1995).

It should be emphasized that algorithms such as CNA, COBRA, Opt-Flux, and other stoichiometric approaches try to reconstruct metabolic fluxes from high throughput data. In the case of genomics, qualitative data is used, i.e. presence or absence of a given gene. In the case of transcriptomics or proteomics, even if relative levels of messenger RNA, protein or post-translational modification are quantitated, there is no dynamic information regarding the rates of processes involved (e.g. enzyme(s)-catalyzed or transport reactions).

The reconstruction of metabolic pathways of an organism of interest can be achieved with the assistance of databases that contain the annotated information about their topology, structure, and links to other pathways. Table 5.1 summarizes the main sources of this type of information. The metabolic pathways so reconstructed can then be subjected to quantitative analysis. A set of feasible solutions can be sought, but in a sufficiently restricted form so as to allow discrimination between desired and undesired outcomes.

All methods described above do not account for the intrinsic spatiotemporal complexity of metabolic and transport networks, and their inherent capacity to self-organize, giving rise to emergent behavior. Later in this chapter, we present and discuss methods for detecting and analyzing nonlinear properties of networks.

Topological Analysis of Networks

The *topological* view of networks is based on network theory and the classical work on random graphs by Erdös and Rényi (1960). This work was further elaborated by Watts and Strogatz (1998) to account for local clustering by adding the two properties of *short paths* and *high clustering*. These

Table 5.1   Databases for genomic, metabolic, and biochemical information

Name of Database	URL	Type of information	Reference
Kyoto Encyclopedia of Genes and Genomes (KEGG)	http://www.genome.ip/kegg/	metabolic pathways, reactions, genes and genomes	(Kanehisa and Goto, 2000)
Encyclopedia of metabolic Pathways (MetaCyc)	http ://metacyg.org/	metabolic pathways, reactions, genes, compounds and organisms	(Caspi *et al.*, 2006)
BRENDA. Comprehensive Enzyme Information System	http://www.brenda-enzymes.org/	Enzyme data, structural and kinetic, and metabolic	(Barthelmes *et al.*, 2007, Schomburg *et al.*, 2002)
ExPASy ENZYME	http://ca.expasy.org/enzyme/	Enzyme data, mainly structural	(Bairoch, 2000, Gasteiger *et al.*, 2003)
ExPASy contains the "Roche Applied Science Biochemical Pathways"	http://ca.expasy.org/tools/pathways/	metabolic pathways with links to structural databases	
BioPath	http://www.molecular-networks.com/ databases/biopath		(Reitz *et al.*, 2004)
Comparative Pathway Analyzer	http://cpa.cebitec.uni-bielefeld.de/	compare metabolic pathways	Oehm, 2008
Biochemical Genetics and Genomics knowledgebase	http://systemsbioloey.ucsd.edu/ In_Silico_Oreanisms/Other_Organisms	metabolic network reconstructions	(Feist *et al.*, 2009)
Plant Metabolic Network	http://www.plantcyc.org/	plant metabolic pathways	(Zhang *et al.*, 2010)

features confer high-speed communication channels between distant parts of the network favoring global coordination as shown for natural, social, and technological networks (Strogatz, 2001; Barabasi, 2009; Bascompte, 2009; Vespignani, 2009). More recently, these initial efforts have been extended to accommodate other characteristics exhibited by real networks. The original ideas on networks by Erdös–Rényi were aimed at describing random connectivity, whereas the new developments have introduced the view of "scale free" networks based on concepts of scaling and inverse power laws. According to this view, most of the nodes in a network will have only a few links, and these will be held together by a small number of nodes exhibiting high connectivity, rather than most of the nodes having the same number of links as in "random" networks (Fig. 1.8). The few nodes having many links represent the network's "hubs" and explain, at least in part, the existence of inverse power law distribution in the connectivity of scale free networks.

A significant finding was that for most large networks the degree distribution significantly deviates from a Poisson distribution. Thus, the statistical distribution of the connectivity degree of scale-free networks (given by the number of links exhibited by the nodes of a network) exhibits a continuous hierarchy of nodes, spanning from rare hubs to numerous tiny nodes, rather than having a single, characteristic scale (see Chap. 1, Fig. 1.8) (Barabasi, 2003). The scale-free topology exhibited by networks can be explained by *growth* and "preferential attachment" among nodes with higher probability of expanding their links, concepts largely derived from studies on network topology of the Internet (Barabasi, 2003). These two concepts introduced a more dynamic view on the development of topology of network connectivity, explaining the origin of hubs and power laws.

Scale-free networks are very robust against failure. This is a key feature since networks must be able to maintain their function in spite of constant perturbation and challenge by fluctuations in variables internal and external to the system. More precisely, the concept of robustness emphasizes the capacity or persistence of topological and functional network properties upon removal of links (Steuer and Zamora Lopez, 2010). Although robust against random attacks, scale-free networks exhibit fragility against selective attacks.

A profuse literature has emerged in the research field of networks in recent years, that is still growing (e.g. Albert, 2005; Albert and Barabasi, 2002; Junker, 2010) and has been applied to various sorts of networks, among them metabolic ones (Almaas *et al.*, 2004; Jeong *et al.*, 2000; Papin *et al.*, 2005; Wagner and Fell, 2001).

Basic concepts on networks and graphs

Topologically, the term network corresponds to a set of elements with connections or interactions between them, which are formally represented with graphs. A graph is a mathematical object consisting of vertices (also called nodes) and edges (links), representing elements and connections, respectively (Schreiber, 2010). A graph $G = (V, E)$ consists of a set of vertices V and a set of edges E, where each edge is assigned to two vertices (Fig. 5.6(A)). An edge where the two end-vertices are the same vertex is called a loop.

In a graph, it should be clearly visible which pairs of vertices are connected by edges and which are not; the layout of a graph is represented by the position of the vertices and lines connecting them. Graphs can be undirected or directed; undirected graphs contain connections between vertices that are without a direction (Fig. 5.6(B)–(D)). Examples of undirected graphs are those used to model protein interaction networks, phylogenetic networks, and correlation networks. In a directed graph, edges have direction which is usually denoted by an arrowhead. Typical examples of biological networks modeled by directed graphs are metabolic networks, gene regulation networks, and food webs.

Mixed graphs contain directed and undirected edges; relevant biological examples of mixed graphs are given by some types of protein interaction networks. A graph where an edge connects more than two elements is called hypergraph, which consists of a set of vertices and a set of hyperedges; these are common in metabolic networks (Fig. 5.6(E)).

Degree distribution. One of the most basic properties of a vertex V_i is its degree k_i, defined as the number of edges (links) adjacent to the vertex (node). Taking all vertices of a network into account, the degree distribution $P(k)$ is the probability $P(k)$ that the degree of a randomly chosen vertex equals k. Thus, the spread in the node degrees is characterized by

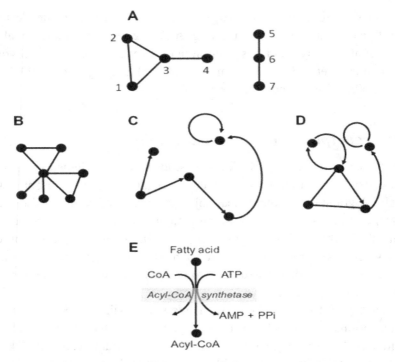

Figure 5.6.   Graphical representation of networks. (A) Graph $G = (V, E)$ with vertex $V = \{1, 2, 3, 4, 5, 6, 7\}$ and edge set $E = \{\{1, 2\}, \{2, 3\}, \{1, 3\}, \{3, 4\}, \{5, 6\}, \{6, 7\}\}$. (B)–(D) Undirected, directed, and mixed graphs, respectively. (E) Metabolic network depicting a hyperedge (Acyl-CoA synthetase), i.e. an edge that connects more than two elements (fatty acid, CoA, ATP, Acyl-CoA, AMP, PP$_i$). The reaction depicted corresponds to the first step in fatty acid degradation (redrawn in part from Schreiber, 2008).

the distribution function $P(k)$, which gives the probability that a randomly selected node has exactly k edges. The degree distribution $P(k)$ is one of the most prominent characteristics of network topology.

A key discovery was that the distribution $P(k)$ of many networks approximately follows a power law $P(k) \sim k^{-\gamma}$, where γ denotes the degree exponent (Albert and Barabasi, 2002; Steuer and Zamora Lopez, 2010). While the vast majority of vertices exhibit a small number of links, a reduced number of vertices are highly connected (hubs) (see Fig. 1.8). Unlike in scale-free networks, in random networks all nodes have the same number of edges (same node degree). Since in a random graph the edges are placed randomly, the majority of nodes have approximately the same degree, close to the average degree $\langle k \rangle$ of the network. The degree distribution of a random graph is a Poisson (exponential) distribution with a peak at $P(\langle k \rangle)$ (see Fig. 1.8).

Although several complex networks such as the World Wide Web, cellular networks, the spreading of viruses (Lloyd and May, 2001), the Internet, some social networks, and the citation network are scale free, not all of them are. For example, the power grid or the neural network of *C. elegans*, appear to be exponential. However, that does not mean that they are random, and are best described as evolving networks, i.e. networks that can develop both power-law and exponential degree distributions (Albert and Barabasi, 2002).

Clustering coefficient. The clustering coefficient C measures the probability that two vertices with a common neighbor are connected. This measure characterizes the internal structure of a network, accounting for its local cohesiveness. The clustering coefficient C_i of the vertex V_i is given by the ratio of the actual number of edges E_i between neighbors to the maximal number E_{max}:

$$C_i = \frac{2\mathbf{E}_i}{\mathbf{k}_i(k_i - 1)} \tag{5.1}$$

The global or mean clustering coefficient $C = \langle C_i \rangle$ of the network is the average cluster coefficient of all vertices. Metabolic networks exhibit high C, indicating local cohesiveness and a tendency of vertices to form clusters or groups (Rosa da Silva *et al.*, 2008; Steuer and Zamora Lopez, 2010).

An alternative, but equivalent, definition of C can be given with respect to the number of triads (triples of vertices where each vertex is connected to both others) within a network (Albert and Barabasi, 2002; Steuer and Zamora Lopez, 2010).

Methods for metabolic network decomposition into relatively independent subsets are essential to better understand the modularity and organization principles of large-scale networks. To this purpose, several methods have been proposed such as elementary flux model analysis (see above) and extreme pathway analysis (Rosa da Silva *et al.*, 2008).

When applying the topological analysis to biological networks one should keep in mind that the concept of networks goes beyond topology, which is essentially an abstract representation. Complex networks are also characterized by emergent features that transcend their topological properties and relate more with the organization and function of the network as a whole. These features will be treated later in this chapter. The structural (morphological) and dynamic view of networks relate to their underlying spatiotemporal organization (see above). For the sake of example, compartmentation in metabolic networks and precise spatial positioning in cellular systems determine that certain links are favored with respect to others. On the other hand, under self-organized spatiotemporal organization, spatially distant elements in the network become synchronous and thus functionally related (Aon *et al.*, 2007 Aon *et al.*, 2008b).

Usefulness of the Analysis of Metabolic Networks in the Context of MCE

There are two levels of information regarding the study of metabolic networks. One of them is qualitative and deals with the structure of the pathways, reactions, and group transfers participating in metabolism. In that respect, the analysis of the type of reactions and general rules for converting compounds revealed some optimality principles in metabolic pathways. Some optimality principles underlying the design of the pentose phosphate pathway have been described (Alon, 2003; Melendez-Hevia, 1990). More recently, an optimality analysis of metabolic pathways design revealed that the paths are the shortest ones resulting in the minimal distances (computed as number of steps) between the 12 key metabolites from which all cell macromolecules can be built as well as from those key metabolites to every macromolecule precursor (Noor *et al.*, 2010). These optimality principles will be useful at the time of (re)designing pathways for specific purposes.

Quantitatively speaking, the usefulness of the analysis of networks is directed toward the evaluation of the feasibility of engineering a process. The elementary flux-mode analysis can provide hypothetical yields in processes leading to a desired product that could point out if there is room for improvement. Also, flux-balance analysis can point to specific hindrance to enhancement of a process due to the topology of the processes involved in the production of a product. Last, but not least, the development of synthetic organisms raises the possibility of MCE strategies aimed at shortening the distance between elements in a network by introducing new activities and evaluating those consequences by metabolic flux analysis (Dong *et al.*, 2011; Lee *et al.*, 2007; Steen *et al.*, 2010).

Kinetic Modeling in Microbial Physiology and Energetics

Earlier attempts of modeling physiological behavior in organisms were based on thermodynamics (Pirt, 1975; Roels, 1983) (see Chap. 2). Thermodynamic models have been extensively used for describing energy transduction in microbial physiology and mitochondrial energetics (Pietrobon *et al.*, 1986; Stucki, 1980; Westerhoff and Van Dam, 1987). The advantage of these earlier attempts was their simplicity and agreement with the fundamental principles ruling biological free-energy transduction, i.e. the thermodynamic laws. However, the lack of mechanistic details accounting for many of the observed behaviors was a serious drawback.

Kinetic modeling provides means to circumvent the limitations of purely thermodynamic and stoichiometric models, by accounting for molecular mechanisms and regulatory interactions (Cortassa *et al.*, 2003; Segel, 1980). Limitations in the availability of quantitative kinetic information (e.g. maximal velocities, affinity constants) are among the drawbacks of the kinetic approach. However, when quantitative data is lacking, the model can be used as a tool for testing hypotheses concerning the mechanism(s) underlying an observed behavior.

Kinetic models can be also phenomenological, i.e. not based on known chemical or transport reaction schemes. However, phenomenological models, by not considering a molecular mechanism specific to the reaction or the process being modeled, cannot be adequately tested and therefore have limited applicability or usefulness.

Modular building of kinetic models

Each module of a kinetic model represents the known or hypothesized kinetic scheme available for a process (Cortassa *et al.*, 2002; Cortassa *et al.*, 2003; Cortassa *et al.*, 2006; Magnus and Keizer, 1997). In the modular approach to computational modeling of networks, a module represents an edge linking at least two nodes. This is analogous to the topological view of networks, conceived as a collection of edges and nodes exhibiting large-scale organization (Almaas et al., 2004; Jeong et al., 2000). A main difference is that the modular approach takes into account functional and regulatory interactions of reconstructed networks, a decisive feature for analyzing the dynamics of complex physiological responses.

According to the level of detail included in the model formulation, the modular approach provides the possibility of zooming in and out in a network of processes. For instance, the TCA cycle, or the glycolytic pathway may be described as set of reactions (8 or 10 reactions, respectively) or as a single aggregated step (Fig. 5.1). Or a biochemical reaction may be written down taking into account the elementary kinetic steps of the catalytic cycle (Hill and Chay, 1979).

Novel behavior in the spatiotemporal organization appears when the modules of a complex system are assembled into a unified scheme. These novel properties are considered emergent because they are unique, and thus could not had been anticipated from the properties of the isolated modules. For example, a model of mitochondrial energetics in which the proton motive force, Δp, is a function of state variables, $\Delta \Psi_m$ and ΔpH, has shown the emergent property of maintaining Δp constant, under conditions in which $\Delta \Psi_m$ and ΔpH markedly vary. However, such constancy is not inherent within the model thus representing an emergent behavior from its inherent dynamics (Wei et al., 2011).

In the following paragraphs we apply the modular approach to building models of biochemical and bioenergetic processes in microbial physiology.

Mathematical models behavior and tools: Reliability criteria

The following set of criteria specify the reliability of a computational model: (i) sound physico-biochemical basis, (ii) ability to reproduce qualitative and quantitative experimental data, (iii) provision of meaningful explanation of the simulated experimental behavior, and (iv) predictive power (Aon and Cortassa, 2005). Along with the buildup and testing of our models, we will emphasize how we account for the criteria above.

Basic biochemical information about the metabolic pathways and transport processes that are being accounted for is required to build kinetic models. This information can be obtained in biochemistry or physiology textbooks and some biochemical databases such as (for a more comprehensive survey, see Table 5.1):

KEGG, Kyoto Encyclopedia of Genes and Genomes,

BRENDA, a comprehensive enzyme information database.

Repository databases of models are available as well. These databases can provide a template for building models, constituting a good starting point for a beginner in the field. The following are some of the many repositories of models that are available:

(1) EMBL-EBI European Bioinformatics Institute (http://www.ebi.ac.uk/biomodels-main/),
(2) IUPS Physiome Proyect repository (http://models.cellml.org),
(3) ModelDB (http://senselab.med.yale.edu/ModelDB/), and
(4) Pathways Logic Models (http://mcs.une.edu.au/~iop/Bionet/index.html).

The models in these databases are stored according to a series of standards in language, annotations (e.g. SMBL, CellML) and are publicly available for download.

In all cases experimental data is required to constrain the modules' kinetics, usually obtained under *in vitro* conditions (see below). Data are also available to constrain the assembled model.

Among the tools required to work with kinetic models are the computational ones that allow us to work with the modules, such as MATLAB (www.mathworks.com), Mathematica (www.wolfran.com), Maple (www. maplesoft.com). These tools may be used in the modular analysis along the process of building up the model, or with the assembled model. Some useful computational modeling packages have been developed for, e.g. MATLAB. Among them the graphical package MatCont (www.matcont.ugent.be) is able to simulate time-dependent behavior and calculate the stability and the parameter sensitivity of the model (Dhooge *et al.*, 2008).

Model building process

A crucial issue is identifying the level of organization accounted for by the model (e.g. molecular, (sub)cellular, (multi)cellular). Such level of organization should suit the question(s) addressed by the model. An important decision is whether the model will take into account spatial coordinates. This choice will determine if the model will be represented by a set of ordinary differential equations (ODEs, the only independent variable is time), or by partial differential equations (with time and spatial coordinates in one, two, or three dimensions, acting as independent variables). We will restrict the scope of this section to the discussion of models of ODEs (Segel, 1980).

The *set of processes of interest along with observables* (*experimental variables*) to be included in the model needs to be identified. For example, if the aim is to model the growth of a microbial population under defined conditions, the concentration of growth-limiting substrate(s) and its consumption kinetics as well as the biomass will be relevant variables. Or if the intent is to model the elimination of a pollutant or xenobiotic from the environment, the significant variables will be those related to the transport and degradation pathway of that compound.

According to the modular approach, the kinetic expressions that best represent the rate of the processes identified previously, including information about the mechanisms underlying the behavior of a particular step, will need to be chosen. Kinetic models of microbial growth processes may encompass, for example, the **exponential rate** equation (see Chap. 3), or the **Fick's laws** of diffusion for molecules following a concentration gradient in a homogeneous medium, if the process is limited by diffusion (Crank, 1975).

Models will not always be mechanistically sound; many processes have not been yet quantitatively and/or kinetically studied to the point of enabling a mechanistic description. In such cases, an "ad hoc" equation may be written down, or a hypothetical functional relation. Savageau had proposed a power-law formalism postulating that each rate expression follows fractal kinetics, where the order of the reaction with respect to each substrate/effector is the fractal dimension of the process (Savageau, 1995). An attempt at a general formalism has been recently proposed, according to which kinetic models may be constrained using perturbation data (Tran *et al.*, 2008).

A modeling package (MatLab, Maple, Mathematica, etc.) will provide the computational environment to represent the kinetic behavior of a particular module as a function of each one of the participating variables in the rate equation (Cortassa *et al.*, 2003; Cortassa *et al.*, 2004; Cortassa *et al.*, 2006; Gunn and Curran, 1971; Segel, 1975). Critical in this procedure is to choose the right set of parameters, which in the case of, e.g. an enzyme following Michaelis–Menten kinetics, correspond to the K_M values for each of the substrates, products or effectors participating in the equation, and the V_{max} value of the enzyme at saturating levels of all substrates and/or effectors. When available, a first approximation about parameter values can be obtained from experimental data (Barthelmes *et al.*, 2007). These values may need some adjustment since the conditions in which they have been measured may not correspond to physiological ones. As an example of a mechanistic rate equation, we consider the ATP hydrolysis associated with the activity of the actomyosin ATPase in cardiac muscle myofibrils (Fig. 5.7). The kinetics of this enzyme was described according to a Michaelis–Menten mechanism (Segel, 1975). The rate expression adopted is able to simulate the experimental data, based on the saturable dependence of the enzyme activity on ATP concentration (represented by the Lineweaver–Burk plot of the experimental data), and the competitive ADP inhibition (Fig. 5.7), thereby justifying its validity for describing this process (see Cortassa *et al.* (2006)).

The sensitivity of the curve relating the rate of a given process versus substrate (or effector) concentration to the values of the kinetic parameters

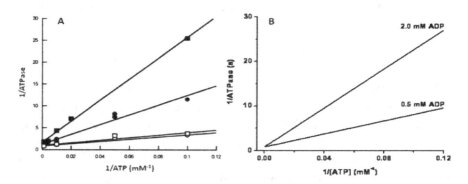

Figure 5.7. Kinetics of the myofibrilar ATP hydrolase in cardiac muscle analyzed with respect to the ATP concentration. The Lineweaver–Burk plots display the experimental (left panel) and modeled behavior of the myofibril ATPase prior to the incorporation of the activity to the ECME model (Cortassa *et al.*, 2006). Taken into account are the ATP dependence as well as the inhibitory effect of ADP. Redrawn from Cortassa *et al.* (2006).

needs to be studied. This will provide information about ways of modifying a module behavior in the fully assembled model when the simulated output differs from experimental data. Studying the sensitivity of the enzyme to parameters or variables also informs about mechanisms through which its activity might be modulated *in vivo*.

The individual behavior of a module should be compared with experimental data obtained *in vitro* from the isolated activity measured either in crude extracts, permeabilized cells, tissues, or partially purified preparation. Eventually, *in vivo* experimental data can be obtained for some biological processes and used to test the equation standing for the description of that process. The agreement of theoretical and experimental data may be qualitative, quantitative or, at best, both.

Assembling the modules

The modules may be assembled once their individual behavior is proved satisfactory according to the criteria above. At this stage, it may be convenient to schematize the relationship between modules in a diagram such as shown in Fig. 5.2 for the model of mitochondrial energetics (Cortassa *et al.*, 2003). The next step is to decide which of the variables participating in the equations will become state variables (i.e. whose values will completely specify the status of a system), and which will be adjustable parameters. For each state variable, an ODE may be written with the following general scheme:

$$\frac{d\mathrm{M}}{dt} = +\mathrm{V}_{production} \pm \mathrm{V}_{transport} - \mathrm{V}_{consumption}, \qquad (5.2)$$

where M represents an intermediary (metabolite) in the system, or in a compartment thereof; $V_{production}$, the sum of the rates of all processes contributing M; $V_{consumption}$, the sum of the rates of all processes consuming M, and $V_{transport}$, the rates of the process that carries M inside or outside the compartment, or system being modeled. Table 1 shows the system of ODEs used in our model of mitochondrial energetics (Fig. 5.2) (Cortassa *et al.*, 2003). Worth of notice at this point is that the rate in ODEs may be subjected to scaling factors to account for volume or buffering effects. In Table 1, the buffering of the concentration of mitochondrial free Ca^{2+} is taken into account by a constant factor, f (Table 1, Eq. 12). In the case that several compartments are considered, the volume ratio of the compartments has to be used to scale the rates if they belong to different compartments.

In order to simulate the behavior of a model, we need to consider the format in which to write the differential equations. The best advice is to take a sample model from the chosen simulation package, and modify it to include the specific ODE from one's own model. Then, parameter files have to be assembled and linked or included, as well as a set of initial values of the state variables (initial conditions).

Another important issue is the choice of the integration algorithm that will be used to simulate the model's behavior. Among matters to be considered for this purpose is the "stiffness" of the model, i.e. if the time scale in which different processes attain steady state is very different ($>10^2$ of the time unit used). These values are usually given in milliseconds for channel gating events, or in seconds for mitochondrial energy transduction processes. If some rapid kinetics is being accounted for, such as fast ion-transport mechanisms, then "stiff" solvers may be required. Most ODEs solver packages have their own integrators. Matlab includes a series of integrators for nonstiff ODEs systems (ode45, ode23, ode113), or for stiff ones (ode15s, ode23s), to name but a few. Fourth-order Runge–Kutta and Euler are commonly used integrators for nonstiff equations.

Running simulations with the model: A combined experimental-computational approach

Simulate the behavior of the assembled model as a function of time; first for a very short time (depending on the temporal resolution of the model may be a few msec, min, or hr) to observe the stability of the model. It is important to watch whether some state variables take negative or very large or small values, or if they exhibit large deviations from physiological values. This step, the most experience-demanding for the modeler, will provide key insights into *which* and *how* (extent and direction of change: increasing or decreasing) parameters may be adjusted to render reasonable simulations of the model behavior.

The next step is to run a model simulation till it reaches a steady state, i.e. where the derivatives of the state variables are less than 10^{-6}–10^{-8} in relative terms ($\Delta x/x$). For the first simulations, this may take a long time, depending on the initial values of the state variables.

At this stage, the modeler should start comparing the model output with experimental data (e.g. steady-state values of variables, or fluxes). Such comparisons will lead the fine-tuning of parameters and criteria utilized to adjusting parameters, which need to be made explicit.

Model simulations should be compared with experimental data from transient behavior when available, especially under parametric conditions which closely reproduce the experiment. As an example of comparison between experimental and theoretical data, Fig. 5.8 shows the temporal profile of bioenergetic variables (NADH, $\Delta\Psi_m$) from isolated mitochondria. The qualitative and quantitative agreement between the experiment and simulations indicates that the model accounts for the mechanisms operating in mitochondria when challenged with substrates, ADP, and uncouplers under *in vitro* conditions. In kinetic modeling, the experimental–theoretical agreement achieved in time-dependent behavior is of utmost importance since it is the performance of the model during transients that provides the most stringent test of its ability to represent physiological mechanisms.

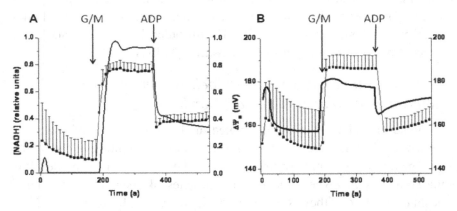

Figure 5.8.   Time-dependent behavior of mitochondrial energetic variables during the state 4 to state 3 transition. Left panel shows the dynamic response of the endogenous fluorescent signal derived from NADH when a suspension of mitochondria was challenged by successive additions of glutamate plus MAL (G/M, 5 mM each) and ADP (1 mM). Right panel shows the corresponding traces of $\Delta\Psi_m$ as obtained from the calibrated ratiometric method with TMRM (Tetramethyl rhodamine methyl ester), and measured simultaneously with NADH. Experimental data are represented by filled squares, whereas the continuous lines stand for the simulated behavior obtained with the mitochondrial model of energy metabolism. Very recently, this model has been extended to account for Na^+, H^+, Pi, and pH as state variables, including the regulation of mitochondrial activities by pH (adapted from Wei *et al.* (2011)).

In the case of the ECME model (Cortassa *et al.*, 2006), the dynamic response of mitochondrial energetics, following changes in workload conditions in heart trabeculae, was a main test of the model's capacity to discriminate between competing hypotheses regarding the mechanisms underlying transient energy supply–demand matching. This is precisely what we meant by "explanatory power," as mentioned in the introduction of this section. Utilizing the ECME model, we tested the involvement of ATP and Ca^{2+} in the coupling between energy supply and demand during transients elicited by a temporary increase in workload, by means of a higher stimulation frequency (Brandes and Bers, 2002; Cortassa *et al.*, 2006) (Fig. 5.9). Under these conditions, the role of ATP was investigated by modifying the rate constant of ATP hydrolysis by the myofibrils, which represents the most conspicuous source of ATP consumption in the working heart. Similarly, in order to investigate the involvement of Ca^{2+}, the maximal transport rate of Ca^{2+} through the mitochondrial Ca^{2+} uniporter was decreased. In each situation, simulations of the temporal profile of mitochondrial NADH and Ca^{2+} were obtained, and conclusions drawn (see caption Fig. 5.9).

Regarding the "predictive power" of a model, it will be necessary to go beyond the traditional interpretation of the word "prediction." Understandably, a model is able to predict if it can simulate a behavior that has not yet been observed. In the model of mitochondrial energetics that incorporates metabolism of reactive oxygen species (ROS), the oscillatory behavior of reduced glutathione, GSH, was first observed in model simulations, and later confirmed experimentally (Cortassa *et al.*, 2004). Broadly speaking, if a model is able to reproduce a behavior whose mechanism was not purposefully built in, even if such behavior was known from experiments performed beforehand then it can be considered that the model has predictive power.

Metabolic Control Analysis of Networks

General concepts of MCA

Metabolic Control Analysis (MCA) is a quantitative methodology that allows answers to an important question which arises when dealing with networks of reactions: What steps control the flux through a metabolic pathway? How is the concentration of the intermediary metabolites controlled?

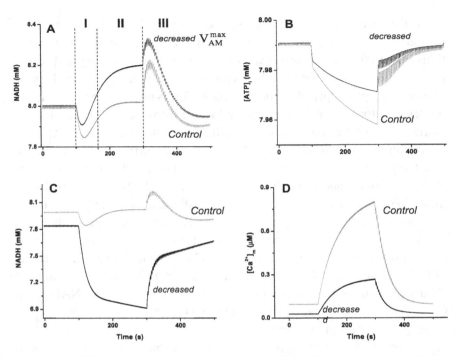

Figure 5.9. Effects on energetic behaviour after modifying the activities of myofibrillar ATPase and the mitochondrial Ca^{2+} uniporter in the ECME model and in response to changes in workload (Cortassa et al., 2006). Panel (A) shows the NADH behavior when the ECME model was studied during transitions in pacing frequency. Before region I, 0.25 Hz pacing was applied, whereas in regions I and II the pacing frequency was kept at 2 Hz and decreased back to 0.25 Hz in region III. The black trace shows the effect of lowering the maximal rate of ATP hydrolysis by the AM-ATPase to one half ($V_{AM}^{max} = 3.6\ 10^{-3}$ mM ms^{-1}) with respect to the control simulation (gray trace); all other parameters being identical in the two simulations. Panel (B) depicts the average profile of cytoplasmic ATP, ATP$_i$, following changes in workload under control (gray trace) or decreased myofibrillar ATPase activity (black trace) according to the same protocol described in panel (A). Panel (C) exhibits the NADH profile obtained when the ECME model was studied with the same pacing protocol described in (A) but decreasing to one-tenth the maximal rate of the mitochondrial Ca^{2+} uniporter (black trace, $V_{max}^{uni} = 2.75\ 10^{-3}$ mM ms^{-1}) with respect to the control (gray trace) simulation. Panel (D) shows the corresponding profile of mitochondrial Ca^{2+} when the uniporter activity was reduced in the simulation depicted in panel (C). Notice the strong reduction in $[Ca^{2+}]_m$ accumulation (black trace) compared to the control simulation (gray trace).

Source: Reproduced from Cortassa S, Aon MA, O'Rourke B, Jacques R, Tseng HJ, Marban E, Winslow RL. (2006) A computational model integrating electrophysiology, contraction, and mitochondrial bioenergetics in the ventricular myocyte. *Biophys J* **91**, 1564–1589, by copyright permission of the Cell Press.

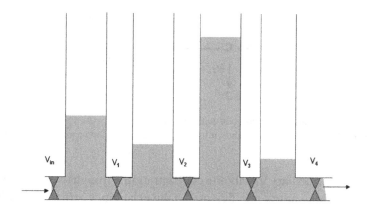

Figure 5.10.   Hydraulic analogy of a metabolic pathway.

This approach has been mainly applied to systems functioning at steady state.

Figure 5.10 shows a metabolic pathway and its hydraulic analogy. The flux of the fluid (e.g. a carbon source, in gray) flowing through communicating containers is regulated by valves (V_1, V_2, etc., i.e. enzymes). The level (e.g. concentration of intermediary metabolite) attained by the fluid in the containers depends on the aperture of the valves. Intuitively, it is easily seen that valve V_3 is the main step controlling the flux, since the fluid accumulates to a higher extent in the container controlled by V_3. Another important feature is that the level attained by the fluid in each container is different and finite, suggesting that each and every valve controls the flux of the fluid to a certain extent.

A hypothetical metabolic pathway and the application of the matrix method of MCA for determining flux and metabolite control coefficients are shown in Fig. 5.11. The metabolic route in Fig. 5.11 represents a network of chemical reactions constituted by branched and linear pathways which share an intermediate B (Appendix). The purpose of this example is to understand in a pathway of medium complexity, the main basis for using MCA. An experimental example of higher complexity is described in Chap. 4 (see Section: *Rational design of microorganisms*).

Considering the hypothetical pathway shown in Fig. 5.11, MCA intends to quantitate two main aspects of its control: of the flux through the pathway,

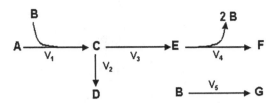

Figure 5.11.   A simplified metabolic pathway. The sketched pathway contains the main elements of the glycolytic model shown in Fig. 4.12 (i.e. they are topologically similar).

and of the intermediary metabolite concentration, through two types of coefficients: the flux control coefficient and the elasticity coefficient. Both type of coefficients represent either global or systemic (flux control coefficient) and local or component (elasticity coefficient) properties of the sequence of reactions represented in Fig. 5.11 (Heinrich and Rapoport, 1974; Kacser and Burns, 1973; Kacser and Burns, 1981; Kacser and Porteous, 1987) (see Fell (1992) and Liao and Delgado (1992), for reviews).

The control coefficients describe how a variable or property of the system, typically a metabolic flux or the concentration of a metabolite, will respond to variation of a parameter, usually enzyme concentration (Fell, 1992):

$$C_{Ek}^{Ji} = \frac{E_k}{J_i} \frac{\partial J_i}{\partial E_k} \qquad (5.3)$$

Figure 5.12 shows different possible relationships between the flux (J_i) of a metabolic route and the activity of a certain enzyme (E_k). In a steady state, the flux control coefficient C_{Ek}^{Ji} is the fractional change in flux for a fractional change in the activity of enzyme, E_k (Heinrich and Rapoport, 1974; Kacser and Burns, 1973).

For each metabolite in the system, there are also concentration control coefficients to quantify for the effect of each enzyme on the level of that metabolite. Thus, similarly a metabolite, M_i, concentration control coefficient, C_{Ek}^{Mi}, can be defined:

$$C_{Ek}^{Mi} = \frac{E_k}{M_i} \frac{\partial M_i}{\partial E_k}. \qquad (5.4)$$

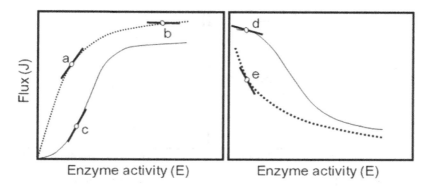

Figure 5.12.   Flux (J_i)-enzymatic activity (E_k) relationships, illustrating positive or negative control coefficients. (A) The flux increases with E_k in a hyperbolic (A, (a), (b)) or sigmoidal (A, (c)) fashion. The flux control coefficient is determined at (a), (b), or (c) points, with positive values in all cases. (a) is slightly lower than 1.0; (b) close to zero; (c) higher than 1.0. (B) Flux control coefficients are either slightly (d) or strongly (e) negative.

Two types of control coefficient have been used, a rate-based (v-type) coefficient and an enzyme concentration-based (e-type) coefficient (Liao and Delgado, 1992). The changes in enzyme activity may arise from changes in the enzyme concentration or modification of their kinetic properties, e.g. k_{cat}.

The link between the properties of an enzyme and its potential for flux control is given by the elasticity coefficient, ε. Figure 5.13 displays possible relationships between the rate, v, of an isolated enzymatic reaction and the substrate concentration, S. The elasticity coefficient ε_{ij} for the effect of intermediary metabolite S on the velocity v_i of enzyme E_i is the fractional change in rate of the isolated enzyme, δv_i, for a fractional change, δ_S, in the amount of substrate S (Heinrich and Rapoport, 1974; Kacser and Burns, 1973):

$$\varepsilon_S^{vi} = \frac{S}{v_i}\frac{\partial v_i}{\partial S}. \tag{5.5}$$

Multiplying by $\frac{E_k}{J_i}$ or $\frac{S}{v_i}$ makes the coefficients dimensionless thereby independent of the units used. All of the terms involved in Eqs. (5.3) and (5.4) are evaluated at steady state.

If an analytic expression for v is available, the elasticity coefficients can be obtained by taking the partial derivative of the rate expression with

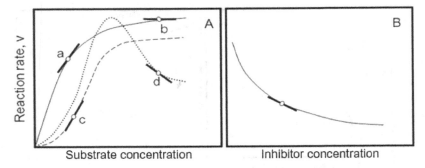

Figure 5.13.   Graphical representation of possible relationships between an isolated enzymatic activity (V) versus substrate (S) concentration. (A) Elasticity coefficients are determined from the slope of the curves, the place (a), (b), (c), (d) depending on the intracellular concentration of S (see Appendix). Points (a) and (b) correspond to different saturation states according to a Michaelis–Menten relationship, whereas (c) and (d) are from a sigmoidal kinetics. Notice that the higher the saturation, the lower the elasticity coefficient. Point (c) corresponds to an elasticity coefficient higher than 1.0, and point (d) to an enzyme inhibited by substrate with a negative elasticity coefficient. (B) Relationship between enzyme rate and inhibitor concentration; the elasticity coefficient is always negative.

respect to the variable or parameter of interest and then evaluating in the steady-state condition (see Appendix).

MCA: Summation and connectivity theorems

Several theorems constitute the main body of MCA. The summation theorem states that the sum of all flux control coefficients with respect to the activities of each of the enzymatic steps involved in the metabolic pathway being considered, is equal to unity (Kacser and Burns, 1973; Kacser and Porteous, 1987).

$$\sum_k C_{Ek}^{Ji} = 1 \tag{5.6}$$

This theorem expresses the concept that the enzymes of a pathway can share the control of the flux. In fact, an important contribution of MCA had been to demonstrate that the control of a metabolic pathway can be shared by multiple enzymes, implying that the traditional concept of a single rate-limiting enzyme (a "bottleneck") is incorrect. This result also illustrates the fact that many mutated loci are needed to increase substantially the flux through a pathway (Kacser and Burns, 1981). Moreover, the summation theorem of MCA highlights the fact that the flux control coefficient of an enzyme is not an intrinsic property of a single enzyme, but a systemic property (Fell, 1998). Otherwise stated, increasing the activity of a rate-controlling enzyme changes its flux control coefficient and also results in alteration of the coefficients of enzymes whose activities have not been changed. The latter accounts for the fact that the summation value for all flux control coefficients remains 1. This is illustrated in the section below entitled: *The TDA approach as applied to the rational design of microorganisms*.

The connectivity theorem relates the elasticities to the control coefficients:

$$\sum_i C_{Ek}^i \varepsilon_k^i = 0. \tag{5.7}$$

This theorem states that the sum over all products of the flux control coefficients (Eq. (5.3)), with respect to the activity of the enzyme catalyzing step i, times the elasticity coefficient of this same enzyme (Eq. (5.5)), is equal to zero. The connectivity theorem is regarded as the most meaningful of the MCA theorems for it provides a precise way to understanding how the kinetics of the enzymes (represented by the elasticities) affect the values of the flux control coefficients (Fell, 1992, p. 212; Sauro *et al.*, 1987; Westerhoff and Kell, 1987).

A corresponding set of theorems exist for the concentration control coefficients. The summation theorem states that the sum of all metabolite control coefficients with respect to the activities of each of the enzymatic steps is equal to zero (Kacser and Burns, 1973; Kacser and Porteous, 1987):

$$\sum_i C_{Xi}^M = 0, \tag{5.8}$$

where M represents any metabolite of a pathway, and X_i stands for the activity of enzyme i.

The connectivity theorem becomes slightly more complex, in that it has one form when the metabolite, the concentration of which is the subject of the control coefficient (say A), is different from the one on the elasticities (say B) (Fell, 1992):

$$\sum_{i=1}^{n} C_{Xi}^{A} \varepsilon_{B}^{i} = 0, \tag{5.9}$$

but the following form when they are the same:

$$\sum_{i=1}^{n} C_{Xi}^{A} \varepsilon_{A}^{i} = -1. \tag{5.10}$$

For unbranched pathways, the summation and connectivity theorems allow the direct calculation of the control coefficients from the elasticities by solving a system of linear algebraic equations (Sauro *et al.*, 1987; Westerhoff and Kell, 1987). In the latter case, the summation theorem and the connectivity theorems for all the metabolites provide exactly the number of simultaneous equations needed for calculating the flux control coefficients of all the enzymes in terms of the elasticities. This is not the case for branched pathways or pathways containing, e.g. substrate cycles (Cortassa and Aon, 1994a; Cortassa and Aon, 1997; Fell, 1992). For systems involving branches and cycles, additional relationships must be used to solve for the control coefficients from the elasticity coefficients (Sauro *et al.*, 1987; Westerhoff and Kell, 1987) (see Liao and Delgado (1992), for a review). These theorems are the basis of the matrix method, and we will take advantage of them to derive the control of ethanol production by *S. cerevisiae* in continuous cultures (see below) (Aon and Cortassa, 1997; Cortassa and Aon, 1994a; Cortassa and Aon, 1994b, Cortassa and Aon, 1997; Sauro *et al.*, 1987; Westerhoff and Kell, 1987).

Control and Regulation

It has been argued that a high control coefficient does not necessarily place control at that step, since an enzyme with low flux control coefficient might be significantly controlled through an effector (Fraenkel, 1992). Thus, control is often confused with regulation.

The problem may be addressed through the following question: Do the flux control coefficients indicate which steps "regulate" the flux? (Liao and Delgado, 1992) Conventionally, enzymes sensitive to feedback inhibition or allosteric effects are said to be the "regulating" enzymes because without these mechanisms the flux will not be responsive to the change of metabolite pools. In this sense, elasticity coefficients rather than the control coefficients will better reveal the kinetic properties around the point of evaluation. Thus, the flux control coefficients contain no information about the "regulating" enzymes in the above sense (Liao and Delgado, 1992). For enzymes that exhibit high Hill coefficients (a common way of "regulation"), the flux control coefficients tend to zero, while a saturated enzyme (no control in the typical biochemical sense) operating in the zero-order region of its substrate concentration at steady state, tends to have a large control coefficient (Fell, 1992; Kacser and Burns, 1973). Therefore, it has been suggested that "control" be used to mean the effect on flux produced by the change of enzyme activity, whereas "regulation" be used to denote the modulation of enzyme activity by effectors (Fell, 1992; Hofmeyr and Cornish-Bowden, 1991; Liao and Delgado, 1992). In a more general and quantitative sense *regulation* refers to how the flux of a pathway is modified through the effect on the rate of an individual step by cellular factors, which may include intermediary metabolite or ion concentrations, post-translational modifications, and is quantified by the *response coefficient*. This coefficient measures the fractional change in flux, e.g. respiration, in response to a fractional change in a parameter (A) (e.g. an effector such as Ca^{2+}) other than enzyme activity. The concept, originally introduced by Kacser and Burns (1973), refers to the effect of a parameter (A) over a flux or an internal metabolite. Formally, it corresponds to the product of the elasticity coefficient of step *i*, with respect to an effector *A*, times the control coefficient that a flux, *J* or an internal metabolite concentration displays with respect to the activity of step *i* (Eq. (5.11)). Therefore, regulation implies the response of a pathway to an effector on two levels: (i) the extent of control exerted on the pathway by the enzyme that is the effector's target, and (ii) the *strength or elasticity* of the effect of *A* on that enzyme (Fell, 1996):

$$R_A^J = C_{Xi}^J \varepsilon_A^i. \tag{5.11}$$

Other regulation modes are related to the localization of a metabolic pathway or an enzyme. Some mechanisms such as enzyme ambiguity, e.g. hexokinase, an enzyme that can function differently depending on its subcellular location, have been shown to influence the *transient* behavior of glycolytic flux and intermediates. On the other hand, the *steady-state* flux of glycolysis was not affected even after large changes in the amount of the mitochondrial-bound form of the kinase (Aon and Cortassa, 1997). After glucose pulses in tumor cells, dynamic ambiguity of HK was able to regulate the transients, although not the steady state control of the glycolytic flux. One important result of the mutual regulation of glycolysis and mitochondrial phosphorylation through HK ambiguity is that it appears as a biochemical device able to regulate transients more effectively than it can control steady-state behavior. Under those conditions, HK ambiguity behaves as a regulatory mechanism exerting feedback on glucose phosphorylation through mitochondrial pools of ATP (Aon and Cortassa, 1997). In connection with these results, it had earlier been stated by Heinrich *et al.* (1977) that the transient control had to be distinguished clearly from the control parameters for the steady state. Enzymes having a great influence on the transition time (i.e. that for interconversion of substrate to product at that step) may have no relevance for the flux control in the steady state and vice versa (Heinrich *et al.*, 1977).

Control of Metabolite Concentrations

In metabolic networks, some metabolites are not only subjected to mass conversion steps but also play a role in information-carrying networks (see Chap. 1, Fig. 1.10 and Chap. 10, for an extensive discussion). Examples are second messengers and allosteric effectors that being produced at a certain metabolic block exert their effects in another. For instance, 3PGA that is synthesized in the Calvin cycle in plants is a positive allosteric effector of ADPglucose pyrophosphorylase (ADPGlcPPase) in the synthetic pathway of starch or glycogen (in plants or cyanobacteria, respectively) (see Chap. 7). Thus, the regulatory role of certain metabolites makes it relevant to know how their intracellular levels are controlled.

One of the main outputs of MCA is the metabolite concentration control coefficient. As a rule, metabolite concentrations are controlled by the rate-controlling steps of the flux, and negatively controlled by the enzyme consuming the metabolite under analysis. Moreover, whenever a metabolite participates in more than one step or pathway, it is additionally controlled by the enzymes catalyzing both or all of those steps (see section below, for an example related to yeast catabolite repression).

In Chap. 7 and in the previous section dealing with control and regulation of metabolic networks, we analyze as an example starch synthesis. In this case, an enzyme such as ADPGlcPPase may effectively function both as mass–energy conversion catalyst and as a transducer device. The latter has important practical applications. When fully activated, an ultrasensitive enzyme such as ADPGlcPPase (Gomez-Casati *et al.*, 2000; Gomez-Casati *et al.*, 2001) is able to become rate-controlling (Aon *et al.*, 2001; Gomez-Casati *et al.*, 2003), thus its overexpression is likely to produce increased levels of starch (Stark *et al.*, 1992).

Control of metabolites in relation to catabolite repression mutants of S. cerevisiae

Metabolite control coefficients were calculated from experimental data in continuous cultures of wild type and catabolite repression yeast mutant. We deduced, by applying MCA, that metabolite concentrations are controlled by the rate-controlling steps of the flux (Cortassa and Aon, 1994a), while the enzyme that consumes the metabolite exerts a negative control (Table 5.2).

The positive control exerted upon the level of a metabolite M1 by a certain step of the pathway means that an increase in substrate consumption results in a rise of M1 at the steady state. On the contrary, a negative control coefficient entrains a decrease of M1 at the steady state. For example, as shown in Table 5.2, the rate-controlling steps of the flux at $D = 0.225\,\text{h}^{-1}$ are: sugar uptake ($C_{IN}^{PEP} = 0.12$), the HK ($C_{HK}^{PEP} = 0.701$) and PFK ($C_{PFK}^{PEP} = 0.231$) that exert a positive control on the PEP concentration, and the ATPase with a negative control on the PEP concentration ($C_{ATPase}^{PEP} = -1.35$).

Another distinctive feature shown in Table 5.2 is that a metabolite concentration with a conversion step involving other metabolites is additionally controlled by the enzymes catalyzing the conversion steps of those other participating metabolites. This is clearly the case for PEP, whose concentration is controlled by aldolase (negatively), alcohol dehydrogenase (negatively), ATPase (negatively), and by the functioning of the TCA cycle, positively.

Table 5.2   Metabolite concentration control coefficients of phosphoenolpyruvate (PEP) and ATP by the different glycolytic steps, in the wild type strain CEN.PK122 and *snf1* mutant growing in aerobic, glucose-limited chemostat cultures

Strain/D	C_{IN}^{PEP}	C_{HK}^{PEP}	C_{PFK}^{PEP}	C_{ALD}^{PEP}	C_{GAPD}^{PEP}	C_{PGK}^{PEP}	C_{PK}^{PEP}	C_{ADH}^{PEP}	C_{TCA}^{PEP}	C_{ATPase}^{PEP}
WT										
0.1	0.132 (0.06)	0.772 (0.35)	1.34 (0.608)	−0.195	−0.008	0	−1.33	−0.415	0.142	−0.433
0.225	0.442 (0.12)	2.60 (0.701)	0.86 (0.231)	−0.115	−0.049	0	−1.54	−1.30	0.447	−1.35
0.3	0.012 (0.008)	0.012 (0.008)	1.51 (0.98)	−0.029	−0.008	0	−1.58	−0.303	1.38	−1.00
snf1										
0.05	0.007 (0.003)	0.011 (0.006)	2.02 (0.003)	−0.046	−0.018	0	−1.45	−0.277	0.095	−0.335
0.15	0.254 (0.13)	0.803 (0.41)	0.96 (0.49)	−0.034	−0.007	0	−1.60	−0.356	0.425	−0.445
0.25	−0.089 (0.021)	−0.081 (0.02)	−3.79 (0.996)	−0.0133	−0.0144	0	−2.044	−0.3055	7.45	−1.10

Strain/D (h^{-1})	C_{IN}^{ATP}	C_{HK}^{ATP}	C_{PFK}^{ATP}	C_{ALD}^{ATP}	C_{GAPD}^{ATP}	C_{PGK}^{ATP}	C_{PK}^{ATP}	C_{ADH}^{ATP}	C_{TCA}^{ATP}	C_{ATPase}^{ATP}
WT										
0.1	0.097	0.571	0.991	0	0	0	0	−0.975	0.335	−1.02
0.225	0.194	1.14	0.378	0	0	0	0	−1.01	0.35	−1.05
0.3	−0.0006	−0.0006	−0.079	0	0	0	0	−0.303	1.38	−1.00
snf1										
0.05	0.005	0.009	1.55	0	0	0	0	−0.837	0.287	−1.01
0.15	0.11	0.35	0.417	0	0	0	0	−0.83	0.99	−1.04
0.25	−0.12	−0.107	−4.99	0	0	0	0	−0.264	6.44	−0.96

The control coefficients were obtained from the **E** matrix inversion, as described in the text. The model of glycolysis and the branch toward the TCA cycle and ethanolic fermentation are depicted in Fig. 4.12.

The *snf*1 mutant when grown in glucose-limited chemostat cultures exhibited low biomass yields and specific rates of ethanol production along with high rates of respiration (Cortassa and Aon, 1998) (Table 5.4). The 3- to 4-fold lower ATP and ADP concentrations and the ability to accumulate large amounts of intracellular G6P with respect to the wild type strain indicated altered energetic metabolism of the *snf*1 mutant (Aon and Cortassa, 1998; Cortassa and Aon, 1997; Cortassa and Aon, 1998) (Table 5.3). This mutant shows strikingly different patterns of control coefficients in two tightly controlled metabolites such as PEP and ATP (Table 5.2), especially at high growth rates. It is remarkable the increase and change of sign (from positive to negative) of the PFK concentration control coefficient on either PEP or ATP. This change may be rationalized as follows: an increase in enzyme activity (PFK) will effect a decrease in a metabolite

Table 5.3   Intracellular metabolite concentrations determined in glucose-limited chemostat cultures of *S. cerevisiae*. Metabolite concentrations are expressed in mmol l^{-1} of intracellular volume

Strain $D\ (h^{-1})$	ATP	ADP	AMP	NAD	G6P + F6P	FdP FdP	DHAP + G3P	DPG	PEP	Pyr
WT										
0.1	3.05	1.249	1.379	0.3650	8.36	0.44	0.44	2.50	2.21	<0.1
0.225	2.00	0.608	0.072	0.0276	4.06	0.51	0.73	3.35	2.87	<0.1
0.3	2.71	0.846	0.174	0.0046	35.65	1.43	1.06	4.50	3.64	5.98
*snf*1										
0.05	0.79	0.719	0.440	0.0092	60.56	0.94	1.15	3.32	2.66	<0.1
0.15	0.88	0.802	0.883	0.1850	5.06	1.20	1.30	3.76	3.64	1.62
0.25	0.95	0.538	1.200	0.0306	19.11	2.41	2.04	7.57	5.47	31.40
*snf*4										
0.1	2.46	0.733	0.347	0.3328	1.90	1.03	1.41	2.95	2.96	0.53
0.2	4.37	1.319	0.556	0.5543	1.21	1.14	1.39	4.22	3.93	0.33
0.3	4.68	1.444	0.320	0.6923	0.24	3.18	1.77	6.80	5.72	24.81
*mig*1										
0.1	2.72	0.940	0.339	0.3039	2.20	1.81	1.02	2.98	2.61	<0.1
0.2	3.74	1.589	0.424	0.4848	0.89	5.41	2.15	5.93	5.10	0.29
0.3	4.32	2.506	0.675	0.2577	1.78	4.79	1.88	6.46	5.00	8.16

Source: Reprinted from *Enzyme and Microbial Technology*, 21, Cortassa and Aon, Distributed control of the glycolytic flux in wild-type cells and catabolite (de)repression mutants of *Saccharomyces cerevisiae* growing in carbon-limited chemostat cultures, pp. 596–602. ©copyright 1997, with permission from Elsevier Science.

Table 5.4   Metabolic fluxes sustained by the wild-type strain and catabolite-repression mutants in glucose-limited chemostat cultures of *S. cerevisiae*

Strain D (h^{-1})	qO2	qCO2	qGlc	QetOH
WT				
0.1	2.12	2.38	1.22	0.0442
0.225	4.31	6.69	3.87	1.7531
0.3	4.85	11.15	7.51	5.7928
Snf 1				
0.05	2.11	2.74	1.04	0.0097
0.15	4.03	5.71	3.78	0.3558
0.25	6.34	11.55	7.23	1.1592
snf 4				
0.1	2.54	2.44	1.37	0.0142
0.2	5.91	10.41	4.13	1.7488
0.3	5.76	16.58	6.99	4.9086
mig 1				
0.1	3.16	3.55	1.48	0.0113
0.2	3.88	5.58	4.21	1.8335
0.3	3.93	13.41	10.72	7.1256

The fluxes are expressed in mmol h^{-1} g^{-1} dw.
Source: Reprinted from Cortassa and Aon (1997) ©copyright 1997, with permission from Elsevier Science.

concentration (PEP or ATP) (a negative control coefficient) leading in turn to higher FDP levels (a PK allosteric activator), and in that way decreasing PEP levels.

Equally remarkable was the increase of the absolute values of the TCA cycle control coefficients on both metabolites. The strong positive effect on the control coefficient of the ATP concentration (almost 7-fold) is due to an increase in the functioning of the TCA cycle that in turn gives higher levels of ATP. The increase of the TCA cycle and the ATP levels decreases ADP ones along with PK activity that allows the increase of PEP levels (almost 20-fold). The drastic increase of the control exerted by the TCA cycle on PEP concentration in the *snf* 1 mutant is due to the basic autocatalytic nature of the glycolytic pathway and, particularly, to the cooperative nature of PK activity. It must be remembered that the fermentative mode of glycolysis is strongly active at high growth rates in *S. cerevisiae*.

MCA of Metabolic and Transport Networks

To analyze control and regulation in a network of processes encompassing over 50 state variables related through 90–100 processes, we applied the stoichiometric matrix method of Reder (Cortassa *et al.*, 2009; Reder, 1988). This method generalizes earlier MCA tools (Kacser and Burns, 1995; Sauro *et al.*, 1987; Westerhoff and Kell, 1987). Calculations of the *control* and *response coefficient* matrices and changes in steady-state fluxes induced by perturbation of the system permit a deeper understanding of the relationship between state variables of different nature: e.g. electrophysiological (involving ion transport through channels or pumps), and mechanical (contractile activity of cytoskeletal components) processes, as they interrelate in a computational model of integrated cardiomyocyte function (Cortassa *et al.*, 2006).

Within the MCA framework, Reder developed a generalized linear algebraic method, using matrices and vectors. This method provides a way of analyzing the sensitivity of metabolic systems to perturbations triggered by either a change in the internal state of the system or by the environment. The departure point of the analysis is the *stoichiometric matrix* obtained from the set of differential equations of the model (see also Fig 5.4 in Section: *Networks Analyzed with Stoichiometric Models*). The stoichiometric matrix defines the structural relationships between the processes and the intermediates participating in the metabolic network under consideration. Otherwise, the stoichiometric matrix contains the mass–energy transformation relations between different nodes of the network. The information contained in the stoichiometric matrix is independent of both the enzyme kinetics and the parameters that rule the dynamic behavior of the metabolic system.

The *elasticity matrix* is the second piece of information required to perform control analysis according to Reder's method. The elasticity matrix is defined by the dependence of each process in the network upon intermediates (e.g. ions or metabolites) included in the model. The elasticity matrix is obtained from the derivatives of the rates of individual processes with respect to each possible effector (see Eq. (5.5)). Each elasticity coefficient reflects the local property of a process, for example, an enzyme activity with respect to its substrate, which is linked to the global behavior of the system through the steady-state levels of metabolites (substrates or effectors) in the network.

By applying matrix operations, the matrices corresponding to control and response coefficients are obtained. Both kinds of matrices quantify the relationships between control and regulation, respectively, in the network of reactions. The regulation exerted by internal or external effectors to a network can be quantified by the response coefficient (Ainscow and Brand, 1999; Kacser and Burns, 1973).

The following matrix relationships were used in the computation of flux and metabolite concentration control coefficients:

$$\mathbf{C} = \mathbf{Id_r} - \mathbf{D_x v L (N_r D_x v L)^{-1} N_r}, \qquad (5.12)$$

$$\mathbf{\Gamma} = -\mathbf{L (N_r D_x v L)^{-1} N_r}, \qquad (5.13)$$

with $\mathbf{C}$ and $\mathbf{\Gamma}$ referring to the matrices of flux- and metabolite concentration control coefficients, respectively; $\mathbf{Id_r}$, the identity matrix of dimension r, the number of processes in the network under study; $\mathbf{D_x v}$ the elasticity matrix; $\mathbf{N_r}$ the reduced stoichiometric matrix and $\mathbf{L}$, the link matrix that relates the reduced- to the full-stoichiometric matrix of the system.

The advantages of the matrix method of Reder (1988) over similar tools developed by others (Sauro *et al.*, 1987; Westerhoff and Kell, 1987) is that: (i) it can be applied to networks of any complexity, including conserved cycles or multiple branches, and hierarchical relationships (Kahn and Westerhoff, 1991); and (ii) it is not based on the compliance of the theorems of MCA and the particular conditions under which the analysis is carried out.

Regarding the limitations of MCA, the tools in this method have been developed to study metabolic systems at *steady state*. This issue raises concerns about the application of metabolic control to analyze time-dependent behavior (see Cortassa *et al.* (2009b)), for a more detailed discussion). However, Ingalls and Sauro (2003) have reported the validity of applying the MCA tools to averages of time-dependent systems such as those undergoing cyclic variations or rhythms (i.e. the value of a state variable during a cycle is averaged over time). These authors showed that the theorems of MCA, such as summation and connectivity theorems, are fulfilled during time-varying behavior.

Analysis and Detection of Complex Nonlinear Behavior in Networks

One way to get from the observations of a system with unknown properties to a better understanding of the dynamics of the underlying system is through nonlinear time series analysis (Aon *et al.*, 2011; Kantz and Schreiber, 2005; Stam, 2005). This "top-down" approach starts with the output of the system such as time series (i.e. successive values in time of an observable, a variable, of such a system) and works back to the state space, attractors, and their properties. Examples of time series include the stock index in the market, temperature in the weather or in the human body, membrane potential or metabolite concentration in a biological system, population size of a species in an ecological system. Analysis of time series represents a major challenge for conjectures on the underlying dynamics of systems as complex as the heart or the brain. With a plethora of methods, nonlinear time series analysis has become the predominant tool for reconstruction of the systems dynamics in state space.

Time series from complex phenomena are often random, but frequently they exhibit correlations or memory over time. A variable that changes randomly in time has no memory, i.e. the probability of occurrence of a positive or negative value is at each time unit one-half independently of its past value. This results in that the probability to have two negatives in a row is $1/2 \times 1/2 = 1/4$; the probability to have three negatives in a row is $1/2 \times 1/2 \times 1/2 = 1/8$, and so on. This defines an exponential distribution, describing the fact that increasing a drawdown by one time unit makes it doubly less probable. This exponential law is also known as the Poisson law and describes processes *without* memory implying that the value of a past observation has no impact on its future value. Any deviation from the Poisson law will suggest some correlation, thus a potential for prediction of future values (Sornette *et al.*, 2004).

Time series with long-term memory exhibit high frequency fluctuations at small scales which are reproduced at low frequency in large scales. The amplitude of the fluctuations is small at high frequencies and large at low frequencies. Time series that exhibit these properties are self-similar and are said to scale (Bassingthwaighte *et al.*, 1994). Scaling of fluctuations in time indicates that the longest and the shortest time scales contributing to the process are tied together, so that what affects one time scale affects them all (West, 1999). The power spectrum of such data is usually proportional to $1/f^\beta$, where f is frequency and β is a constant (Bak, 1996; Yates, 1992). Phenomena whose power spectra are homogeneous power functions lack inherent time and frequency scales; they are *scale free*, meaning that there is no characteristic time or frequency (Schroeder, 1991). Consequently, no one scale in space or time is able to characterize by itself the evolution of the system. The scale-free property (also named scale-invariance or scale-independence) is present in a wide class of complex systems (Gisiger, 2001; West, 1999; Yates, 1992; Yates, 1993). Long-ranged spatiotemporal correlations may develop when a system exhibits scale-free behavior, and so is able to support fluctuations over scales ranging from those of intermolecular forces to those billions of times longer (Aon and Cortassa, 1997; Ball, 2004).

The self-similarity and scale-free properties of time series correspond to fractal dynamics, i.e. originating from processes in which the statistical behavior of their dynamics is self-similar over increasing time intervals. Fractals may be of geometrical, statistical, and dynamic nature (Aon and Cortassa, 2009; West, 1990). Geometrically, fractals can be looked at as structures exhibiting *scaling* in *space* since their mass as a function of size, or their density as a function of distance, behave as a power law (Mandelbrot, 1983; Vicsek, 2001). The self-similarity of the scaling implies that the object remains invariant at several length scales. The fractal dimension, D_f, characterizes the invariant scaling of objects exhibiting this type of geometry. Mandelbrot and Hudson (2004) and West (1990) among others, expanded the concept of fractals from geometric objects to dynamic processes exhibiting fractal statistics.

Considering the topological properties of networks in the context of dynamics of cell function, the relationship between structure and function in complex networks is currently unknown. Recent findings show that the time series resulting from mitochondrial function at the subcellular (heart) and cellular as well as cell population levels (yeast) exhibit fractal dynamics (Aon *et al.*, 2007a; Aon *et al.*, 2008a). The scale-free dynamic organization is emergent since visible self-organized spatiotemporal behavior is present in both cellular systems (Aon *et al.*, 2003; Aon *et al.*, 2006a; Roussel and Lloyd, 2007). Whether the resulting scale-free dynamics stems from topological scale-free networks is still as open question.

Methods of Time Series Analysis

Mathematically, the fractal, scale-free nature of complex dynamic behavior results in inverse power laws, in the form of $1/f$ noise. An inverse power law in time of the correlation function (i.e. the correlation between points in time decreases with increasing time separation) and in frequency of the spectrum (i.e. the Fourier transform of the correlation function) are a signature of fractal dynamics (West, 1999).

The double nature of fractals as geometric (spatial) objects and dynamic processes has allowed their characterization and quantification through the use of common techniques. For instance, *lacunarity* utilized as a measure of the nonuniformity (heterogeneity) of structures is quantified through mass-related distributions. These distributions can be characterized by the coefficient of variation or relative dispersion, RD (= standard deviation, SD/mean) that is a highly dependent function of scale (Bassingthwaighte *et al.*, 1994; Mandelbrot, 1983; Smith and Lange, 1996). Relative dispersion as applied to the temporal domain suggests that events in different time scales are tied together through fractal statistics, provided that the scaling holds for at least two orders of magnitude.

Relative Dispersional Analysis

The method consists in sampling the data set at successively and progressively larger scales and to compare the level of dispersion on the resulting data set at each of the resultant scales (West, 1999). Adding adjacent points in the time series at 2, 4, 8, 16, and 32 successive values of the data set to calculate the RD is performed. The RD is repeatedly calculated while sampling (coarse-graining) the data set at successively larger time scales and for each grouping. A log–log plot of the RD value against the aggregation number, m, is represented. The slope of this relation provides information on the extent of long-term correlation (memory) as well as the statistical fractal nature of the dynamics in the time series.

For self-similar time series, relative dispersional analysis (RDA) shows that RD is constant with scale (i.e. the object looks the same at all scales) (Bassingthwaighte *et al.*, 1994; West, 1999). Since RDA implies the binning of the data set at successively larger time scales, for a system exhibiting completely random fluctuations in time, its RD drops off more rapidly as compared with a system that shows long-term memory. As a result, an inverse power law with a slope corresponding to a D_f close to 1.0 suggests high correlation among the components of a system exhibiting dynamic processes with long-term memory (Aon *et al.*, 2006a).

Power Spectral Analysis

In the temporal domain, the existence of dynamic fractals has also been investigated using power spectral analysis (PSA). The power spectrum of the time series was analyzed after Fast Fourier Transform (FFT). As in geometrically fractal objects, the dynamics of processes with long-term memory correspond to self-similar scaling in frequency. Thus, correlations in time series are also revealed using PSA; the power spectrum of such data usually follows an inverse power law proportional to $1/f^{\beta}$.

Double log plots of amplitude versus frequency indicated a decrease of power proportional to $1/f^{\beta}$, where f is frequency and β is the spectral exponent. This power law, more generally known as $1/f$ noise, describes colored noise depending on the value of β or the spectral exponent (e.g. $\beta = 1, 2$, or 3, for pink, brown, or black noise, respectively). The spectral exponent equals zero for white noise; thus, the spectrum of white noise is independent of frequency. If we integrate over time a series that display white noise, we get "brown" noise that has a power spectrum that is proportional to f^2 over an extended frequency range.

Applications to Time Series Analysis from Cardiomyocytes and Spontaneously Synchronized Continuous Yeast Cell Cultures

Cardiomyocytes

Extended time series were obtained from isolated cardiomyocytes loaded with TMRM, a mitochondrial membrane potential, $\Delta\Psi_m$, probe (Fig. 5.14(B)) or either of two different ROS probes, CMH_2-DCF and

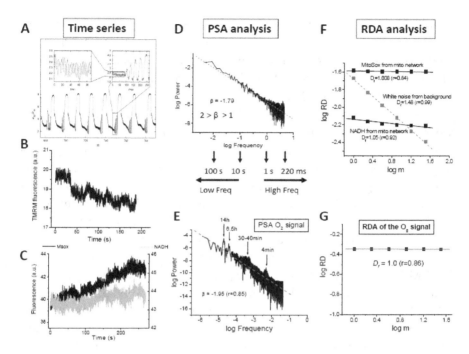

Figure 5.14. RDA and PSA analyses from time series of cardiomyocytes and *S. cerevisiae*. In yeast the time series correspond to multioscillatory behavior in dissolved O_2 in spontaneously synchronized continuous cultures ((A), (E), (G)) as described in (Roussel and Lloyd, 2007). In cardiomyocytes, the time series correspond to $\Delta\Psi_m$, NADH, and ROS signals from the mitochondrial network ((B)–(D), (F)) (Aon *et al.*, 2008a). (A) In the yeast time series (47,200 time points with temporal resolution of 12 s representing 118 hr of continuous culture), periods of 13 hr, 40 min and 4 min (see the sub-panels) in dissolved O_2 were detected by membrane-inlet mass spectrometry. (E) Power spectrum of the yeast time series after FFT. (G) Log–log plots of RD versus the aggregation number, *m*, after pooling the data set at successively larger time scales by adding adjacent points in the time series ((B), (C)). Time series from isolated cardiomyocytes loaded with TMRM, a $\Delta\Psi_m$ sensor, or the superoxide probe MitoSOX, and autofluorescence (NADH) were obtained by imaging with two photon laser scanning fluorescence microscopy at temporal resolution of 110 ms to 120 ms. The time series of MitoSOX and NADH in panel (C) were obtained simultaneously. (D) Power spectrum of the TMRM signal time series after FFT. (F) RDA analysis of the MitoSOX and NADH signals time series after binning the data set as in panel (G). A control corresponding to uncorrelated white noise from the background of the same stack of images is included (Aon *et al.*, 2006a; Aon *et al.*, 2008a).

Source: Redrawn from Aon *et al.* (2006) and Aon *et al.* (2008a).

MitoSOX (H_2O_2 or superoxide, O_2^-, sensor, respectively) (Fig. 5.14(C)). In cardiomyocytes, the time series of the fluorescencent images, consisted of 2,000 to 4,000 frames of 2-photon excited specimens loaded with TMRM plus CMH_2DCF or emission from MitoSOX plus autofluorescence (Aon *et al.*, 2006b; Aon *et al.*, 2008b).

Collectively, cardiac mitochondria behaved as a highly correlated network of oscillators (Aon *et al.*, 2006b; Aon *et al.*, 2008b). On RDA, the fluorescence time series exhibited long-term memory quantitatively characterized by an inverse power law with a fractal dimension, D_f, of approximately 1.0 (Fig. 5.14(F)). This measuring, a signature of self-similar fractal processes, is distinct from processes without memory that show completely random behavior (white or brown noise), which are characterized by an exponential (Poisson) law with a slope corresponding to $D_f = 1.5$.

Self-similar scaling was also revealed by PSA after applying FFT to the TMRM fluorescence time series. The power spectrum followed a homogenous inverse power law of the form $1/f^\beta$ with $\beta \sim 1.7$ (Fig. 5.14(D)). Both, CMH_2DCF and MitoSox exhibit scale-free dynamic behavior, as expected from a network of coupled oscillators (Aon *et al.*, 2006b; Aon *et al.*, 2008b). These results revealed that mitochondrial oscillations exhibit a broad frequency distribution spanning at least three orders of magnitude (from milliseconds to a few minutes). Thus, statistically, the collective behavior of the mitochondrial network is fractal, self-similar, and characterized by a large number of frequencies in multiple time scales, rather than an inherent "characteristic" frequency. It has been proposed that in the physiological state these mitochondrial oscillators are only weakly coupled by low levels of mitochondrial ROS. However, an increase in ROS levels under metabolic stress can reach a threshold that results in strong coupling through mitochondrial ROS-induced ROS release (Zorov *et al.*, 2000), and organization of the network into a synchronized cluster spanning the whole cell (Aon *et al.*, 2004a). A dominant low-frequency high-amplitude oscillation ensues (Aon *et al.*, 2003; Aon *et al.*, 2004a; Aon *et al.*, 2006a; Aon *et al.*, 2008a).

Yeast cultures

A time series recorded simultaneously from yeast cultures of dissolved oxygen, O_2, and carbon dioxide, CO_2 (47,200 time points with temporal resolution of 12 s representing 118 hr of continuous culture) were subjected to RDA and PSA analyses (Aon *et al.*, 2008a; Roussel and Lloyd, 2007). Yeast can produce multiple frequencies when grown continuously under precisely controlled conditions (Fig. 5.14(A); see also Chap. 6). The analysis of O_2 and CO_2 time series using RDA and PSA revealed that the observed multioscillatory dynamics correspond to statistical fractals, as can be judged by the perfect correlation between oscillators in the 13 hr, 40 min and 4 min time domains (Fig. 5.14(G)). An inverse power relationship with a fractal dimension, D_f $(= 1.0)$ was obtained, implying that RD is constant with scale (i.e. the time series looks statistically similar at all scales). The inverse power law behavior found is consistent with long-term memory in each of the data sets, and suggests fractal dynamics of processes on different time scales (from seconds to several hours) (Aon *et al.*, 2008a; Bassingthwaighte *et al.*, 1994; West, 1999).

PSA indicated an inverse power law proportional to $1/f^\beta$, as expected for a time series exhibiting self-similar scaling by RDA (Fig. 5.14(E)). The value of $\beta = 1.95$ obtained for the O_2 signal is close to that characteristic of colored noise, and this is as expected for chaotic time series (Aon *et al.*, 2008a; Roussel and Lloyd, 2007).

Complex Qualitative Behavior in Networks

Collective behavior of particles, mitochondria, agents, molecules, in networks present in, e.g. cells, markets, the atmosphere, is taking center stage in the analysis of complex systems (Aon *et al.*, 2009; Aon *et al.*, 2011; Aon *et al.*, 2006). Cells, organisms, social-, economic-, and eco-systems are complex because they form networks. These networks consist of a large number of usually nonlinearly interacting parts that operate in multiple spatial and temporal scales, and extend beyond their complex connectivity highlighting the inseparability of the spatial and temporal aspects. In the case of metabolic or signaling networks, nonlinear control mechanisms governing the function of the biochemical circuitry produce, under certain conditions, emergent macroscopic response. Manifesting themselves as novel, and unexpected, macroscopic spatiotemporal patterns, emergent behavior is the single most distinguishing feature of organized complexity in a system. Emergent organization cannot be anticipated in any way from the behavior of the isolated components of a system, and does not result from the existence of a "central controller." Emergence in complex systems arises from self-organizing principles resulting from nonlinear mechanisms and the continuous exchange of energy, matter, and information with the environment (Aon and Cortassa, 1997; Aon and Cortassa, 2009; Haken, 1983; Lloyd *et al.*, 2001; Nicolis and Prigogine, 1977). Self-organization is one of the main mechanisms that generate organized complexity, and this is expressed as disparate types of nonlinear outputs (e.g. oscillations, chaos) in a wide range of systems (Ball, 2004), including biological ones (Aon and Cortassa, 1997; Lloyd, 2009).

Nonlinear dynamic stability and bifurcation analyses of computational models, represented by coupled differential equations, allow the exploration of the qualitative behavior of complex networks (Aon and Cortassa, 2011; Cortassa *et al.*, 2003; Cortassa *et al.*, 2004). Mitochondria represent a good example of metabolic networks in several meanings of the word; on the one hand, they are spatially organized as a network with physical continuity, as in the case of neurons or cancer cells, or with chemical communication but without apparent physical connection, they behave dynamically as a network in heart cells (Aon *et al.*, 2003; Aon *et al.*, 2004a). On the other hand, since mitochondria are at the convergence between catabolic and anabolic networks they appear as "hubs" in central catabolic pathways (Aon *et al.*, 2007a).

Stability and Bifurcation Analyses

The concept of *phase space* representation is the hallmark of both non-linear dynamical time series and stability and bifurcation analyses (Kantz and Schreiber, 2005; Williams, 2003). Phase space or state space is an abstract mathematical space in which x- y- (2-dimensional) or x- y- z- (3-dimensional) coordinates represent the variables needed to specify the phase (or state) of a dynamical system (Fig. 5.15). The phase space includes all the instantaneous states the system can have, and as such is a complement to the time series plot with the additional power to reveal hidden patterns or structure (Williams, 2003). The succession of instantaneous states of a dynamic system describe a *trajectory* (time path) in phase space, and a family of

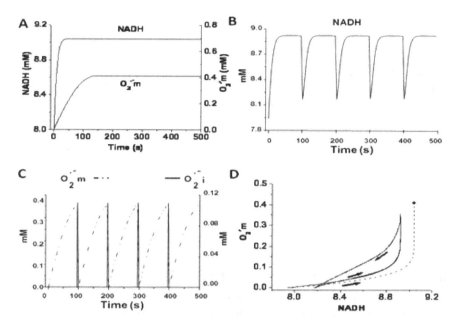

Figure 5.15. Time series, phase space, phase portrait, and bifurcation diagram. Simulation results obtained with an integrated model of mitochondrial energetics, ROS production, and scavenging (Aon *et al.*, 2006a; Cortassa *et al.*, 2004) are presented. Shown are the evolution of NADH and mitochondrial superoxide, $O_2^{\cdot-}$m, toward (A) steady (i.e. a fixed-point attractor) or (B), (C) oscillatory (i.e. limit-cycle) states, respectively. Panel (D) depicts the phase space plot of NADH and $O_2^{\cdot-}$m for the steady (dashed) and oscillatory (continuous) solutions. The change from a fixed-point attractor to limit-cycle behavior was achieved by increasing only the concentration of respiratory electron carriers (Cortassa *et al.*, 2004). (E) Phase portrait plot showing the relationship between $\Delta\Psi_m$ and the superoxide anion released to the periplasmic space, $O_2^{\cdot-}$c, during the depolarization phase of the oscillation. Plotted are the trajectories followed by several limit cycles corresponding to oscillatory periods ranging from 70 to 200 ms after sequentially changing the rate of ROS scavenging. (F) Bifurcation diagram of the state variable NADH as a function of parameters related with ROS production (fractional ROS production) and scavenging (superoxide dismutase, SOD, concentration). The results concerning stability and type of steady states exhibited by the model were represented in a plot, which in this case consisted of an upper branch, with predominantly reduced NADH, and a lower branch, in which NADH was mainly oxidized. The transitions at the borders of the steady states marked by arrowheads and numbers indicate change in stability of the solution; 1 and 2 in the upper branch indicate Hopf bifurcations delimiting the oscillatory region (thin line); 3 and 4 denote limit points.

Source: Modified from Cortassa, S, Aon, MA, Winslow, RL, O'Rourke, B. (2004) Biophys. J. 87, 2060–2073.

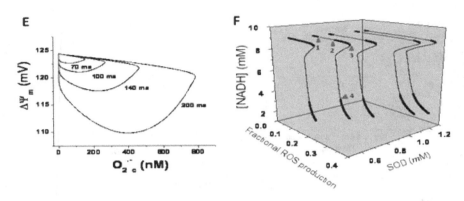

Figure 5.15. (*Continued*)

different trajectories that can evolve for the same system constitutes the *phase space portrait* or phase portrait (see Chap. 6).

Trajectories will converge to a subspace of the phase space, if we observe a system dynamics long enough (i.e. after transients die out); this subspace is called the *attractor* of the system (intuitively, because it "attracts" trajectories from all possible initial conditions). Unlike linear systems that can only exhibit *point* attractors, nonlinear systems display a wide repertoire of them in phase space, including *limit cycles* or closed loops (from oscillations), *torus* of "doughnut" shape (quasiperiodic dynamics from superposition of different periods), and *strange* with fractal geometry (deterministic chaos) (Aon *et al.*, 2011; Stam, 2005).

Changes among different types of attractor happen through bifurcations. Bifurcations are abrupt changes in the stability of a system dynamics as reflected by a radically different qualitative behavior (e.g. steady, oscillatory, chaotic) at critical values of the control (bifurcation) parameter (Aon and Cortassa, 1997) (Fig. 5.15; see also Chaps. 1 and 6). Oscillatory and chaotic behaviors constitute typical dynamic outcomes of nonlinear systems. A substantial difference between oscillatory and chaotic dynamics is that the former do not exhibit sensitive-dependence on initial conditions. Following two different initial conditions, trajectories in nonchaotic regimes get closer together describing limit cycles (periodic attractors) (Fig. 5.15(D) and (E)), or remain equidistant describing toroidal attractors (these combine two or more limit cycles and can be periodic or quasiperiodic). Oscillatory systems will evolve to the same attractor if changes in initial conditions belong to the basin of attraction of the limit cycle. Unlike oscillatory systems, chaotic ones will exhibit extreme sensitivity to initial conditions since even a slight change will make trajectories diverge widely (exponentially), although bounded to a strange attractor in phase space.

Emergent Spatiotemporal Behavior in Networks

Under metabolically stressful conditions such as substrate deprivation or oxidative stress, the role of mitochondrial function becomes a key arbiter of life and death at the cellular and organ level. While under normal physiological conditions, the availability of energy is fine tuned to match changes in energy demand but under stress this is not the case. Most myocardial ATP production occurs in the mitochondria through oxidative phosphorylation.

The remarkable nonlinear properties of the mitochondrial network, and of the heart itself, make them prone to the appearance of critical phenomena and bifurcations leading to self-organized, emergent, behavior. A dramatic example of the latter is the succession of failures shown to escalate from the mitochondrial network to the whole heart resulting in reperfusion-related arrhythmias after ischemic injury, and eventually the death of the organism (Akar *et al.*, 2005; Aon *et al.*, 2006a; O'Rourke *et al.*, 2005). The transition between the physiological and pathophysiological domains of mitochondrial oscillation is a clear example of emergent self-organization in both time (limit cycle oscillation) and space (synchronization across the mitochondrial network) (Aon *et al.*, 2006a).

Stability analysis of a computational model of mitochondrial energetics, Ca^{2+} dynamics, and ROS showed the presence of Hopf bifurcations, a signature of the existence of limit cycles, visualized as sustained oscillations (Fig. 5.15(B), (C), and (F)). The bifurcation diagrams obtained have the typical S-shape that describes the behavior of systems exhibiting bistability and show an unstable region (thin lines) between the upper (polarized) and lower (depolarized) branches of steady states (thick lines in Fig. 5.15(F)). This diagram illustrates the importance of the balance between mitochondrial ROS generation and ROS buffering since the oscillatory domain, flanked by Hopf bifurcations, appears within the upper branch of the curve as the concentration of ROS scavenger increases. This domain gradually expands at higher SOD concentrations (from left to right in Fig. 5.15(F)) until the limit between the upper or lower branch, where the model behavior changes precipitously and $\Delta\Psi_m$ jumps from polarized to depolarized steady states or vice versa.

At each oscillatory cycle, the computational model predicts a burst of respiration triggered by the rapid membrane uncoupling and the concomitant oxidation of the mitochondrial NADH pool together with the rapid release of O_2^- to the mitochondrial intermembrane space (Fig. 5.15(C)).

Mitochondria from heart cells act as a network of coupled oscillators capable of producing frequency- and/or amplitude-encoded ROS signals under physiological conditions (Aon et al., 2006b; Aon et al., 2007a; Aon et al., 2008a). This intrinsic property of the mitochondria can lead to a mitochondrial "critical" state, i.e. an emergent macroscopic response manifested as a generalized $\Delta\Psi_m$ collapse followed by synchronized oscillation in the mitochondrial network under stress (Fig. 5.16(A)–(C)) (Aon et al., 2004a). With respect to the spatial organization, we have applied percolation theory to explain how a local interaction among mitochondria organized in a lattice can lead to a widespread change in the state of energization of the system. The mitochondria in the cardiomyocyte are particularly amenable to this analysis because they are organized almost as a cubic lattice, with a spacing of $\sim$0.2–1 μm between elements. In two-dimensional images of a single focal plane of the myocyte, we measured $\Delta\Psi_m$ and ROS accumulation after a local laser flash and determined the percentage of the mitochondria showing a ROS signal above a certain threshold level (20% above baseline fluorescence) as a function of time after the flash (Fig. 5.16(B)). At the point in time just before the first cell-wide collapse of $\Delta\Psi_m$, which we referred to as "criticality," we found that $\sim$60% of the mitochondrial network had ROS levels at or near threshold (Fig. 5.16(B)), and this cluster of mitochondria spanned the whole cell (i.e. a "spanning cluster" was evident) (Fig. 5.16(C)) (Aon et al., 2004a). This number is very close to the theoretical prediction for systems at a percolation threshold ($p_c = 0.59$) (Feder, 1988; Stauffer and Aharony, 1994). Moreover, fractal analysis of the shape of the spanning cluster was also consistent with percolation theory (Aon et al., 2004b).

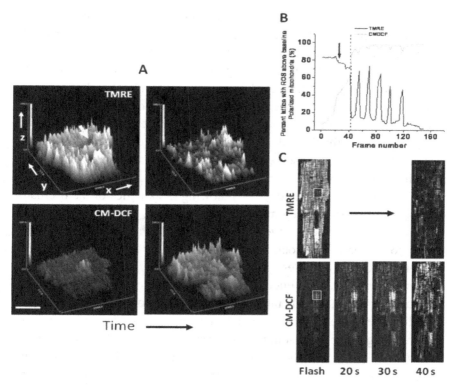

Figure 5.16. Emergent, self-organized whole-cell oscillations and time course of cluster formation in mitochondrial metabolism triggered by a local perturbation. (A) Top panels, surface plot of mitochondrial membrane potential ($\Delta\Psi_m$) in a TMRE-loaded cardiomyocyte during the flash (white arrow, top left panel) showing the local $\Delta\Psi_m$ depolarization, and the cell-wide $\Delta\Psi_m$ collapse (top right panel) ~40s (on average) after the flash. Bottom panels, surface plot of the fluorescence intensity of the ROS-sensitive fluorophore CM-H$_2$DCF during the flash showing the local (bottom left panel) and the cell-wide (bottom right panel) increase in ROS generation associated with the $\Delta\Psi_m$ collapse. x, y, z, in the axes of the surface plot correspond to the transversal and longitudinal axes of the cell, respectively, and the fluorescence intensity of TMRE (top panels) or CM-DCF (bottom panels). Bar ~20 μm. (B) Sustained oscillations in $\Delta\Psi_m$ happening beyond criticality are depicted. Arrow indicates the timing of the flash. (C) At criticality, a cell-wide $\Delta\Psi_m$ depolarization (TMRE, top right) happens in a cluster possessing about 60% of the mitochondria with ROS levels above baseline (CM-DCF, bottom).

Source: Adapted from Aon *et al.* (2004) and Aon *et al.* (2008b).

The significance of this characterization of the mitochondrial network as a percolation process is that it helps to explain *why* $\Delta\Psi_m$ depolarization occurs in a synchronized manner throughout the cell and *where* it will occur. Only those mitochondria belonging to the spanning cluster take part in the global limit cycle oscillation, explaining why some mitochondria appear to resist oscillation even though they are surrounded by depolarized mitochondria. Furthermore, it explains why there does not seem to be a single point in the cluster (e.g. the flashed area) from which each cycle originates — the system at criticality is susceptible to small perturbations anywhere in the spanning cluster to initiate a transition.

With faster time resolution, or by slowing the kinetics of the oscillator (e.g. with submaximal oligomycin concentrations), the $\Delta\Psi_m$ depolarization wave can often be resolved. This wave travels at a speed of $22\ \mu m/s$ which corresponds to a total time of about $4\,s$ for global depolarization (see Fig. 5.16(B)) (considering an average myocyte length of $100\ \mu m$), in agreement with the experimental data. Although the $\Delta\Psi_m$ depolarization is global (Fig. 5.16(A)), the interaction between mitochondria is local. The short lifetime of the O_2^- radical and the presence of fast scavenger systems prevents long range diffusion, but in our mechanistic model, O_2^- liberated from a mitochondrion in the spanning cluster must only diffuse to its neighbors. The second criterion is that the neighbor must also be a member of the spanning cluster (i.e. be close to the threshold for depolarization). According to percolation theory, a p_c of 0.59 corresponds to a coordination number of four, again emphasizing that each mitochondrion is influencing its nearest neighbors in the network. These results also explain why the speed of the depolarization wave is faster than it would be if sustained by diffusion of the chemical messenger (O_2^-) alone.

Recent computational and experimental studies further support the interpretation that ROS, and more specifically the superoxide anion, propagates by reaction-diffusion among mitochondria in the network, giving rise to the observed emergent macroscopic properties such as widespread $\Delta\Psi_m$ depolarization and ROS waves. The implementation of the mitochondrial oscillator computational model into Reaction-Diffusion (RD) in one- and two-dimensional mitochondrial network models (Zhou *et al.*, 2010), allowed the first mechanistic assessment of ROS-induced ROS release (RIRR). In a two-dimensional network composed of 500 mitochondria, model simulations revealed $\Delta\Psi_m$ depolarization waves similar to those observed when isolated guinea pig cardiomyocytes are subjected to a localized laser flash (Aon *et al.*, 2003; Aon *et al.*, 2004a), antioxidant depletion (Aon *et al.*, 2004a; Aon *et al.*, 2008b; Cortassa *et al.*, 2004). The sensitivity of the propagation rate of the depolarization wave to O_2^- diffusion, production, and scavenging in the RD-RIRR model is similar to that observed experimentally. These results indicate that local gradients of cytoplasmic O_2^-, determined by diffusion and scavenger capacity, play a significant role in the rate of propagation of the $\Delta\Psi_m$ depolarization and repolarization waves (Zhou *et al.*, 2010). The results obtained with the RD-RIRR model illustrate how local neighbor–neighbor interactions (1–2 μm distance) can lead to long distance ($>100\,\mu$m in cells, and $>4000\,\mu$m in whole hearts) spatiotemporal patterns in cells (Aon *et al.*, 2003; Aon *et al.*, 2004a; Brady *et al.*, 2006) and in whole hearts (Aon *et al.*, 2009; Brown *et al.*, 2010; Lyon *et al.*, 2010; Slodzinski *et al.*, 2008).

Appendix

A simplified mathematical model to illustrate the Sauro–Westerhoff method of MCA

The simplified model depicted in Fig. 5.11 is used to show the main steps taken to implement the Sauro–Westerhoff method of MCA. How the latter is experimentally achieved, is described under the Sections: "Metabolic Control Analysis" and "Rational Design of Microorganisms." Our attempt here is to show the construction of the matrix **E** of elasticity coefficients. Mathematically, the terms of the matrix **E** are numbers that are obtained from derivation of the rate expressions (e.g. V_1, V_2, V_3) with respect to the substrate (A for V_1) and substrate or effector (B for V_1). Thus, the matrix comprises five reactions (i.e. the columns) and three intermediates B, C, E (i.e. the rows) (Fig. 5.11):

$$Matrix\ E = \begin{bmatrix} 1 & 1 & 1 & 1 & 1 \\ ElB1 & ElB2 & ElB3 & ElB4 & ElB5 \\ ElC1 & ElC2 & ElC3 & ElC4 & ElC5 \\ ElE1 & ElE2 & ElE3 & ElE4 & ElE5 \\ 0 & Vr3 & Vr2 & Vr4 & Vr5 \end{bmatrix} \qquad (5.14)$$

The rate equations for each one of the steps in the hypothetic pathway in Fig 5.11 are as follows:

$$V_1 = \frac{k1 \; A \; B}{A \; B + Ka \; B + Kb \; A + Ka \; Kb} \tag{5.15}$$

$$V_2 = \frac{k2 \; C}{C + Kc} \tag{5.16}$$

$$V_3 = \frac{k3 \; C}{C + Kcd} \tag{5.17}$$

$$V_4 = \frac{k4 \; E}{E + Ke} \tag{5.18}$$

$$V_5 = \frac{k5 \; B}{B + Kb5} \tag{5.19}$$

The derivatives of V_1 with respect to B (*E1B1*), and of the branch at C (V_{r3}), are shown as an example.

$$E1B1 = \frac{Kb}{B + Kb} \tag{5.20}$$

$$V_{r3} = \frac{k3 \; C \; (A \; B + Ka \; B + Kb \; A + Ka \; Kb)}{k1 \; A \; B \; (Kcd + C)} \tag{5.21}$$

The experimentally determined values of metabolite concentrations are fed into the derivative of the corresponding rate equations (e.g. B for *E1B1*), in order to obtain the elasticity coefficients, which are replaced into the explicit expressions of the following **E** matrix:

$$
\begin{bmatrix}
1 & 1 & 1 & 1 & 1 \\
\frac{Kb}{B+Kb} & 0 & 0 & 0 & \frac{Kb5}{B+Kb5} \\
0 & \frac{Kc}{C+Kc} & \frac{Kcd}{C+Kcd} & 0 & 0 \\
0 & 0 & 0 & \frac{Ke}{E+Ke} & 0 \\
0 & \frac{k3 \; C \; Z}{(C+Kcd) \; k1 \; A \; B} & -\frac{k2 \; C \; Z}{(C+Kc) \; k1 \; A \; B} & -\frac{k4 \; E \; Z}{(E+Ke) \; k1 \; A \; B} & -\frac{k4 \; E \; (Kb5+B)}{(E+Ke) \; k5 \; B}
\end{bmatrix} \tag{5.22}
$$

with $Z = (A \; B + Ka \; B + Kb \; A + Ka \; Kb)$

The matrix **E** is inversed using standard mathematical software (e.g. Maple, Matlab, Mathematica).

Chapter 6

Dynamic Aspects of Bioprocess Behavior

Transient and Oscillatory States of Continuous Culture

Continuous culture, the subject of many laboratory experiments, is not only used as research tool, but also is increasingly utilized as a unit process in industry. Examples include the continuous production of beer (Branyik *et al.*, 2005; Dunbar *et al.*, 1998). In order to ensure the efficient operation of these processes, we must understand their unsteady states (dynamic behaviour) as well as their steady states. This is necessary so that it may be possible to design effective control systems. Because fluctuations are unavoidable in input variables, the dynamics of chemostat behavior can be unpredictable, but in many cases understanding can help to reduce perturbation to a minimum. It may also be possible to use unsteady states to improve productivity.

Mathematical model building

Mathematical model building provides a theoretical framework for understanding mechanisms that underlie the behavior of continuous culture systems, e.g. the importance of hydrodynamic and biological lags in determining the kinetics of responses to altered environmental conditions. Predicted responses made on the basis of idealized models (often derived from chemical reactor studies) can provide new insights into important state variables.

Steps in setting up such a model usually involve:

(1) Studies of attainable steady states under specific operating conditions.
(2) Determination of whether the behavior of a culture represents a true steady state, or whether it is characterized by oscillations.
(3) Studies of the responses of the system to disturbances.

The simplest case consists of a single-stage, well-stirred, continuous-flow system (CSTR) without recycle or feedback. In practice, continuous cultures are often operated with some type of superimposed control, e.g. of turbidity, or of substrate level, by manipulation of dilution-rate. These controls necessarily introduce additional lags, which may impose instabilities not present in the open-loop response. Even though the aim of control strategy is to have more stable operation, simple control systems of the on/off type give rise to overshoot, damped oscillations, and off-set problems. More precise control requires more elaborate correction systems (e.g. PID, a proportional integral derivative system). No single model can possibly be adequate to fit the complexity inherent in any microbial system. There is thus always a need to interpret a model system cautiously and verify the major output trends experimentally, and inevitably a model is constructed on the basis of generalized variables. Simplifications arise because the parameters used represent average properties; variations in individual organisms are usually not accounted for. Similarly, the possible presence of subpopulations is neglected. Growth is for the most part treated as a deterministic rather than as a stochastic process. Such assumptions are to some extent valid for large populations of microorganisms, because random deviations from mean values that average out these models are sometimes referred to as "unsegregated" and actually represent an abstraction from, and simplification of, real biological behavior. The following example is based on the treatment proposed by Harrison and Topiwala (1974).

Biological systems are thereby regarded as (and described as) homogeneous chemical reactions in terms of vector space-state whose elements (X_i) represent various component concentrations, physical parameters, temperature, pressure, stoichiometric constants, etc. The biological rate expression for any material component (y_i) can be expressed as a continuous function:

$$R_j = f(y_1, \ldots, y_k \ldots T, pressure, mass - transfer\ terms, pH, ..),$$

$$j = 1, \ldots k, \tag{6.1}$$

where, R_j is the rate of overall biochemical reaction producing or consuming component y_k per unit volume, per unit time.

A mass balance on any single component y_j the continuous culture yields the equation:

$$\frac{dy_j}{dt} = R_j + D(y_j^1 - y_j), \tag{6.2}$$

provided R_j is assumed to be uniformly distributed through the culture, and the growth process has negligible effect on fluid density. Equation (6.2) can be modified to include components present in more than one physical phase to include transport terms. The detail of this kinetic model (i.e. the nature of the R_j terms), determines the theoretical stability of the system. Because microbial growth is so complex, the equations are semi-empirical, and their formulation involves simplification of the real behavior; R_j represents a macroscopic process, and does not describe control systems within an individual cell. Therefore it is difficult to decide the level of complexity of the rate expression necessary for the correct interpretation of the dynamics of the continuous culture system.

In general, the model of continuous culture system as defined by Eqs. (6.1) and (6.2) is represented by a set of first-order, nonlinear differential equations and can be written in matrix notation as

$$\frac{dx}{dt} = \underline{f}(\underline{x}) \tag{6.3}$$

where x and f are n-vectors.

The equilibrium condition of the state–space dynamic model is given by solution of

$$\underline{f}(\underline{x}) = 0 \tag{6.4}$$

At any point x which satisfies Eq. (6.4) is termed a singular point or steady state (a solution may not exist, or more than one solution may satisfy the equilibrium conditions: e.g. where substrate inhibition occurs (Andrews, 1968)). The stability of the steady state can be found by the Liapounov indirect method (see MacFarlane (1973)). This establishes stability (or instability) of the nonlinear differential equations by examination of stability of the singular point for the locally approximating set of linear equations. To linearise, the model Eq. (6.3) is expanded to a Taylor series about the singular point to obtain:

$$\frac{dx}{dt'} = \underline{\mathbf{J}}\mathbf{x}' + \underline{\mathbf{N}}(\mathbf{x}') \tag{6.5}$$

where x' is the perturbation variable, J the Liapounov first approximation matrix and $N(x')$ a matrix which represents first order terms. The stability of the steady state to small perturbations is determined by the eigenvalue of the matrix J. A necessary and sufficient condition for a sufficiently small perturbation to die away is that all the eigenvalues of J have negative real parts. This represents an overdamped system (Fig. 6.1(a)).

If an eigenvalue has a positive real part, the steady state will be unstable, and a small perturbation will lead the system away from the steady state (Fig. 6.1(b)). If J has any pure imaginary eigenvalues, the stability will be determined by higher-order terms in $N(x^1)$. This approach to the steady state does not require exact solution of the nonlinear differential Eq. (6.3). For example, in the Monod model of continuous culture:

$$\frac{dx}{dt} = \mu(s)x - Dx \tag{6.6}$$

$$\frac{ds}{dt} = D(S_R - s) - \frac{\mu(s)\mathbf{x}}{Y}, \tag{6.7}$$

where, x is the biomass, s is the substrate concentration in the culture, S_R is substrate concentration in the feed stream (which contains no biomass), μ an arbitrary function of s, is defined as the specific growth-rate, and Y is the constant yield factor.

If x and s represent steady-state values, the J matrix for the linearized form of the above differential equations is given by

$$\underline{J} = \begin{vmatrix} \mu(\bar{s}) - D & \bar{x}\left(\dfrac{d\mu}{ds}\right)_{\bar{s}} \\ -\dfrac{\mu(\bar{s})}{Y} & -\left[\dfrac{\bar{x}}{Y}\left(\dfrac{d\mu}{ds}\right)_{\bar{s}} + D\right] \end{vmatrix}. \tag{6.8}$$

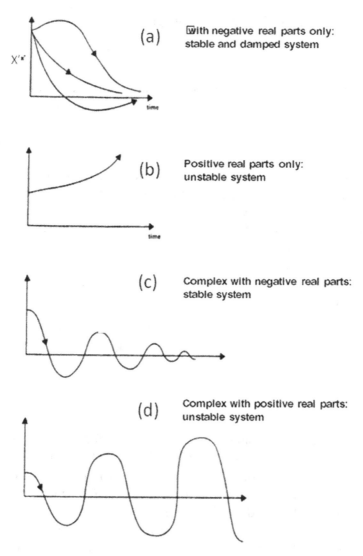

Figure 6.1.   Stability of the nonlinear system to a small perturbation of the steady state. Theoretical response as determined by the eigenvalues of the linearized model equations. *Source*: Reproduced from Harrison and Topiwala (1974) ©copyright Springer-Verlag, with permission.

The two eigenvalues of the J matrix are given by

$$\lambda_1 = -D \quad \text{and} \quad \lambda_2 = \frac{\bar{x}}{Y}\frac{d\mu}{ds_{\bar{s}}}. \tag{6.9}$$

Examination of λ_1, and λ_2 yields the condition that the nontrivial steady state $(x > 0, s < S_R)$ will be stable to small perturbations if

$$\frac{d\mu}{ds_{\bar{s}}} > 0, \tag{6.10}$$

i.e. the steady state will be stable to small perturbation provided that the rate of change of specific growth rate with respect to the substrate is positive, the rate of change being evaluated at the steady state.

For the Monod Model: where μ, the specific growth-rate is a simple hyperbolic function.

The nonwashout state will always be stable.

The eigenvalues of the linearized Monod model can be obtained from Eq. (7.9) as

$$\lambda_1 = -D \quad \text{and} \quad \lambda_2 = -\frac{(\mu_m - D)\left[S_R(\mu_m - D) - K_S D\right]}{\mu_m K_S}, \tag{6.11}$$

Where,

$$\mu = \frac{\mu_m s}{K_S + s}. \tag{6.12}$$

Since λ_1 and λ_2 consist only of negative real parts for the nontrivial steady state, this model will not give rise to oscillations when subjected to small disturbances. The speed of response of the system as it returns to the steady state will be characterized by two exponentially decaying modes which will be associated with λ_1 and λ_2, respectively (Fig. 6.2).

Two widely differing modes of response could exist depending on operative dilution rate; λ_1 (representing mixing lag) is much smaller than λ_2 at low dilution rates, but this trend reverses as D approaches μ_m. Important examples of industrial processes which involve inhibition by the substrate can be considered (e.g. waste-water treatment of effluent from a chemical plant which contains toxic xenobiotics).

Stability criteria of Eq. (6.10) are not met when substrate concentration is in excess of S_1 (Fig. 6.3).

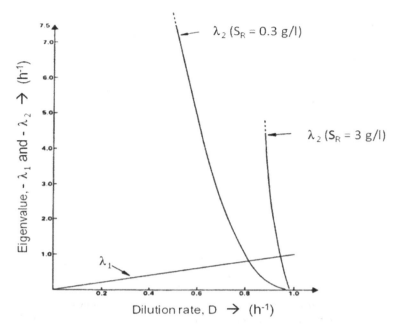

Figure 6.2. Relative importance of the two eigenvalues which determine the speed of response for the Monod chemostat model. Computed numerical values of the eigenvalues plotted as a function of the operating dilution-rate. The values were obtained using the parameter combinations: $S_R = 3\,\mathrm{g}\,\mathrm{l}^{-1}$; $K_S = 0.012\,\mathrm{g}\,\mathrm{l}^{-1}$; $\mu_m = 1.0\,\mathrm{h}^{-1}$.
Source: Reproduced from Harrison and Topiwala (1974) ©copyright Springer-Verlag, with permission.

When D does not give washout and feed substrate $S_R > S_1$, this model gives two possible steady states for every value of dilution rate $(D = \mu)$, but only that steady state corresponding to the lower substrate concentration and to the left of S, is stable.

Transfer-function analysis and transient-response techniques

The Liapounov direct method cannot be applied in order to determine responses to finite disturbances from a steady state. A common example is that of a system at start-up from the batch operation to that in continuous mode. A generally applicable method for the determination of which of many possible steady states will be attained is not available.

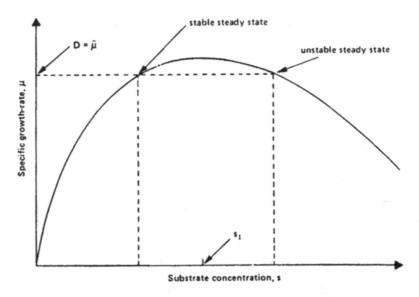

Figure 6.3.　Effect on available steady states of substrate/specific growth-rate relationship based on Monod-type hyperbolic model, including a substrate-inhibition function. *Source*: Reproduced from Harrison and Topiwala (1974) ©copyright Springer-Verlag, with permission.

System dynamics can be investigated experimentally by frequency-forcing or pulse testing: e.g. pH, temperature, inlet substrate concentration dilution rate. Lags (mixing and kinetic) limit the usefulness of this approach. The use of sinusoidal variations in input variables is of limited applicability. Pulse-testing is usually more convenient and useful.

Theoretical transient response and approach to steady state

Continuous culture systems behave in a nonlinear manner; so for large perturbations nonlinear differential equations must be solved. Analytical solutions are rarely feasible and recourse to numerical (computer) solutions is usually necessary (see Chap. 5). Figure 6.3 shows generalized transient responses of the Monod model to stepwise increases in S_R or D. The result is always an overdamped response, provided that the system that was in steady state prior to disturbances shows a single overshoot or undershoot, depending on whether S_R is increased or decreased. After disturbance, the system returns to a unique steady state. The initial condition of the system does not affect the final steady state. More structured models (i.e. those that account for the "physiological state" of the culture) give more complex (often oscillatory) behavior.

Substrate-inhibition model

To study the response of the cell and substrate concentrations after a change in operating conditions, the solution of the following differential equations can be considered:

$$\frac{dx}{dt} = \frac{\mu_m xs}{(K_s + s)\left(1 + \dfrac{s}{K_i}\right)} - Dx, \qquad (6.13)$$

$$\frac{ds}{dt} = D(S_R - s) - \frac{\mu_m xs}{Y(K_s + s)\left(1 + \dfrac{s}{K_i}\right)}. \qquad (6.14)$$

A computer solution of the response of the system starting from two different steady states shows that washout occurs in the case where system is originally in the unstable steady state, but not when system is in the stable steady state prior to the stepwise change.

Phase plane analysis

Many models of continuous culture have been formulated in terms of only two dependent variables (e.g. average cell and substrate concentrations). State variables are plotted against one another through a series of states from initial to final states (Fig. 6.4).

Only trajectories associated with a single steady state or single limit cycle are shown. Cases where multiple states/cycles are also possible occur in more complex examples. Limit cycle behavior indicates that continuous oscillation in state variables occurs, even though temperature, pH, medium flow rate are kept constant. The dynamics of populations in chemostat cultures were simulated in early studies (Tsuchiya, 1983). Soon afterward, data analysis and computer control of biochemical processes (Nyiri, 1972)

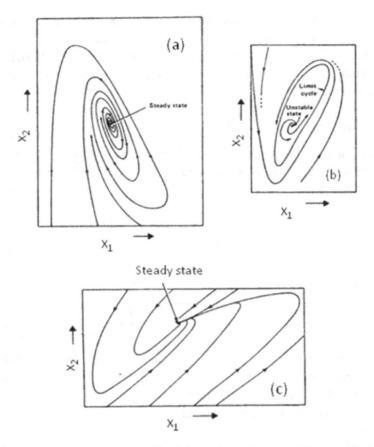

Figure 6.4.   (a–c) Examples of generalized phase-planes of state-variables x_1 and x_2 around the steady state. (a) A system displaying damped oscillations as a unique steady state is approached (Process Analysis and Simulations, Himmelblau and Bischoff ©copyright 1968 John Wiley & Sons, Inc. Reprinted by permission of Wiley-Liss, Inc., a subsidiary of John Wiley & Sons, Inc.). (b) A system displaying limit-cycle behavior. (c) An overdamped system: no oscillations are obtained as the unique steady state is approached (reproduced from Elementary Chemical Reactor Analysis, Aris ©copyright 1969 Prentice Hall, Inc., reprinted with permission).

enabled the development of online balancing of carbon conversion processes (fermentation and respiration) with the growth of yeast, using off-gas analysis by mass spectrometry (Cooney *et al.*, 1977). Optimization of this process so as to produce high biomass with minimal carbon overflow into ethanol, the major fermentation product, was made possible by use of a simple stoichiometric model. This principle then became widely adopted to improve biomass yields (e.g. when producing heterologous gene products in bacteria cultures). Much more complex models are necessary when dealing problems of integration of the massive information flow from assays now available. Early approaches to the "reconstruction" of the growth of single *E. coli* "model bacterium," from nutrient uptake to energy metabolism, macromolecular synthesis and cell cycle traverse, so as to give direct output of doubling times as a function of glucose concentration (Domach and Shuler, 1984; Domach *et al.*, 1984), led on to simulations of mixed *N*-source utilization (Shu and Shuler, 1989), and prediction of the effects of amino acid supplementation on growth rates (Shu and Shuler, 1991). The regulation of plasmid replication in engineered organisms is an important factor in the development of efficient systems for the production of recombinant proteins. Using information on molecular interactions (e.g. of the affinities DNA-binding regulatory proteins), Lee and Bailey (1984) were able to build a detailed mathematical model which predicted the behavior of the model organism. This "genetically structured model" enables the mapping from nucleotide sequence, through transcriptional control to the initiation of plasmid replication; it is this which in turn determines plasmid copy number. The principles of Metabolic Control Analysis (see Chap. 5) for the determination of the control properties of individual steps in a pathway and the consequences for metabolic fluxes (Heinrich and Rapoport, 1974; Heinrich *et al.*, 1977; Kacser and Burns, 1973) have been extensively applied to problems in biochemical engineering (Cortassa and Aon, 1994b, 1994c; Cortassa and Aon, 1997; Cortassa *et al.*, 1995; Stephanopoulos *et al.*, 1998; Varma and Palsson, 1994). Information from *in vivo* NMR analyses of stable isotope incorporation kinetics into metabolic intermediates currently enables rapid advances based on noninvasively obtained details of metabolic fluxes. Nonlinear dynamics control theory can be incorporated into models in order to predict regions of operation which are characterized by bistability, limit cycle behavior, quasiperiodic or chaotic outputs (Hatzimanikatis and Bailey, 1997; Lloyd and Lloyd, 1995; Schuster, 1988). Control of complex dynamic behavior offers new approaches to the predictable operation of fermenters driven into their nonlinear operating states (Ott *et al.*, 1994). Forced glycolytic oscillations may be used to increase ethanol yields (Zimmerman, 2005).

Transient Responses of Microbial Cultures to Perturbations of the Steady State

Dilution rate

Stepwise alteration in the dilution rate, D, is the most commonly employed means of perturbation; this is achieved by raising or lowering the medium feed rate. This imposes on the organism a change in growth rate.

Implicit in the unstructured Monod model (Eqs. (6.6), (6.7), and (6.11)) are the principles that growth is regulated only by the concentration of growth-limiting substrate, and that microbes possess all the constituents necessary for growth at maximum growth rate and can accelerate to maximum growth rate instantaneously when substrate concentration in the culture fluid is raised. These assumptions are not entirely reasonable, because growth processes (metabolism and cell division) are regulated by highly complex control systems. Thus RNA content increases markedly with growth rate.

Work with Mg^{2+}, K^+, or Pi-limited culture (Tempest *et al.*, 1965) shows that ribosome content regulates protein synthesis rates and hence growth rates. The organism requires a finite adjustment time for the biosynthesis of more RNA before faster growth can occur. In general, only small increases in D can be accommodated almost immediately and without lag. Larger changes in substrate concentration give finite lag times, especially where a substrate is potentially toxic at high concentrations (e.g. methanol). In that case, a sequence of reactions occurs whereby a lag leads to a build-up of substrate, giving growth inhibition and "washout."

Responses to decreased D will follow the predictions of the unstructured model with regard to cell concentration (because lower growth rate can be achieved instantaneously). However, a finite time will be required for precise readjustment of intracellular conditions (and this depends on turnover times of constituents).

Feed substrate concentration

This should have immediate effect on growth rate according to the simple unstructured Monod-type growth model. Effect of sudden increase of substrate (or pulse) is qualitatively similar to increase in D and gives rise to a transient state due to unbalanced growth (Vaseghi *et al.*, 1999). Individual cellular components changing at different rates give rise to change in cellular composition; the RNA content increasing faster than total cell mass. The term "unbalanced growth" often applied in this situation is actually a misnomer: this state is really controlled change between two equilibria states. A small change gives rise to fast accommodation to the new conditions, whereas a large change will give a lag (usually of about 1 to 2 h for a small (1 liter) laboratory fermenter.

Growth with two substrates

Cultures containing a mixture of carbon sources have been extensively studied (e.g. glucose and xylose in *E. coli* cultures). In batch culture diauxie occurs with glucose used preferentially before xylose, the substrate that requires an inducible system. In continuous culture, the switch over from glucose to xylose leads to a decreased cell concentration, and to an accumulation of xylose during a transient phase. Then the xylose is metabolized and the cell concentration is re-established. If the nutrient feed is switched back to glucose the result is a smooth transition with no loss of cell density and no accumulation of glucose (i.e. operation of the "constitutive system"). The long recovery times observed (measured in days) indicates that a slow response is required for induction of xylose-metabolizing enzymes (which have been previously subjected to glucose repression). In a recombinant strain of the high-secreting yeast, *Pichia pastoris*, angiostatin production can be enhanced by manipulation of a two-substrate system (glycerol/methanol), (Xie *et al.*, 2003).

We can conclude that modeling transient responses in chemostats with two substrates must take account of cell physiology. Relaxation times may be expected to be much longer than those observed for single-substrate systems.

Temperature

Change in temperature produces altered growth rates, yield coefficients, and affinity for substrate. The content of lipid, carbohydrate, and RNA in cells is temperature sensitive. Within limits, responses of steady-state growth rates to temperature follows simple Arrhenius relationship. So it would be expected that a sudden increase of temperature applied to a growing culture would give an immediate increase in the maximum growth rate, but in practice a lag is observed before the growth rate is accelerated to the new value. This is because intracellular controls have to be adjusted, new cellular machinery must by synthesized; levels of RNA, regulatory molecules, enzymes, and membrane structure and function all have to be modulated. Decrease in temperature in general produces a smoother adjustment process and consequently a smaller lag. Here again, transient responses indicate complex intracellular regulation.

Dissolved Oxygen

The oxidases (e.g. cytochrome c-oxidase in yeast, or several different bacterial oxidases) have high affinities for O_2 in organisms that have been grown aerobically. Apparent K_m values for O_2 of about 0.1 μM are typical. Thus, O_2 the terminal electron acceptor, saturates the oxidase above a low threshold, and over a wide range of concentrations and does not limit respiration. So over this range (e.g. 5–280 μM for bakers' yeast) the culture is usually completely insensitive to the concentration of O_2. At very high (especially hyperbaric) O_2, almost all microbial species are inhibited, due to formation of reactive oxygen species. At very low dissolved O_2, dynamic responses can be very complex, especially in facultatively anaerobic species; e.g. *Klebsiella aerogenes* (Harrison and Pirt, 1967) or *Saccharomyces cerevisiae* (Lloyd, 1974). Changes occur on differing time scales. Rapid responses (on time scales of minutes) represent metabolic feedback in allosteric enzymes. These are followed by slower changes (measured in hours) that require new enzyme synthesis.

The meaning of steady-state performance in chemostat culture

So we may ask "What is meant by the steady-state?" How long after a perturbation to a continuous culture must we allow to elapse before we can assume that the culture is again at steady state? There is no absolute answer to this question; a true steady state probably never exists in a biological system (otherwise there could be no process of evolution). In most studies, the term "steady state" culture is an operational one, and refers to a population of cells kept constant over several generations. There may be multiple possibilities for steady states. This is especially so in mixed cultures of microorganisms. For instance, this is a very important principle in waste-water treatment. The actual steady state attained depends on past history of the culture and on the previous perturbations it has experienced. Transitory responses provide the best means of defining regulatory mechanisms involved in cell metabolism e.g. the Pasteur Effect in yeast, mammalian cells, and bacteria. When O_2 is decreased, glucose-6- phosphate decreases and then recovers, while fructose 1, 6-diphosphate and triose phosphate increase, before decreasing again. This "cross-over point" between those metabolites, which show transient decreased concentrations and those that increase, indicates an important site of glycolytic control by phosphofructokinase (Ghosh and Chance, 1964).

Oscillatory Phenomena in Continuous Cultures

Oscillations as a consequence of equipment artifacts

Poor feedback control of parameters, e.g. pH, stirring speed, foam or volume control, can lead to instabilities. Even small oscillations in these environmental factors are amplified to give large fluctuations in other culture parameters. Thus in a continuous culture of *Pseudomonas extorquens* (e.g. at pH 7.0, pH fluctuations of < 0.1 unit), the usual feedback system used for pH control (linked to the output of a pH electrode) has been known to give fluctuation of ±6% total respiration; this in turn can give a fluctuation in dissolved O_2 of as much as ±25% of air saturation (Harrison and Loveless, 1971). Sensitivity of the cells to small changes in pH varies across the pH range; sometimes when pH is altered even by small amounts, acid fermentation products (acetic and formic acids) replace nonionisable products (ethanol and acetoin) and the pH of the culture then falls precipitously.

(1) *Temperature*: especially at upper or lower end of growth range, small changes in temperature can give large changes in physiological activities and growth rates.

(2) *Stirring rate fluctuations*: affect mixing and especially gas transfer functions (e.g. K_{La}).

(3) *Foaming*: also affects K_{La}; antifoam agents (surface active agents) affect membrane functions (e.g. cell respiration).

(4) *Discontinuous substrate feed*: Pulse-feeding of substrate can give oscillations in respiration rate, dissolved O_2 and pH. Also the yield coefficient may vary with pulse frequency even though the average D is kept constant. This observation indicates poor regulation of energy metabolism in these organisms — in the sense that excess substrate is wastefully oxidized. Some feed pumps do not provide steady continuous flow; a peristaltic pump is not as good as a syringe pump activated by stepper motor in this respect. In largescale processes, components of a single-feed stream may not be completely mixed.

(5) *Spatial heterogeneities in reactors*: All the preceding factors can contribute to the important problem of heterogeneities in the bulk phase: to live with this situation is often necessary, and attenuation of the consequences has been reviewed (Lara *et al.*, 2006). The same group has investigated the fast (within 2 s) dynamic response of fermentative metabolism in a 3.5 L reactor (Lara *et al.*, 2009).

Oscillations derived from feedback between cells and environmental parameters

Here, feedback between cells and environment has an effect on the environment, which in turn feeds back on the cells. Thus when the pH of a weakly buffered culture is not controlled, or if the culture growth produces acid, the pH falls, and the metabolism slows or stops. Then the pH increases again due to influx of fresh medium; metabolic activity increases again, giving damped or continuous oscillations depending on lags in the system. This sequence of events thereby leads to oscillations in respiration and dissolved O_2. Thus it was shown that in *K. aerogenes* the oscillatory state requires three conditions.

(i) At low O_2, respiration rate increases (Degn and Harrison, 1969); whereas, at higher levels of dissolved O_2 ($<20\,\mu M$ O_2), respiration rate is independent of O_2.

(ii) Some metabolic intermediate, a substrate for respiration, builds up and then becomes depleted.
(iii) Oxygen transfer from gas–liquid phase is limited by a low value for K_{La}. Increased stirring rate (i.e. the lowering of the diffusional barrier to O_2) in this case prevents the oscillations.

Growth of *E. coli* in a chemostat can give oscillating pyruvic acid production, and this was observed as spikes which occurred at hourly intervals. Pyruvate initially was produced at a high rate, and then rapidly oxidized. Production repressed at high pyruvate concentration but, derepressed at low pyruvate concentration. Because of time lags in the system, feedback control continuously produced overshoots and this led to undamped oscillations in the concentration of pyruvic acid in the culture.

Oscillations derived from intracellular feedback regulation

Rhythmic phenomena are ubiquitous in biological systems due to physiological control processes. These oscillatory states, may represent "sloppy" regulation, but more importantly serve a variety of functions including time-keeping or signaling cell–cell and intracellular (Goldbeter, 1996). Where advantageous they are highly conserved during biological evolution as biological rhythms and clocks (Edmunds, 1988; Lloyd, 1992). Complex metabolic interactions in recombinant *E. coli* fermentations indicate the nonlinearity of network behavior and the often unpredictable behavior of cultures (Andersen *et al.*, 2001; Stowers *et al.*, 2009) have shown how autonomous oscillations can be exploited to maximize product yield (e.g. in this case ethanol). Modeling the effects of high product and substrate inhibition predicts the onset of complex oscillatory behavior (Lenbury *et al.*, 1999).

Glycolytic oscillations

These are the most frequently studied example of oscillatory states. They were first reported by Ghosh and Chance (1964) who observed NADH oscillations in yeast and measured the fluctuating pool sizes of glycolytic intermediates. Mechanisms were predicted (Higgins, 1967; Sel'kov, 1968) and experimentally validated (Chance *et al.*, 1967). The first demonstration of a sustained glycolytic oscillation in a suspension of intact yeast was that of Von Klitzing and Betz (1970).

Respiratory oscillations

K. aerogenes grown in continuous cultures sometimes shows NADH and dissolved O_2 oscillations at high frequency, and low amplitude, i.e. <1% total respiration rate (Harrison, 1970; Harrison *et al.*, 1969). These oscillations (Fig. 6.5) have been shown to be insensitive to changes in pH, temperature, O_2 tension, or medium supply rates, and they occurred at high O_2 (i.e. well above "critical" dissolved O_2).

They were not due to cell–environment interactions, but arose from intracellular feedback loops initiated by an anaerobic shock. Damping was decreased by repeated anaerobic shock, although sometimes these oscillations arose spontaneously. It was concluded that there must be some cell–cell communication in order to produce and maintain population synchrony although the putative synchronizing substance has never been identified in this bacterial system.

The induction and elimination of oscillation in continuous cultures of *S. cerevisiae* (Parulekar *et al.*, 1986) by varying dilution rate, agitation speed, and dissolved O_2, did not provide a mechanistic explanation.

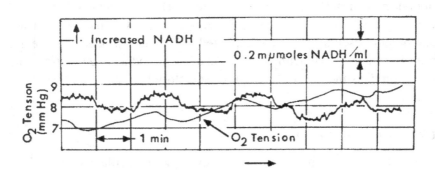

Figure 6.5.   Oscillations of NADH fluorescence and dissolved oxygen tension obtained in a glucose-limited chemostat culture of *K. aerogenes* at $D = 0.2\,h^{-1}$ and pH = 6.0.
Source: Reproduced from Harrison (1970), with permission.

Growth rate oscillations

Dean and Moss (1970) showed that barbiturate inhibition of *K. aerogenes* induced oscillations in growth rates in a turbidostat culture. Uptake of the drug was dependent on growth rate and this phenomenon gave a highly damped oscillation with a period of about 2 generation times.

Oscillations derived from interactions between different species in continuous culture

Protozoa and bacteria grown together in continuous cultures often show the oscillatory predator/prey population relationships under constant environmental conditions, e.g. with glucose as the limiting nutrient (Curds and Cockburn, 1971). In this system variations in the temperature conditions or residence times gave a reasonable fit to a model, except that damping eventually occurred (perhaps due to wall-growth which would be expected to stabilize the system).

This approach to growth dynamics of mixed cultures has proved useful for the modeling of microbial interactions in natural environments, e.g. in sludge-treatment plant, soil, sediments, and in medical microbiology. Developments include the modeling of the spread of antibiotic resistance through and between populations and an understanding of epidemiology (e.g. the spread of human disease; AIDS, malaria, and measles).

Oscillations due to synchronous growth and division

To study synchronized population growth and division in continuous culture, periodic starvation and reseeding regimes have been most valuable, e.g. in *Candida utilis* and in *E. coli* (Dawson, 1985). "Spontaneous" synchrony described by Kuenzi and Fiechter (1969), and Von Meyenburg (1973), has been analyzed in continuous cultures of *S. cerevisiae* by a number of groups including those in Zurich and Milan (Porro *et al.*, 1988; Strassle *et al.*, 1988/1989). In the former, investigations of a number of variables were simultaneously monitored (Fig. 6.6) and it was shown that oscillatory behavior of O_2 consumption, CO_2 production, ethanol production, NADH fluorescence intensity and biomass results from spontaneously generated synchronization of cell division cycles in the entire population of organisms.

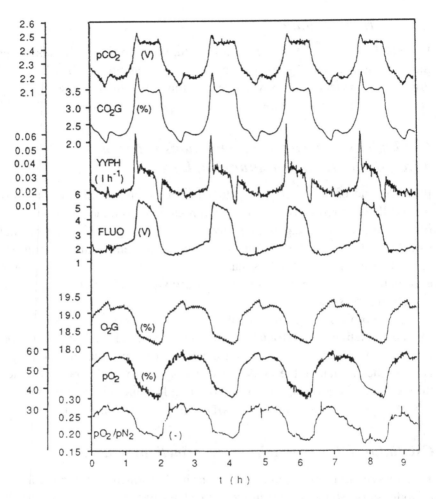

Figure 6.6. Synchronous growth of *S. cerevisiae* in continuous culture at a dilution rate of $0.182\,h^{-1}$: ratio of intensities (MS-membrane inlet) of CO_2 and N_2 in liquid phase (pCO2/pN2)' Col partial pressure in liquid phase (pCO2, Ingold electrode). CO_2 content in exhaust gas (CO2 G). Flow rate of pH controlling agent (YYPH). Culture fluorescence (FLUO, arbitrary voltage units). Ratio of intensities (MS-membrane inlet) of ETOH and N_2 in liquid phase (pETOH/pN2)' O_2 content in exhaust gas (O2 G)' O_2 partial pressure in liquid phase (PO2, Ingold electrode). Ratio of intensities (MS-membrane inlet) of O_2 and N_2 in liquid phase (PO2/pN2). The ripple in the O_2 G-signal is due to a badly tuned thermostat in the oxygen analyzer.

Source: Reprinted from Strassle *et al.* (1989) ©copyright 1989, with permission from Elsevier Science.

More recent analyses have shown that in "forced" synchrony cultures (Fig. 6.7) the period and degree of synchrony of spontaneous oscillations can be manipulated by repetitive small pulses of carbon and energy source (Munch *et al.*, 1992).

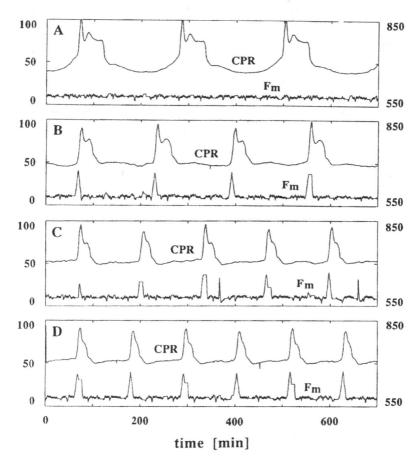

Figure 6.7. The subplots A–D show a spontaneously synchronous (A) and three forced synchronous cultures (B–D) of *S. cerevisiae* at the constant dilution rate $D = 0.13\,h^{-1}$ at pH 4.0. The forcing period varies from 162 (subplot B), 132 (C) down to 112 (D) min. The spontaneous oscillation period is 215 min F, is the medium flux (measurement value), CPR is the carbon dioxide production rate. The amount of the pulses in the forcing function was $60\,mg\,l^{-1}$ glucose.

Source: Reprinted from Münch *et al.* (1992) ©copyright 1992, with permission from Elsevier Science.

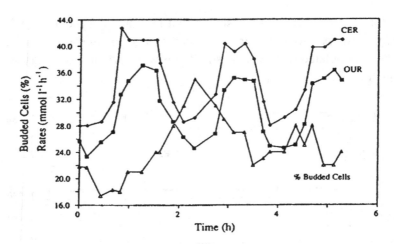

Figure 6.8.   Cyclic carbon dioxide production, oxygen uptake and percentage budded cells for synchrony oscillations at $0.14\,h^{-1}$ in the calorimeter. The budded cell curve is out of phase with the CER and OUR.
Source: Reprinted from Auberson *et al.* (1993) ©copyright 1993, with permission from Elsevier Science.

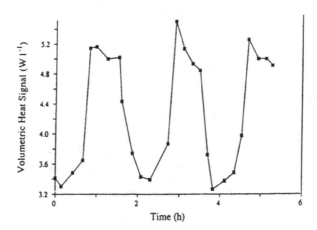

Figure 6.9.   Cyclic volumetric heat production rates for synchrony oscillations at $0.14\,h^{-1}$ in the calorimeter. With this smaller time scale, it can be seen that the period for one cycle is $1.8\,h$.
Source: Reprinted from Auberson *et al.* (1993) ©copyright 1993, with permission from Elsevier Science.

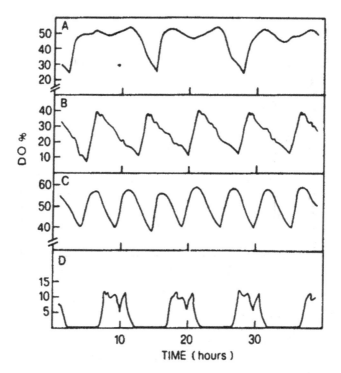

Figure 6.10.   Oscillation of dissolved oxygen (DO%) in continuous cultures of *S. cerevisiae* at different dilution rates: (A) D = $0.07 \, h^{-1}$, agitation speed 200 rpm; (B) $D = 0.106 \, h^{-1}$, 200 rpm; (C) $D = 0.15 \, h^{-1}$, 300 rpm; and (D) $D = 0.166 \, h^{-1}$, 200 rpm.
Source: Porro *et al.* (1988) ©copyright 1988 John Wiley & Sons, Inc. Reprinted by permission of Wiley-Liss, Inc., a subsidiary of John Wiley & Sons, Inc.

The periodic nature of respiro-fermentative metabolism have been further analyzed by calorimetry (Figs. 6.8 and 6.9) (Auberson *et al.*, 1993), and by cell component analysis (Duboc *et al.*, 1996).

Systematic analysis of oscillations as a function of two important controlling parameters, dilution rate and dissolved O_2 (Figs. 6.10 and 6.11) have been made by Porro *et al.* (1988).

In a newer model for spontaneous oscillations in yeast cultures alternate growth on limiting glucose and limiting glucose plus ethanol is invoked (Martegani *et al.*, 1990). More complex behavior in continuous cultures controlled by ac impedance feedback includes fluctuations in growth rate (Davey *et al.*, 1996). A variety of time series

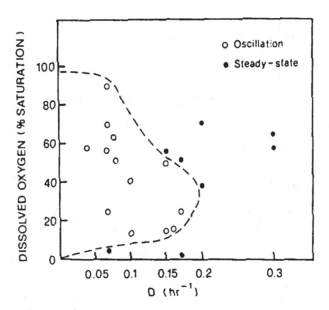

Figure 6.11.   Existence of oscillations as a function of D and DO% in glucose-limited continuous cultures of *S. cerevisiae*.
Source: Porro *et al.* (1998) ©copyright 1988 John Wiley & Sons, Inc. Reprinted by permission of Wiley-Liss, Inc., a subsidiary of John Wiley & Sons, Inc.

analyses (Fourier transformations, determination of Hurst, Lyapunov and embedding dimensions as well as nonlinear forecasting techniques) were employed to demonstrate the presence of a chaotic attractor in the dynamics.

Self-synchronized continuous cultures of yeast

A robust autonomous respiratory oscillation ($\tau = 30$–120 min) also occurs in certain strains of acid-tolerant *S. cerevisiae* grown under continuous aerobic culture conditions (Satroutdinov *et al.*, 1992). The oscillation occurs independently of glycolysis, the cell cycle (Keulers *et al.*, 1996b) and no difference in oscillations was observed when cultivation was carried out in light or darkness (Murray *et al.*, 1998). The oscillation is dependent on pH (Satroutdinov *et al.*, 1992), aeration, and carbon dioxide (Keulers *et al.*, 1996a). Oscillation occurs when glucose, ethanol, or acetaldehyde is used as a carbon source. For the oscillation to occur, ethanol has to be present.

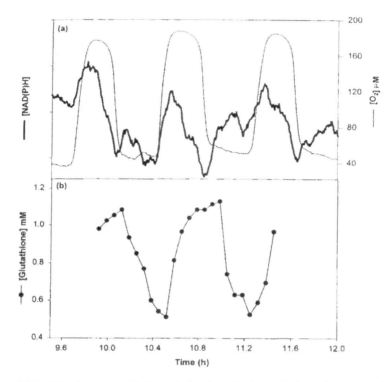

Figure 6.12. Respiratory oscillations of the changing intracellular redox states of yeast. (a) Dissolved O_2 and NAD(P)H fluorescence (365 $\rightarrow$ 450 nm) measured continuously online. (b) Total intracellular glutathione.
Source: Reprinted from Murray *et al.* (1998) ©copyright 1998, with permission from Elsevier Science.

It has been suggested that a stage of ethanol metabolism may be a locus for both population synchrony and intracellular regulation giving oscillatory dynamics (Keulers *et al.*, 1996b).

With ethanol, continuous monitoring indicates an oscillation of NAD(P)H with a complex waveform, and with the predominant period of 45 min identical with that in dissolved O_2 (Murray *et al.*, 1998). Intracellular GSH also oscillates with the same period (Fig. 6.12).

These respiratory oscillations are extremely sensitive to perturbation by pulse addition of Na nitroprusside (Fig. 6.13) and this effect appears to be specifically mediated by nitrosonium ions (NO^+), as $NO^{\cdot}$ (gas) or

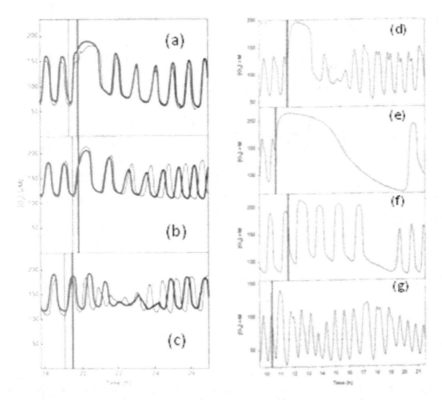

Figure 6.13.   Perturbation of dissolved oxygen oscillation during continuous aerobic yeast culture, with GSH (50 μM) (a), GSSG (25 μM) (b), NF (5-nitro-2 furaldehyde, an inhibitor of GSH reductase) (50 μM) (c), 100 μM glutathione (d), 5 μM Na nitroprusside (e), 8 μM NaNO$_2$ (f), and 10 μM S-nitrosoglutathione (g). In (a)–(c), the thick lines represent injection at high dissolved oxygen, the fine lines represent injection at low dissolved oxygen, and the vertical bars represent the time of addition. In (d)–(g), vertical bars indicate time of the additions specified.

Source: Panels (a)–(c) are reproduced from Murray *et al.* (1999) by permission of The Society for General Microbiology. Panels (d)–(g) are reproduced from Murray *et al.* (1998); ©copyright 1998, with permission from Elsevier Science.

NO-donors are not effective (Murray *et al.*, 1998). Preferred target sites are probably either thiols or protein metal centers.

Reduced glutathione (GSH) itself also produces a marked effect on the respiratory oscillations (Murray *et al.*, 1999) leading to an interruption of oscillatory behavior due to respiratory inhibition (Fig. 6.13). No evidence has been found for cell division cycle synchrony in these cultures;

the mean budding time under the conditions employed is about 12 h. Our interpretation of these data suggests the operation of a respiratory switch during cycling between high and low respiratory activity, concomitant with alternation of redox states.

This continuously oscillating yeast system provides a convenient model, which may be analogous to a longer-period redox cycling system, the Circadian Clock (Lloyd and Murray, 2007). Temperature compensation of the period (Murray *et al.*, 2001) indicates a timekeeping function. This ultradian rhythm was first discovered in *Schizosaccharomyces pombe* (Poole *et al.*, 1973), then in a variety of yeasts and protozoa (Lloyd, 1992). Its timekeeping characteristics were first shown in the small soil amoeba, *Acanthamoeba castellanii* (Edwards and Lloyd, 1978; Edwards and Lloyd, 1980; Lloyd *et al.*, 1982). This Ultradian Clock, most evident as an easily monitored respiratory oscillation in the *S. cerevisiae* system, as originally described by Kuriyama's group, underpins the coherence of all cellular functions. Thus it acts as the time base for metabolism, biosynthesis, and assembly of membranes and organelles, and cell division, as well as all the necessary attendant signaling and control functions: at the core of this 40 min clock is a nicotinamide nucleotide driven redox cycle (Fig. 6.14) (Lloyd and

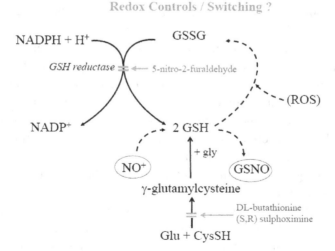

Figure 6.14. The redox core of the ultradian clock.

(a)   (b)

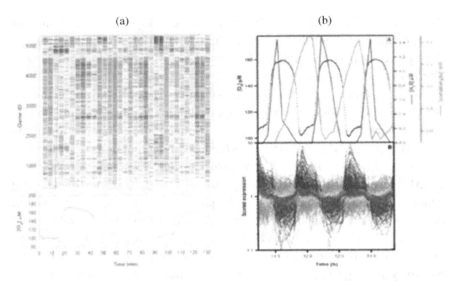

Figure 6.15.   The transcriptome of *S. cerevisiae* in continuous culture. (a) Transcripts of 5,329 genes were separated from samples taken every 5 min through 3 ultradian cycles. (b) Dissolved O_2, H_2 S, and acetaldehyde in the culture up- and down-transcript expression: those in phase with the oxidative state are dark gray, and with the reductive phase, light gray. *Source*: Klevecz *et al.*. (2004).

Murray, 2000; Lloyd and Murray, 2005; Lloyd and Murray, 2006; Lloyd and Murray, 2007)).

The cellular network of transcription is subject to its temporal control (Klevecz *et al.*, 2004) (Fig. 6.15), as is the corresponding expression of the metabolome (Murray *et al.*, 2007a). Higher order functions (e.g. the inner membrane transmembrane electrochemical potential) in mitochondria (Lloyd *et al.*, 2002) and the coordination of mitochondrial function with the cell division cycle (Lloyd, 2003) are also subject to Ultradian Clock timekeeping (Figs. 6.16 and 6.17).

Underlying clock function is a tuneable metabolic attractor which provides robust and yet flexible operation (Murray and Lloyd, 2007), and shows periodic, quasiperiodic, and chaotic outputs. Real-time monitoring of the self-synchronized continuous culture of yeast for dissolved O_2, CO_2, and H_2S reveals multioscillator operation, with periodicities of 13 h, 36 min, and 4 min (Fig. 6.18(a)) (Roussel and Lloyd, 2007). Computation of the

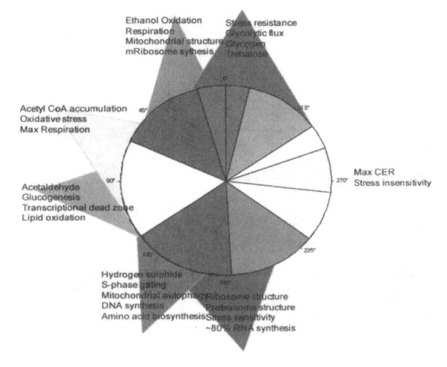

Figure 6.16. Ultradian clock cycle maps. The 40 min cycle in yeast is described by sequential outputs of physiological and biochemical events as the traverse is accomplished in the anticlockwise direction.
Source: Courtesy of D.B. Murray, in Lloyd and Rossi (2008).

leading exponent reveals the dynamics on the metabolic attractor to be chaotic (Fig. 6.18(b)).

The 4 min component is likely to arise from intracellular mitochondrial function as the autofluorescence of individual yeasts arising from NAD(P)H as observed by two-photon excitation (Aon *et al.*, 2007b) shows similar periodicity. The 13 h period may represent a circadian subharmonic (the experiments were conducted at constant temperature and in natural daylight). Under similar conditions the 40 min Ultradian Clock, monitored as the RQ of the off-gas from the fermentor, has been shown to lead to the gradual appearance of a Circadian cycle (Murray *et al.*, 2007b). The fractal scaling and inverse power-law nature of the yeast multioscillator is revealed

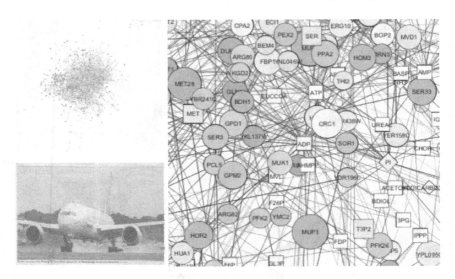

Figure 6.17.   The yeast reactome simplified to a network of 16,723 edges and 4912 nodes: its complexity is comparable to the control wiring of a large airliner, and can only be meaningfully illustrated as a complex of subgraphs.
Source: Murray *et al.* (2007a).

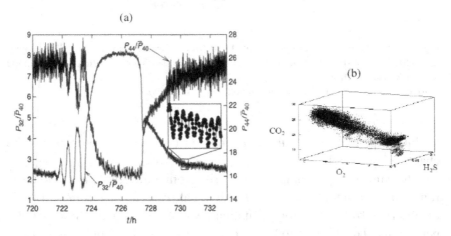

Figure 6.18.   Dissolved O_2 (P32/P40) and CO_2 (P44/P40) in the self-synchronized yeast culture. (a) Membrane inlet mass spectrometric monitoring at m/z 32, 40, and 44 for O_2, Argon, and CO_2, respectively showing 13 h, 40 min and 4 min periodicities for both respiratory gases relative to the inert control signal for Argon (Roussel and Lloyd, 2008). (b) A plot of dissolved O_2, CO_2, and H_2S continuously monitored by membrane inlet mass spectrometry shows the chaos inherent in the dynamics of the self-synchronized yeast culture. The plot has 36,374 data points sampled every 15 s over several months (Roussel and Lloyd, 2007).

by relative dispersional and power spectral analyses; robust yet flexible responses to environmental influences is characteristic of such networks. Coherent operation of intracellular temporal behavior is underpinned by heterarchical and delocalized control, coordinated through multiple time-domains (Aon *et al.*, 2008a).

Chapter 7

Bioprocess Development with Plant Cells

Microorganisms were the first biological systems employed in MCE. The relatively ease of manipulation of many bacteria and of a few fungi (typically *Saccharomyces cerevisiae*), including short generation times, as well as the information available on their genetics and physiology, have contributed to the preferred choice of microorganisms for MCE. This is well mirrored in the previous chapters of this book, where mathematical approaches, experimental methods, and designs are exemplified and derived from studies mainly performed with microorganisms.

Unlike microorganisms, plant cells are more complex in structure, physiology, and metabolism. In addition, for many years, biochemical studies on metabolism were concentrated on bacteria and animal cells; with the idea that what happens in a plant cell should be very similar to other organisms. This is clearly observed in most textbooks on biochemistry, where metabolism of animals and microorganisms are exhaustively considered; whereas distinctive fundamental aspects of plant biochemistry receive no mention. Despite this, several features of plants determine that they are currently being considered as the most promising organisms to be used in MCE. In fact, the production of transgenic plants with increased resistance to herbicides, insects, or viruses shows that, in this area, the development of plant biotechnology is more advanced than in animals. Also, photosynthetic microorganisms like green algae, are of high value for applications in bioindustries producing cosmetics, foods, fuels, and plastic-like polymers. For these reasons, it is clear that a complete characterization of plant metabolism and physiology, followed by accurate studies of matter and energy balance and cellular process quantitation, are highly relevant issues at the present time.

In this chapter, we consider the use of plants as biological systems for MCE. We focus on the current and potential resources in the field and discuss the distinctive problems arising from the complexity of plant cells for the rational design of metabolism and metabolic engineering. The analysis involves both the plant cell and the whole organism and presents a challenge for MCE at higher levels of complexity.

MCE in Plants: Realities and Potentialities

Many distinctive characteristics make plants a unique target system for MCE. Mostly based on autotrophy toward the assimilation of inorganic CO_2, photosynthetic organisms are the most important renewable resource commodity on Earth (Owen and Pen, 1996). Plants have been historically used to satisfy demands for food, as well as to obtain raw materials for nonfood industries (including chemicals and pharmaceuticals) (Somerville and Bonetta, 2001). The emerging techniques of genetic transformation have opened relevant possibilities for their application in plants not only in order to improve the productivity of traditional crops but also to obtain organisms synthesizing novel products. Thus, a vision of plants as natural bioreactors, having the advantage of producing high quality biomass at relatively low cost, is increasingly being considered. Many goals have already been attained in this field, and the future of this discipline is highly promising (Birch, 1997; Collins and Shepherd, 1996; DellaPenna, 2001; Iglesias, 2004; Ohlrogge, 1999; Owen and Pen, 1996; Sharma and Sharma, 2009; Somerville and Bonetta, 2001; Willmitzer, 1999). A number of genetically modified (GM) plant varieties (mainly represented by the crops soybean, maize, canola, and cotton) have already been implemented as an alternative to classical breeding. Commercial production of the transgenic utilized more than 110 million hectares in 2007, expanding in developed and developing countries. Application of this technology is modifying scenarios and is posing key questions related with food security, as well as socioeconomic and environmental issues (Lemaux, 2008; Lemaux, 2009).

Approaches to plant metabolic engineering are multifaceted. A schematic division of purposes and objectives of plant transformation by genetic engineering is illustrated in Table 7.1. Manipulation of genetic information in plant cells is a very important tool that can primarily be utilized for basic or applied research (Birch, 1997; DellaPenna, 2001; Iglesias, 2004; Somerville and Bonetta, 2001). Obviously, applied strategies are dependent on genetics; mostly because the use of molecular biology tools results in a better understanding of physiological processes; this fundamental work feeds potential new applications. In fact, a good knowledge of plant metabolism, including metabolic fluxes and their control and regulation, is essential for a rational design of plant transformation in MCE.

Plant transformation for studies on metabolism and physiology

The possibility of up- or down-modulating specific genes in plants contributes to the study of their metabolism and physiology in new and powerful ways than those allowed by other biochemical methods. Table 7.1 shows examples of key plant processes that are being studied utilizing this tool. These studies have provided key information for potential application to the improvement of plant products. First, the analysis of the enzymatic role in the allocation and partitioning of photoassimilates is relevant for the evaluation of factors determining the harvestable yield of a crop; this is critical for rational manipulation and improvement (Herbers and Sonnewald, 1996; Somerville and Bonetta, 2001; Stitt and Fernie, 2003). Second, the characterization of specific enzymatic and hormonal contributions to plant development is important for determining postharvest altering traits for fruits and flowers maintenance (Mol *et al.*, 1995; Theologis, 1994). Finally, the study of plant-microbe interactions and their dependence on cellular signals has allowed gaining key information about the functioning of natural plant resistance to pathogens (Chisholm *et al.*, 2006; Keen *et al.*, 2000; Staskawicz, 2009; Staskawicz *et al.*, 1995).

Emerging scenarios from -omics projects and systems biology

The emerging "omics" technologies, together with the improvement of analytical techniques, have provided valuable tools for obtaining complete sets of data necessary to understand functional regulation of metabolism in different cells. The subject is very appropriate for application in plants

Table 7.1　Plant genetic transformation: approaches, potential, and actual achievements

Plant transformation (main objective)	Possibilities	Examples	References
1. As a tool for basic research in plant biochemistry/ physiology	Expression, over- or under-expression or deletion of specific genes coding selected proteins (enzymes, translocators, hormones, receptors, etc.). Development of the RNAi strategy.	Studies on carbon allocation and partitioning. Characterization of enzymes and hormones involved in plant development. Studies on cellular signals for plant-microbe interaction	(Hebert *et al.*, 2008; Herbers and Sonnewald, 1996; Mol *et al.*, 1995; Staskawicz, 2009; Staskawicz *et al.*, 1995; Stitt and Fernie, 2003; Theologis, 1994)
2. As a tool for the improvement of plant uses	a. Introduction of new traits to plants for increasing resistance to chemicals, pathogens, and abiotic stress	Transgenic plants with resistance to glyphosate, pests, osmotic, and water stress	(Collins and Shepherd, 1996)
	b. Manipulation to improve the quantity and/or quality of natural products derived from plants	Transgenic tomato with longer postharvest life. Transgenic plants producing higher amounts or structurally modified fats, carbohydrates, proteins, vitamins. Transformed plants producing thermoplastic polymers (polyhydroxyalkanoate-based). Plant transformation to improve production of secondary metabolites	(Capell and Christou, 2004; Carrari *et al.*, 2003; Collins and Shepherd, 1996; DellaPenna, 2001; Heyer *et al.*, 1999; Iglesias, 2004; Ohlrogge, 1999; Slater *et al.*, 1999; Somerville and Bonetta, 2001; Weselake *et al.*, 2009)

(Continued)

Table 7.1 (*Continued*)

Plant transformation (main objective)	Possibilities	Examples	References
	c. Production of heterologous proteins for pharmaceuticals and other industries	Transgenic plants producing cholera toxins and human enzymes and antibodies.	(Basaran and Rodriguez-Cerezo, 2008; Collins and Shepherd, 1996; Drake and Thangaraj, 2010; Lienard *et al.*, 2007)

and other photosynthetic organisms. Expectedly, genomics is leading with the major number of projects and data feeding of bioinformatics registers. *Arabidopsis thaliana* was the first higher plant genome sequenced and was completed more than a decade ago (Arabidopsis Genome Initiative, 2000). At present, the website http://www.ncbi.nlm.nih.gov/genomes/leuks.cgi shows 141 projects involving plants at different states of progress. From these, 15 correspond to green algae and 136 to land plants. Among the genome projects already completed are those from higher plants *Arabidopsis thaliana* Columbia-O, *Oryza sativa* Japonica Group Nipponbare, and *Zea mays* B73, and green algae *Ostreococcus lucimarinus* CCCE9901, *Ostreococcus tauri* OTH95, and *Micromonas* sp. RCC299. Many of the genome projects correspond to crops, although only the one on rice has been finished (Feuillet *et al.*, 2011). Information gained from genome sequencing constitutes the basis for the current era of plant genomics research. In such a way, genomic information is helping the experimental design aimed at obtaining more data. The information gathered is utilized for addressing biological questions of speciation, evolution as well as to select crops and other kind of photosynthetic organisms for sequencing in the future.

Based in the "omics" strategy, many studies related with functional genomics are generating valuable as well as profuse information about function, regulation, and coordination between different metabolic pathways in plants. All this information needs to be integrated into modeling of networks with a Systems Biology approach (Fukushima *et al.*, 2009). Integration of high-throughput data derived from studies on transcriptomics, proteomics (including posttranslational modifications), and metabolomics, has spurred the development of intelligent bioinformatic software. A better software quality favors the elucidation of the actual function of a gene of interest, associating modes of regulation and synthesis, posttranslational modification, interaction with other partners, and effect of the final product on the cell metabolic network (Pitzschke and Hirt, 2010). The approach of metabolomics is also of great value for determining functionality of genome information and for devising biotechnological applications (Saito and Matsuda, 2010). Furthermore, integrative projects using robotic/computational platforms for measurement of multiple molecular/functional profiles (e.g. enzymatic activities and kinetic parameters, together with metabolites and proteomics) are useful tools to reconstruct metabolic networks and to rationalize variations associated with growth and development as well as with different environmental conditions (Kopka *et al.*, 2005; Stitt *et al.*, 2010). Studies are also relevant in supplying evidence about the coordination of assimilates supply and sustained plant growth under a changing environment (Smith and Stitt, 2007).

Improving Plants through Genetic Engineering

Table 7.1 illustrates the use of genetic engineering as a practical tool for improving plants, considered as chemical factories that use sunlight and atmospheric CO_2 as sources of energy and feedstock, respectively. Plant transformation is performed with three main objectives: (i) incorporation of new traits, endowing plants with increased resistance to chemicals, pathogens, or different stresses; (ii) the improvement of products naturally synthesized by plants through the increase of their quantity and/or quality; and (iii) the incorporation of genes to plants for the production of heterologous proteins or novel compounds with pharmaceutical or other industrial applications. We will consider examples and possibilities in each of these three categories as outlined in Table 7.1.

Improving plant resistance to chemicals, pathogens, and stress

Development of plants with higher resistance to different biotic and abiotic stresses is a main goal that needs to be reached in order to solve the increasing demand for food production with a maximal economy in the use of land and water (Somerville and Briscoe, 2001). The genetic transformation of plants is a key tool to produce economically relevant species with improved resistance to chemicals, pathogens, and other stresses.

The modification of many crops in order to incorporate new traits for higher resistance to chemicals and pathogens, or conferring longer postharvest life, constitutes the first generation (or phase) of GM plants (Iglesias, 2004; Ohlrogge, 1999; Willmitzer, 1999). Production of plants with resistance to herbicides is a common example of plant transformation by genetic engineering for increasing the yield of many crops. Herbicides are chemicals that act inhibiting target enzymes involved in photosynthesis or amino acids metabolism that are essential to plants (Lein *et al.*, 2004). The widely utilized herbicide glyphosate is an analogue of phospho-*enol*-pyruvate (PEP). This key metabolite is a biosynthetic precursor of three essential amino acids: tryptophan, phenylalanine, and tyrosine. Glyphosate inhibits EPSP (5-enolpyruvylshikimate-3-phosphate) synthase, an enzyme involved in the anabolic pathway of aromatic amino acids (Herrmann and Weaver, 1999). The effect of glyphosate on plants is nonselective and of broad spectrum; thus, this compound is toxic not only for weeds but also for crops, by affecting protein synthesis (as a consequence of the effect on aromatic amino acids anabolism), and cell growth. In transgenic plants, tolerance to the herbicide is achieved by overexpression of the normal EPSP synthase, or by expression of a glyphosate-insensitive enzyme, naturally occurring in *Agrobacterium* spp. Alternatively, transformation includes the expression in plants of a bacterial glyphosate oxido-reductase, an enzyme that degrades the herbicide to nontoxic compounds (Duke, 2010).

The increase in plant resistance to a number of pests has been obtained by transforming different crops with genes encoding for endotoxins from *Bacillus thuringiensis* (*Bt*) (Collins and Shepherd, 1996; Iglesias, 2004). This aerobic, Gram-positive bacterium (with soil habitat), produces a number of insect toxins. More important are δ-endotoxins (protein crystals formed during sporulation), with activity against a number of caterpillars.

This approach represents a good alternative to chemical insecticides for controlling many species of pest insects (DellaPenna, 2001; McPherson and MacRae, 2009; Somerville and Bonetta, 2001). A strategy followed for successful plant transformation comprises the introduction of *Bt* toxin genes linked to a constitutive promoter allowing the expression of toxic proteins in all plant tissues. Advances have also been made to incorporate the transgene in plastidic genomes (Chakrabarti *et al.*, 2006; Liu *et al.*, 2008). Using resistance conferred by *Bt* genes has the advantage of being environmentally safe while exhibiting high efficiency. Disadvantages include high cost of production and low persistence, as well as the appearance of resistant insects.

Plant transformation using genes encoding for proteinase inhibitors (proteins naturally produced by certain plants that inhibit proteinase action) was also successfully applied to improve resistance to pests. The accumulation of the expressed proteinase inhibitor becomes toxic for herbivorous insects, thus resulting in an effective and broad spectrum strategy of pest control. Transgenic plants overexpressing hydrolytic enzymes (i.e. chitinase) and exhibiting enhanced resistance to fungal infections have also been developed (DellaPenna, 2001; McPherson and MacRae, 2009; Somerville and Bonetta, 2001).

Many environmental stresses (e.g. drought, high salinity, temperature extremes, toxicity by metals and other pollutants, UV-B radiation) significantly affect crop productivity (Al Khateeb and Schroeder, 2009; Farinati *et al.*, 2009; Smirnoff, 1998). Different genetic engineering approaches have been proposed to obtain economically important plants with increased resistance to a certain stress. In this sense, a very promising field is the transformation of plants for osmotic stress resistance (Apse and Blumwald, 2002; Munns and Tester, 2008). Various osmoprotectants (or compatible solutes) were found in plants and bacteria. Different plant species naturally synthesize and accumulate osmotically compatible metabolites. Remarkably, some species produce sugar-alcohols (mannitol, sorbitol), and these metabolites may represent a major product of photosynthesis together with sucrose and starch (Figueroa and Iglesias, 2010; Figueroa *et al.*, 2011). A strategy for plant transformation is to express genes in species that do not produce these compounds, or to accurately manipulate their level and timing of expression in plants that normally synthesize them. Convenient engineering of carbon metabolism may thus allow not only the production of plants that can grow under osmotic stress but also plants with improved resistance to other abiotic or biotic stressors (Figueroa *et al.*, 2011). Interestingly, it has been shown that some osmoprotectants help plants to cope with oxidative stress and pathogen attack (Smirnoff, 1998). In fact, any stressful condition ultimately induces oxidative damage, and understanding the different factors involved in maintaining the redox balance (Arias *et al.*, 2011) is critical for designing plants with built-in resistance to adverse environments.

Improving quality and quantity of plant products

Transgenic tomatoes, in which fruit ripening was conveniently modified (Flavr Savr tomato), was the first genetically engineered whole food sold commercially (Collins and Shepherd, 1996). The transgenic fruit exhibits a longer postharvest life. This genetic modification allows both better shipping and handling and complete fruit ripening on the plant, resulting in a product with improved flavor. Transformation of tomato plants was performed utilizing antisense technology to reduce the expression of polygalacturonase (PGase, the enzyme degrading pectin in the cell wall) by near 90%. The PGase antisense gene was constructed by fusing a cDNA clone of the enzyme gene in reverse orientation to a constitutive promoter, and then introducing it into tomato plants. During gene transcription, transgenic cells produce normal as well as antisense (complementary) mRNA molecules. The latter interact by binding to normal mRNA molecules thus inhibiting translation into PGase protein. In order to improve manipulation of fruit ripening, an alternative approach based on the regulation of ethylene synthesis (a phytohormone involved in modulating biochemical processes related with ripening), is being evaluated. For this purpose, the enzymes producing or degrading **a**mino**c**yclopropane-1-**c**arboxylate (**ACC** synthase and deaminase, respectively) are key targets (Collins and Shepherd, 1996).

Engineering of plant fatty acid composition is one of the most developed areas in the field of genetic transformation used to enhance the quality of plant food products (Collins and Shepherd, 1996; Ohlrogge, 1999; Somerville and Bonetta, 2001; Thelen and Ohlrogge, 2002). Triacylglycerols metabolism in plants became biotechnologically relevant not only as a food source but also for biofuels production. Outstandingly, fat products constitute the naturally most abundant, energy-rich forms of reduced carbon (Carlsson, 2009; Durrett *et al.*, 2008). A main objective in these studies is to modify chain length and degree of saturation of the fatty acids present in triacylglycerols, the plant storage oils. Successful results were obtained by manipulating up or down the expression of genes coding for specific thioesterases (which catalyze the hydrolytic release of fatty acids bound to acyl-carrier protein) or desaturases (which specifically introduce unsaturation in the fatty acid chain). Transgenic plants may synthesize modified or novel oil compounds with relevance for the production of healthier food. Additionally, utilization of fatty acids of plant origin in industrial processes represents an advantage as a renewable, biodegradable resource (Collins and Shepherd, 1996; Durrett *et al.*, 2008; Thelen and Ohlrogge, 2002). Attempts to modify oilseed fatty acid composition by genetic engineering to produce other chemical structures (i.e. conjugated double bonds, epoxy functions) are also main objectives in the field (Ohlrogge, 1999; Thelen and Ohlrogge, 2002).

An important advance related with the transformation of plants to produce biopolymers as an alternative to petrochemical plastics was achieved in 1999 (Slater *et al.*, 1999). In this case, the transgenic product resulted from the combined modification of fats (keto acids) and amino acids metabolism in plants. By this mean, *Arabidopsis thaliana* and *Brassica napa* were engineered with four distinct transgenes that divert metabolic pools of acetyl-CoA and threonine to synthesize the copolymer poly 3-hydroxybutyrate-*co*-3-hydroxyvalerate. Synthesis of the polymer takes place in plastids of leaves (*Arabidopsis*) or seeds (*Brassica*) of transformed plants. The commercial relevance of the copolymer is based on its biodegradability as a thermoplastic, thus representing an early example of the applicability of plants to biorefinery strategies.

Manipulation of the content and composition of carbohydrates is also a primary challenge for plant genetic engineering (Collins and Shepherd, 1996; DellaPenna, 2001; Iglesias, 2004). Considering that carbohydrates (mainly starch and sucrose) are major photosynthetic products, enhancement of CO_2 assimilation would straightforwardly lead to biomass increase (Figueroa *et al.*, 2011; Iglesias and Podestá, 2005). The importance of the latter is both ways, as a source of foodstuff and nonfood carbohydrates (mostly biofuel production). Although definite success has not been achieved yet, important partial goals have already been claimed, especially in terms of understanding carbohydrate allocation and partitioning in plants (Collins and Shepherd, 1996; Iglesias, 2004; Iglesias and Podestá, 2005; Zeeman *et al.*, 2010).

Starch is the main storage polysaccharide found in plants. Its biosynthesis plays different roles during plant growth and development, critically determining crop yield (e.g. corn, wheat, rice, potato, yam, cassava). Starch has multiple uses, as it constitutes the main source of food for humans and is utilized in a wide number of industrial processes, including those associated with nanotechnology (Le Corre *et al.*, 2010; Morell and Myers, 2005; Tharanathan, 2005). Starch synthesis takes place in the chloroplast (in photosynthetic cells) or in the amyloplast (in storage tissues), and it involves three enzymes: ADP-glucose pyrophosphorylase (ADPGlcPPase), starch synthase (SS) and branching enzyme (BE). ADPGlcPPase catalyzes the production of ADP-glucose (ADPGlc) from glucose-1-P (Glc-1-P) and ATP, and it is the key regulatory step in the biosynthetic pathway of the storage polysaccharide. Because of its role in starch synthesis, ADPGlcPPase is given as an example for the understanding of control and regulation of a metabolic flux (see below). ADPGlc is the glucosyl donor molecule utilized by SS to elongate an α-1,4-glucan chain, followed by the BE, which introduces α-1,6-branches. As such starch is a mixture of two polysaccharides: amylose (a linear α-1,4-glucan) and amylopectin (the branched polymer) (for further details see reviews by Ballicora *et al.* (2003), Ballicora *et al.* (2004), Hannah and James (2008), Iglesias and Podestá (2005), Kotting *et al.* (2010), Zeeman *et al.* (2010)).

Production of transgenic potatoes accumulating increased amounts of starch represents a remarkable success in the use of plant metabolic engineering for enhancing food quality traits (Collins and Shepherd, 1996; DellaPenna, 2001; Stark *et al.*, 1992). In this example, the strategy followed was to express a mutant (unregulated) ADPGlcPPase from *Escherichia coli* into potato plants with a transit peptide sequence that targeted it to amyloplast stroma (Stark *et al.*, 1992). The use of a mutated nonregulated enzyme circumvented the necessity for the presence of specific metabolites, the natural allosteric regulators of ADPGlcPPase (see below in this Chapter). By this procedure synthesis of ADPGlc within amyloplasts was enhanced (constantly and independently of regulatory metabolites) resulting in higher amounts of starch (Collins and Shepherd, 1996; Stark *et al.*, 1992). The transgenic product possesses a series of advantages, e.g. it is more suitable for storage and gives healthier fried food derivatives. Using a similar strategy, high starch tomato and canola plants were also obtained (Collins and Shepherd, 1996).

The success in the manipulation of the amount of starch synthesized by plants strongly supports the prospect of important achievable aims in the field of biodegradable polymers (Mooney, 2009). The possibility of modifying the quality of starch would be a highly relevant result, and a necessary next step in plant biotechnology. The relative content of amylose and amylopectin greatly influences the physicochemical properties of starch, and this determines its use in different industrial processes (Heyer *et al.*, 1999; Le Corre *et al.*, 2010; Morell and Myers, 2005). Approaches to modifying the ratio of polysaccharides include manipulation of SS isoenzymes and branching protein isoforms, as well as a proper ADPGlcPPase. To this respect, it has been shown that alterations in the supply of ADPGlc can affect the production of amylose (Heyer *et al.*, 1999; Morell and Myers, 2005; Stark *et al.*, 1992). The degree of phosphorylation of the polysaccharide molecule is another critical factor affecting the quality of starch for industrial use. The biochemical steps involved in starch phosphorylation have been elucidated in the last decade, starting with the identification of a gene responsible for glucan phosphorylation, which was utilized to genetically engineer potato starch (Heyer, 1999). This finding was followed by α-glucan water dikinase characterization and demonstration of the occurrence of isoforms of this enzyme that specifically phosphorylate C6- and C3-position of glucosyl residues in starch (Hejazi *et al.*, 2009; Mikkelsen *et al.*, 2004; Ritte *et al.*, 2006). All this progress opened new avenues for producing modified polymeric carbohydrates in the near future.

Plant secondary metabolites comprise a myriad of phytochemicals including alkaloids, polyphenols, steroids, terpenoids, anthocyanins, and anthraquinones (Allen *et al.*, 2009; Pichersky and Gang, 2000). They are specific to special groups of plants and, in some cases, only produced in certain tissues of a given species. These compounds have a high commercial value due to their extensive industrial use mainly as drugs, perfumes, pigments, agrochemicals, and food flavors. Good examples of valuable pharmaceuticals are the anticancer drugs vinblastine and taxol (Hughes and Shanks, 2002; McChesney *et al.*, 2007). Also relevant is the potential to manipulating volatile secondary metabolites for enhancing plant defenses along with the quality of flowers and fruits (Dudareva and Pichersky, 2008). Efforts are being made to use MCE to improve the performance of whole biochemical pathways, thereby enhancing the *in vivo* production of secondary metabolites (usually synthesized in low amounts) (Allen *et al.*, 2009). Important advances have been registered in the development of plant cell culture techniques for producing phytochemicals (Collins and Shepherd, 1996; Dudareva and Pichersky, 2008).

Plant proteins are deficient in certain amino acids that are essential for humans and animals, resulting in a shortfall for the nutritional quality of vegetarian diets. Typically, cereal proteins are poor in lysine and tryptophan, whereas legume and vegetables contain low amounts of methionine and cysteine. In order to overcome this deficiency, genetic engineering emerges as a very convenient tool with remarkable advantage over traditional breeding regimes. Identification of seed storage proteins rich in sulfur amino acids in plant members of the Brazil nut family, followed by the engineering of their expression in target plants, represent an important approach aimed at obtaining methionine-rich transgenic seeds (Collins and Shepherd, 1996). Also, promising advances in the enrichment of maize in lysine and tryptophan, have launched the stage of improving the content in essential amino acids of rice and other major grain crops (Wenefrida *et al.*, 2009).

Also worth mentioning is the accomplishment of improving the content of certain vitamins in plants through genetic engineering (DellaPenna, 2001). One specific example is the expression of the enzyme γ-tocopherol methyltransferase in *Arabidopsis* seeds, which significantly increases the amount of vitamin E. The latter represents a very promising strategy to manipulate levels of the antioxidant vitamin E in other plant seeds (e.g. soybean, maize, canola), with the aim of obtaining better quality oils for the human diet (Chen *et al.*, 2006). Another relevant example is given by the remarkable product called "golden rice" because of the yellow endosperm exhibited by rice after introducing the simultaneous expression of three enzymes (two from plants and one from bacteria) involved in the carotenoid biosynthetic pathway (Ye *et al.*, 2000). The final product (golden rice) contains high levels of β-carotene (precursor of vitamin A), thus representing a very important tool to cope with a serious vitamin deficiency, a recurrent problem in the developing world. Additionally, golden rice exhibits the strategic advantage of being a major commodity in developing countries. This is one of the many current approaches directed to seeking nutritionally fortified and enriched food from GM plants, as a strategy to cope with problems that critically affect poorer countries (http://www.the-scientist.com/article/display/55926/). Variation in the level of other carotenes (e.g. lycopene) was also explored, with the aim of modifying the content of dietary antioxidants (Rosati *et al.*, 2000). Recently, efforts have been directed to manipulate the content of vitamin B_6 in plants, whose relevance has increased due to existing evidence which shows an association between the deficiency of this vitamin with impaired cognitive functions, Alzheimer's disease, cardiovascular alterations, and different types of cancer (Chen and Xiong, 2009).

Using genetic engineering to produce heterologous proteins in plants

Plants represent a very convenient biological system for the production of recombinant proteins. This is increasingly being used for medicinal applications, ranging from antibodies to diagnostic proteins, and to nutraceuticals, as well as in other industries, including the potential use to produce enzymes and polymers for biofuels and biorefineries (Obembe *et al.*, 2011). The use of plants for "molecular farming" to produce eukaryotic proteins has many advantages (Basaran and Rodriguez-Cerezo, 2008; Drake and Thangaraj, 2010; Lienard *et al.*, 2007; Obembe *et al.*, 2011). Plants represent a less risky alternative due to reduced technical, ethical, and safety issues, including costs when compared with the use of mammalian-cell cultures or transgenic animals. Genetically engineered plants can synthesize fully folded functional proteins, even complex, glycosylated proteins. Plants possess similar post-translational and co-translational modification mechanisms as in mammalians. In addition, the photosynthetic capacity of plants, combined with the advances of modern agriculture, enables low-cost production of large amounts of recombinant macromolecules. The possibility of expressing heterologous proteins in a certain tissue (e.g. seeds) represents an additional asset, since it improves the capacity for storage of active biomolecules. Successful achievements deserving mention include the expression of cholera toxin subunits, human enzymes, vaccines, coagulation factors, and antibodies in plants (Basaran and Rodriguez-Cerezo, 2008; Collins and Shepherd, 1996; Obembe *et al.*, 2011).

Tools for the Manipulation and Transformation of Plants

Specific methods have been developed for manipulating and introducing foreign DNA into plant cells. Basically, the strategy for obtaining GM plants (mainly applied to crops) requires resourceful protocols for introducing foreign functional DNA into a host cell, and the capability of regenerating adult-transformed individuals exhibiting the desired trait (Shewry *et al.*, 2008). We will briefly consider the methods successfully employed for plant genetic engineering in the following order: (i) *in vitro* plant tissue culture techniques, and (ii) plant transformation systems.

(i) *In vitro plant tissue culture.* Tissue culture of higher organisms is based on the theory developed in 1838 by Schwan and Schleiden. This theory established that, because of its totipotency, a single plant cell is self-sufficiently able to regenerate a whole plant (Vasil, 2008). In 1939, Nobécourt, Gautheret and White were the first to successfully obtain plant tissue culture by growing tomato root tips using an artificial medium. It was after the discovery of phytohormones and their action as growth phytoregulators that appropriate techniques for the culture of plant cells were rapidly developed (Pierik, 1987). The recovery of plants from GM somatic cells is made using two main strategies: somatic embryogenesis or organogenesis (Pierik, 1987; Shewry *et al.*, 2008) (see Fig. 7.1).

Plants can be grown *in vitro* under sterile culture conditions in a medium containing a carbon source, minerals, growth factors, and regulators (Pierik, 1987; Shewry *et al.*, 2008; Vasil, 2008). Different types of culture methods comprise the use of single cells or part of a plant to initiate growth. The latter has been extensively utilized to develop micropropagation techniques, i.e. generation of plants by asexual means under laboratory conditions. This technology is being successfully applied to different commercial plants, including ornamentals (orchids, ferns), woody species (*Eucalyptus*, *Pinus*,

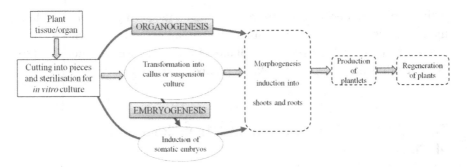

Figure 7.1.   Different strategies for plant tissue culture. Tissues can be cultured to generate shoots and roots (organogenesis) or somatic embryos (embryogenesis). In both cases the procedure can be made directly or indirectly via the previous transformation into callus or suspension culture.

Sequoia), and several crops (banana, carrot, cassava, celery). The culture of single cells is mainly utilized for the production of secondary metabolites under controlled conditions.

Intact whole plants can be cultured from seeds that are aseptically germinated or from pieces of different tissues. A variant to this approach is the culture of embryos (or embryo-like structures) previously isolated from the seeds or obtained by somatic embriogenesis from other plant tissues or organs. The explant (i.e. the initial piece of the plant introduced *in vitro*) is conveniently selected from meristems, shoot tips, axillary buds, leaves, roots, stems, petioles, or flower parts; then disinfected (to clean out of contaminating surface microorganisms) and cultured in defined medium to promote morphogenesis. Figure 7.1 shows that the generation of a plant can be made via organogenesis (direct morphogenesis into shoots and roots) or through somatic embryogenesis (somatic embryos are induced prior to morphogenesis). Moreover, both procedures, morphogenesis and organogenesis, can be achieved directly or indirectly via the initial transformation of a sterile piece of tissue into a callus (a mass of unorganized, undifferentiated cells) or suspension culture (Fig. 7.1).

On the other hand, single cells are obtained enzymatically or mechanically from a callus of tissue culture in suspension, and then maintained continually growing in aerated liquid media. An interesting variant is the culture of plant protoplasts (plant cells without the walls), which can be prepared from cell suspensions after incubation in media containing mannitol at hypertonic levels (to detach the cytoplasmic membrane from the cell wall) followed by enzymatic digestion of the wall with pectinase.

Alternative techniques to reduce problems associated with tissue culture are methods known as *"in planta."* In this case, delivery of foreign DNA is performed in germ-line tissues. First achievements using this approach were obtained in transforming *Arabidopsis* by *Agrobacterium* vacuum infiltration and floral dip methods. Later, success was documented *in planta* transformation of the legume *Medicago trunculata* via *Agrobacterium* inoculation of flowering plants or younger seedlings. Moreover, male germ-line transformation was achieved in tobacco plants using biolistics (Shewry *et al.*, 2008).

(ii) *Plant transformation systems.* Systems for the successful genetic transformation of plants necessarily have to overcome specific limitations and problems created by intrinsic characteristics of plant cells and tissues (Birch, 1997; Owen and Pen, 1996). For instance, the cell wall represents a primary barrier limiting the introduction of foreign DNA into any kind of plant cell. Additionally, there are difficulties inherent to differences found between plant tissues and species, which determine dissimilar responses to a transformation protocol. Various methods have been developed for the transfer of foreign DNA into plants, which include the use of electroporation, microinjection, polyethylene glycol, and laser-mediated uptake. Although the use of these techniques allowed different degrees of success, the predominant tools utilized for the transformation of crop species are *Agrobacterium*-mediated or particle bombardment (Shewry *et al.*, 2008).

Overall, depending on the way by which the genetic material is transferred, methods to produce transgenic plants can be conveniently divided into vector-mediated or vector-free systems.

Vector-mediated gene transfer systems for plants are based upon the use of bacteria or viruses. Two species of the genus *Agrobacterium*: *A. tumefaciens* and *A. rhizogenes* are the main bacterial vectors for plant transformation (Shewry *et al.*, 2008; Tzfira and Citovsky, 2006; Zupan *et al.*, 2000). *A. tumefaciens*, a Gram-negative soil bacterium causing a tumor (crown gall disease) in leguminous dicot plants, is the best known and more commonly utilized vector. Bacterial infection initiates at sites of mechanical wounding in the plant through chemotaxis induced by phenolic compounds released from damaged cells. Plant infection and disease formation by *Agrobacterium* involve gene transfer between cells, resulting in the integration of a DNA sequence from the invading pathogenic bacterium into the DNA of the host plant cell. The latter induces cell proliferation with unusual biochemical characteristics, as they produce opines (amino acids utilized by *Agrobacterium* as a source of carbon and nitrogen) and exhibit hormone-independence (auxins and cytokinins) for growth.

The genetic material (named T-DNA, from "**T**ransferred **DNA**") causing the tumor is carried on a 200 kbp plasmid (the **Ti** plasmid from "**T**umor-**i**nducing") that is separate from the *Agrobacterium* main chromosome. During infection, T-DNA is transferred from the bacterium into plant cells where it enters the nucleus and integrates into the plant genome. In addition to the T-DNA (T-region), Ti plasmid also has the virulence region (coding for proteins involved in the T-DNA transfer) and the opine catabolic region (allowing the use of these molecules by the microorganism). Transformation systems for plant genetic engineering using *Agrobacterium*, take advantage of the fact that deletion of genes of the virulence and T-region, from the Ti plasmid, produces no adverse effect on the transfer and integration of T-DNA. Thus, with different strategies, Ti plasmid is engineered to construct the transformant vector by deleting the coding region of opine biosynthetic genes and addition of the foreign genes. Transformation using Ti plasmid as a vector is useful for plants susceptible to infection by *Agrobacterium* (mostly dicotyledonous species) and is generally applied to cultures of callus or leaf discs, where cut edges provide the wound region that initiates the bacterial infection (Owen and Pen, 1996; Shewry *et al.*, 2008; Tzfira and Citovsky, 2006; Zupan *et al.*, 2000).

Certain plant viruses have been utilized as vectors to transform different species. Most of them are single-stranded RNA viruses and are of relevance because of their potential for high levels of expression of transgenes; a wide range of plant species are susceptible to virus infection (Owen and Pen, 1996). The virus-mediated plant transformation system is particularly important for the production of vaccine epitopes (e.g. the use of tobacco mosaic tobamovirus, cowpea mosaic comovirus, and johnsongrass mosaic potyvirus) (Collins and Shepherd, 1996). Virus-induced gene silencing is an approach of great potential for plant reverse genetics that utilizes RNA-mediated antiviral defense mechanism (Godge *et al.*, 2008).

Vector-free systems for the transfer of naked-DNA into plant cells, comprise the use of chemical, physical (mechanical), and electrical methods (Owen and Pen, 1996; Rakoczy-Trojanowska, 2002; Shewry *et al.*, 2008). In any of these methods, the transfer is preferably performed on protoplasts, in order to overcome the barrier of the plant cell wall. Chemical methods include techniques using polyethylene glycol to facilitate the uptake of DNA by cells or to fuse liposomes (encapsulating the DNA) with protoplast. Among mechanical methods, DNA microinjection into the nucleus of individual cells through fine glass needles is worth mentioning; long needle-like crystals (whiskers) of silicon carbide are formed, which penetrate the cells allowing the entry of DNA (Rakoczy-Trojanowska, 2002). Also relevant is the bombardment of cells with microprojectiles of tungsten or gold coated with DNA (Shewry *et al.*, 2008). Finally, electroporation based on the increased permeability of the plasma membrane to hydrophilic molecules upon application of well-defined electrical pulses to protoplasts or cells suspensions containing the transformant DNA (Owen and Pen, 1996; Rakoczy-Trojanowska, 2002).

Among other strategies utilized for plant genetic engineering is the methodology based on RNA interference (RNAi). Unlike transformation protocols pursuing the overexpression of a certain protein, RNAi interference seeks to down-regulate native cellular proteins (Hebert *et al.*, 2008). This methodology employs DNA constructs designed to generate double-stranded RNA thus giving sequence-specific silencing of genes. The discovery and application of RNAi constitute an outstanding contribution to biology, as it has been recognized by the 2006 Nobel Prize in Physiology and Medicine awarded to Andrew Z. Fire and Craig C. Mello, who performed decisive advances in the field. Interestingly, pioneer work with RNAi were made in plants; however, as stated by Bots *et al.* (2006), these studies were given low consideration by the Nobel committee in their decision to assigning the award.

Plant Metabolism: Matter and Energy Flows and the Prospect of MCA

In the first edition of this book, we remarked the relatively low understanding existing at that time on the subject of control of matter and energy fluxes in plant cells. In fact, a main conclusion was that, by then, the scarce research made on MCA resulted in a qualitative rather than quantitative knowledge on the control of plant metabolism. Fortunately, the landscape of advances achieved during the last decade allowed a more quantitative analysis of the functioning of plant metabolic networks. Nevertheless, this research field still represents a challenge for plant scientists, constituting a hot topic of research. Current efforts focus on gaining key information for developing tools in order to design MCE strategies as applied to plant systems. Approaches based on the use of robotic/computational platforms for determining multiple variables (including level of metabolites, enzyme activities and kinetic parameters, as well as genotype-to-phenotype relationships) are promising advances (Kopka *et al.*, 2005; Stitt *et al.*, 2010). Key progress has been achieved after gathering massive data which allowed the analysis of fluxomes and reconstruction of metabolic networks described by mathematical models (Stitt *et al.*, 2010). These relevant tools for applying MCE are described in Chap. 5.

A number of features make plants outstanding complex organisms in which to perform MCA. Plants are able to grow and reproduce under quite a diverse range of environmental conditions. This property is bestowed to plants by a flexible metabolism, represented by multiple, inter- and trans-connected metabolic pathways. A variety of regulatory processes complement the complex physiological and biochemical scenario exhibited by plant cells. On the other hand, the highly compartmentalized metabolism of plants results in complete or partially duplicated metabolic pathways, operating in different subcellular locations (Lunn, 2007).

Metabolic Compartmentation in Plant Cells

In addition to the nucleus, mitochondria and other organelles found in fungi and animal cells, photosynthetic eukaryotes possess plastids (Archibald, 2009; Lunn, 2007). These are specific self-replicating organelles surrounded by a double membrane, which occur in a broad range of types, sizes, shapes, and colors. Plastids are derived from the so-called primary endosymbiosis (where a free-living cyanobacterium was engulfed by an eukaryote) and they are found in the different algae (red, glaucophyte, and green algae)

(Archibald, 2009). Green algae, the unicellular lineage leading to land plants, constitute highly amenable organisms for studying and quantifying metabolic fluxes in photosynthetic cells. Thus, because of its phylogenetic and genetic properties, the species *Chlamydomonas reinhardtii* has been called "green yeast," to emphasize the potential relevance of this organism for biological studies (Goodenough, 1992). The different plastids exhibit distinctive metabolic capacities. The most typical is the chloroplast that contains chlorophyll pigments and is the site of photosynthesis, or the amyloplast, found in nongreen tissues and that functions as a storage organelle for starch.

The extensive compartmentation and duplication of metabolic pathways in plants makes difficult studies of flux analysis. According to the quantitative methodology utilized, metabolite concentrations and the measurement of *in situ* enzyme activities (which should correspond to the level of metabolite and to the activity of the isoenzyme in each specific intracellular compartment) may be needed (Archibald, 2009; Lunn, 2007). Glycolysis is a good example of this complexity; in plants, glycolysis comprises two different sets of enzymes, one in the cytoplasm and another in the plastid, operating simultaneously and largely independently (Iglesias and Podestá, 2005; Plaxton, 1996). An example of this complex network is shown in Fig. 7.2 that depicts reactions operating in different intracellular compartments from green algae or higher plant. The interaction between reductive/anabolic pathways (e.g. carbon assimilation) and oxidative/catabolic routes (e.g. respiration), happens between these pathways taking place in different intracellular compartments (cytoplasm, chloroplast, vacuole, and mitochondria).

Another layer of complexity is given by autotrophic or heterotrophic modes of growth existing in algae and plant cells, which involve opposite fluxes of carbon and energy (Iglesias and Podestá, 2005). In higher plants, heterotrophic growth occurs during dark conditions within photosynthetic cells (e.g. leaf cell), or constantly in tissues having a nonautotrophic nature (e.g. root or seed cells). Many algae are capable of growth under autotrophic, heterotrophic or mixotrophic conditions depending on the source of carbon, nutrients, and the availability of light. Examples of organisms exhibiting multifaceted intermediary metabolism are protists, including the green alga *Chlamydomonas reinhardtii* and the brown alga (diatom) *Phaeodactylum*

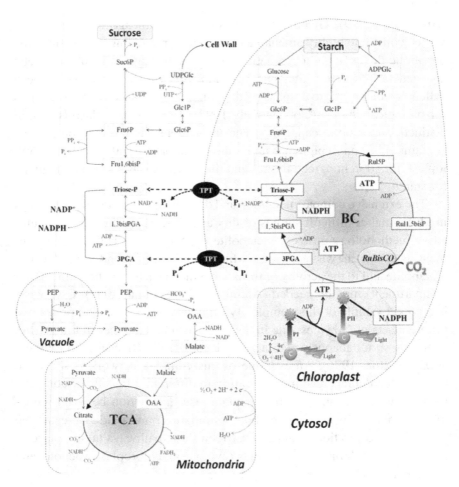

Figure 7.2. Scheme of the flux of carbon in photosynthetic cells from green algae or higher plants. The figure describes interactions between photosynthesis and respiration in an autotrophically carbon-assimilating cell. Abbreviations: BC, Benson-Calvin cycle; OAA, oxaloacetate; TPT, triose-P/Pi translocator.

tricornutum (Ginger *et al.*, 2010). The multiple metabolic pathways occurring in these eukaryotic organisms make them potentially useful for MCE. To this respect, genetic manipulation of the diatom *P. tricornutum* aimed at modifying its capacity to grow on glucose has been reported (Zaslavskaia *et al.*, 2001). The brown alga was genetically engineered by the introduction of a gene encoding a glucose transporter resulting in cells that can thrive on exogenous glucose in the absence of light. The work by Zaslavskaia *et al.* (2001) clearly demonstrated that the introduction of a single gene in an organism could produce substantial changes in its metabolism, despite the overall complexity involved (Fig. 7.2). The advantage of this microalga is given by its dual ability to develop photo- or hetero-trophically. This versatility makes this organism potentially amenable for fermentative technology as well as for studies on carbon and energy fluxes under different throphic conditions, as recently shown in *C. reinhardtii* (see Chap. 4 and Boyle and Morgan, 2009).

Carbon Assimilation, Partitioning, and Allocation

In higher plants, the existence of distinct photosynthetic (e.g. leaves) and heterotrophic (e.g. roots and seeds) tissues makes necessary the analysis of carbon and energy fluxes in the different cells (Emes *et al.*, 1999; Iglesias and Podestá, 2005). In the leaves, atmospheric CO_2 is fixed into carbohydrates by photosynthesis. A major product of the process, sucrose, is a mobile carbohydrate that is transported to the different parts of the plant where it provides carbon skeletons for nonphotosynthetic cells. In heterotrophic tissues, sucrose is the starting feeding substrate for different and specific metabolic pathways, many of them occurring in distinctive plastids (i.e. starch synthesis in amyloplasts of reserve tissues such as endosperm, or fatty acids production in leucoplasts of certain seeds). For the sake of clarity, we will analyze carbon assimilation and partitioning separately, including metabolic variants existing in different plants for these processes. The integrated understanding of the whole network of reactions associated with carbon photoassimilation, intra- and inter-cellular partitioning, and utilization of photosynthates is critical for MCE in photosynthetic organisms, particularly in higher plants. Knowledge about carbon assimilation, partitioning, and allocation ultimately determines the way in which plants produce carbohydrates, oils, sugar-alcohols, biomass, and other high-value compounds.

Carbon Fixation in Higher Plants

The common route for inorganic carbon fixation in oxygenic photosynthetic organisms was elucidated by studies performed with green algae by Benson, Calvin and coworkers during the mid-20th century. The metabolic pathway was established to be cyclic, interconverting phosphorylated trioses, tetroses, pentoses, hexoses, and heptoses. Most of the reactions of the Benson–Calvin cycle (BCC) are common with those of the pentose-P pathway taking place in almost all organisms, but these routes operate in opposite ways. Thus, the pentose-P path is functionally oxidative, and so is called **O**xidative **P**entose **P**hosphate **P**athway (OPPP); NADPH and ribose-5P are obtained form Glc-6P oxidation. On the contrary, the BCC is basically reductive (afterward named **R**eductive **P**entose **P**hosphate **P**athway, RPPP), and utilizes NADPH and ATP generated photosynthetically (Iglesias and Podestá, 2005; Iglesias and Podestá, 2008). As schematized in Fig. 7.3, three different phases can be identified in the RPPP: carboxylative, reductive, and regenerative. The reaction of CO_2 fixation in the BCC is the only one involved in the phase 1 of carboxylation, and is

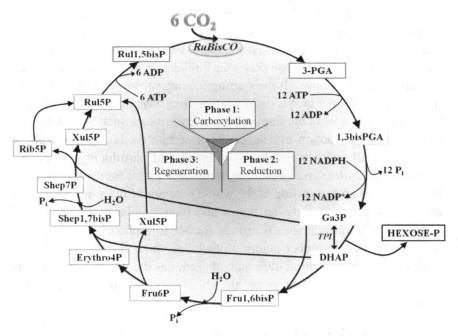

Figure 7.3. The RPPP or BCC, showing the net production of one hexose-P from the photoassimilation of six CO_2 molecules. Specific abbreviations for the figure are: DHAP, dihydroxyacetone-P; Ga3P, glyceraldehyde-3P; TPI, triose-P isomerase; Erythro-4P, erythrose-4P; Shep, sedoheptulose.

catalyzed by ribulose-1,5-bisP carboxylase/oxygenase, Rubisco (Fig. 7.3). The result of ribulose-1,5-bisP carboxylation is the production of two molecules of 3P-glycerate (3PGA). This metabolite is then converted to triose-P [glyceraldehyde-3P (Ga3P) and dihydroxyacetone-P (DHAP)] in two consecutive reactions consuming ATP and NADPH (produced by photosynthetic electron transport at the thylakoid membrane). The latter two reactions conform the phase 2 of reduction in the cycle, being the only one that consumes NADPH (Fig. 7.3). The pathway continues in the series of reactions conforming phase 3, that interconvert compounds of three, four, five, six, and seven carbon atoms and regenerate the initial acceptor of CO_2 (ribulose-1,5-bisP, Rul-1,5-bisP) with the additional consumption of ATP. Thus, the RPPP is autocatalytic and produces the assimilation of one inorganic carbon molecule into one organic metabolite per turn. The net synthesis of one hexose-P molecule by the RPPP needs fixation of 6 CO_2 (Fig. 7.3), utilizing 18 ATP and 12 NADPH molecules (Iglesias and Podestá, 2005; Iglesias and Podestá, 2008).

As indicated by its name, Rubisco utilizes Rul-1,5-bisP and is able to catalyze the reaction as a carboxylase:

$$Rul\text{-}1,5\text{-}bisP + CO_2 \rightarrow 3P\text{-glycerate} + 3P\text{-glycerate},$$

and as an oxygenase:

$$Rul\text{-}1,5\text{-}bisP + O_2 \rightarrow 3P\text{-glycerate} + 2P\text{-glycolate}.$$

Both reactions take place within the chloroplast and the relative extent of their occurrence is mainly determined by the $CO_2:O_2$ ratio within the plastid (Iglesias and Podestá, 2008). The oxygenase activity of Rubisco initiates a metabolic pathway known as photorespiration, which provides precursors to different intracellular compartments (chloroplast, cytosol, peroxysome, mitochondria) (Iglesias and Podestá, 2008; Slatyer and Tolbert, 1971). As shown in Fig. 7.4, photorespiration consumes O_2 and releases CO_2, clearly functioning in an opposite way to carbon assimilation.

Photorespiration not only adds a degree of complexity to carbon metabolism and partitioning in photosynthetic cells but also affects the carbon fixation yield in higher plants. Under certain environmental conditions, photorespiration can strongly affect the efficiency of photosynthesis. As an example, at high temperatures, O_2 solubility decreases but relatively

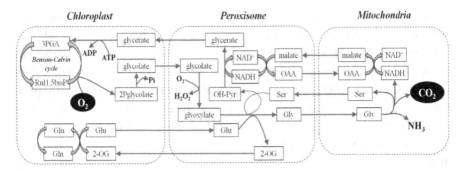

Figure 7.4. Photorespiration, the pathway followed by carbon between different compartments of a photosynthetic cell (chloroplast, peroxisome, and mitochondria) as a consequence of the oxygenase activity of Rubisco. Specific abbreviations for the figure are: 2-OG, 2-oxoglutarate; Gln, glutamine; Glu, glutamate; Gly, glycine; OH-Pyr, hydroxypyruvate; Ser, serine.

much less than that of CO_2, thus resulting in a lower $CO_2:O_2$ intracellular ratio. Under these conditions, net carbon assimilation decreases significantly (Iglesias and Podestá, 2008). Since photorespiration arises from the oxygenase activity of Rubisco, numerous efforts were conducted to obtain an enzyme having only (or substantially increased) carboxylase activity. However, this problem seems to have no solution, because the reaction mechanism of Rubisco is such that it is not possible to affect one of the activities without changing the other, a particular trait of this enzyme that otherwise seems to have been perfectly optimized in terms of kinetics during evolution (Iglesias and Podestá, 2008; Tcherkez *et al.*, 2006).

The evolutionary path to decrease or eliminate photorespiration appears to have been different. It comprised the addition of a series of metabolic steps to the process of photosynthetic CO_2 fixation. The RPPP cycle is common to all autotrophic cells of higher plants, being the obligate primary route for carbon assimilation. Because the first product of CO_2 fixation via RPPP is a 3-carbon compound (3P-glycerate, see Fig. 7.3), this route is also known as the C_3-cycle. C_3-plant is the name given to those species possessing only this route for carbon assimilation. Certain plants have an additional (not alternative) pathway that operates separately (in space or time) with RPPP. In such additional route, the reaction for fixing atmospheric CO_2 is catalyzed by phosphoenolpyruvate carboxylase (PEPCase), an enzyme that cannot use O_2 as a substrate:

$$PEP + HCO_3^- \rightarrow \text{oxaloacetate} + Pi.$$

This reaction starts a cyclic metabolic pathway named the C_4-route because of the 4-carbon compounds (oxaloacetate, malate, aspartate) produced during CO_2 fixation. In this cycle, oxaloacetate is reduced to malate or transaminated to aspartate and, after a series of steps, a 4-carbon metabolite is decarboxylated at the chloroplast. The CO_2 released serves as a substrate for Rubisco that ultimately reprocesses it via RPPP. The C_4-route is cyclic because the C_3-carbon compound resulting from the decarboxylation is then utilized to regenerate PEP, the substrate of PEPCase.

Figure 7.5 schematizes the operation of the C_4-route in a C_4-plant, where the whole carbon assimilation process involves two different photosynthetic cells: mesophyllic and bundle-sheath. C_4-plants exhibit a characteristic cellular anatomy in their photosynthetic tissues that is of key physiological relevance for carbon assimilation (Iglesias and Podestá, 2008). As depicted in Fig. 7.5, atmospheric CO_2 fixation by PEPCase occurs in the cytosol of mesophyll cells. The C_4-carbon product is then metabolized in the chloroplast or cytoplasm of the same cell followed by transport to the bundle-sheath cell where it is decarboxylated, after being delivered to the chloroplast. There, the RPPP is operative for the assimilation of the released CO_2. The C_3-metabolite produced by decarboxylation is recycled as it is metabolized and finally regenerating PEP, a reaction occurring at the chloroplast of mesophyllic cells.

In Fig. 7.5, it can be visualized that in a C_4-plant, the additional C_4-pathway physiologically operates as a biochemical device for pumping atmospheric CO_2 to the cellular location where the RPPP operates. In this way, the $CO_2:O_2$ ratio at the chloroplast of bundle-sheath cells increases, thus favoring the occurrence of the carboxylase (over the oxygenase) activity of Rubisco (Iglesias and Podestá, 2008). Consequently, under certain environmental conditions, C_4-plants exhibit a better performance than C_3-species in terms of photosynthetic efficiency. A main reason for the greater efficiency is given by the markedly diminished or the elimination of photorespiration in C_4-plants. Species of C_4-plants include some crops, such as maize, sugar cane, sorghum, and millet.

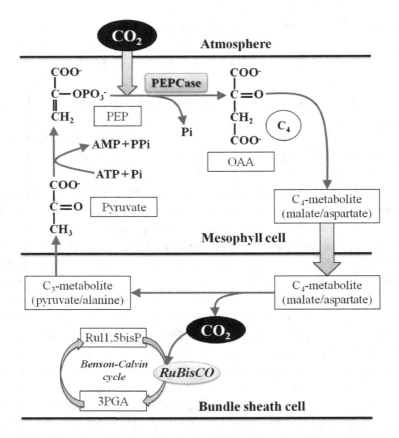

Figure 7.5. The carbon flux in a C_4-plant, which involves two different photosynthetic cells: mesophyll and bundle sheath. C_3 and C_4 indicate 3- or 4-carbon compounds involved in the metabolic pathway.

A variant of the C_4-pathway is that found in plants where both, the C_4-pathway and the C_3-cycle function in a single photosynthetic cell but they operate during different time periods. These plants usually grow in desertic regions and are named CAM-plants (CAM after **C**rassulacean **A**cid **M**etabolism) because the first species characterized as performing this metabolism belonged to the Crassulaceae family (Borland *et al.*, 2009; Iglesias and Podestá, 2008). In CAM-plants, atmospheric CO_2 is fixed during the night via PEPCase. The oxaloacetate thus produced is reduced to malate, which accumulates in the vacuole to attain very high levels (from this derives the term Acid in the pathway's name). During the day, malate is delivered to the chloroplast, where it is decarboxylated, the CO_2 released is refixed by Rubisco, and finally assimilated via RPPP. In this way, CAM-plants not only concentrate CO_2 in the chloroplast during the day (increasing the $CO_2:O_2$ ratio and thus reducing photorespiration) but also by exchanging gases with the environment only during the night period; they avoid excessive water loss. The latter is critical for survival of organisms under extreme conditions, such as those existing in desertic or marginal lands, where water economy has to be maximized (Borland *et al.*, 2009).

Strategies followed by nature to "metabolically engineer" plants during evolution for optimizing carbon assimilation under different environmental conditions have inspired efforts directed to apply similar approaches to modifying plants for augmenting photosynthetic efficiency. Thus, the possibility of converting economically valuable C_3-plants (e.g. wheat, rice, potato, soybean) into C_4-plants may theoretically increase their productivity through reduction of photorespiration. Similarly, and emulating CAM-plants, it should be possible to increase the cultivable area of the planet after fine-tuned genetic engineering of certain species, allowing them to grow in more arid habitats. Another technological approach which takes advantage of the carbon fixation characteristics, is to exploit the potential of plants performing CAM for biofuel production in current nonproductive land regions (Borland *et al.*, 2009).

However, it has to be considered that carbon metabolism is very complex in plants. At this point, it would be convenient to review Fig. 7.2 and to consider what needs to be integrated to it from Figs. 7.3, 7.4, and 7.5. Furthermore, a coordination of reactions in time and space for the desired pattern of operation will also be necessary. Despite this complexity, some initial results suggest that important goals could be obtained by engineering carbon assimilation in plants. In example, a C_3-crop like rice has been successfully transformed with an *Agrobacterium*-mediated system to introduce the intact gene of maize PEPCase (Ku *et al.*, 1999). Transgenic plants showed: (i) high-level expression of the maize gene, with high PEPCase activity in leaves (where the enzyme reached up to 12% of the total soluble protein), and (ii) reduced O_2 inhibition of photosynthesis. The latter result is very relevant; it shows that relatively simple changes in enzyme activity can result in the successful improvement of carbon assimilation after the effective operation of complementary metabolic pathways in a single photosynthetic cell. Apparently, efficient carbon fixation is not strictly dependent on the existence of the complex cellular anatomy occurring in C_4-plants or the physiological chronology found in CAM species, but it can be attained by increasing levels of certain enzymes in a single cell. This was substantially reinforced by the discovery that certain plant species naturally perform a distinctive mechanism of C_4-photosynthesis characterized by the operation of the C_4-pathway during the day in only one photosynthetic cell (Voznesenskaya *et al.*, 2001).

Carbon Partitioning within One Cell and between Source and Sink Tissues

The high degree of compartmentalization of metabolism in plants and the occurrence of tissues with different trophic conditions determines the distribution of enzyme activities and the partial or total sequestration of pathways. This happens between different cells and/or subcellular compartments, and implying as well the intra- and inter-cellular transfer of different metabolites. In this way, carbon assimilated by photosynthesis in leaves is partitioned both intracellularly and between different cell types in distinct plant tissues. Thus interactive source–sink relationships for carbon metabolism are established between different tissues (Iglesias and Podestá, 2005; Paul and Foyer, 2001; Roitsch, 1999). In any case, the overall picture of carbon allocation leading to the accumulation of carbohydrates, oils, or other metabolic products should be integrated to the whole physiology and metabolism of a plant. More specifically, this picture has to include steps occurring in source cells, partitioning and transport of photoassimilates, and metabolism in sink tissues. The understanding of carbon, energy and redox fluxes in a whole plant thus requires control and regulation analyses in at least three different metabolic processes. Each one of these requires a particular type and distribution of enzymes as well as physiological regulation with singular characteristics. Deciphering the operation and modulation of carbon partitioning is critical for applying biotechnological approaches using plants as sources of starch and other polysaccharides, including lignocellulosic biomass, sugar-alcohols, and oils.

Plants as a source of soluble and storage carbohydrates and sugar-alcohols

As pointed out by Emes *et al.* (1999), nowhere is carbon partitioning better exemplified than in the field of plant carbohydrate metabolism. We show the complexity of carbohydrate synthesis and utilization in source and sink tissues of plants in Figs. 7.6 and 7.7, respectively. Carbon assimilation and partitioning of photosynthates within the leaf cell is illustrated in Fig. 7.6. Carbon assimilation occurs in the chloroplast through the BCC (Iglesias and Podestá, 2008). From hexose-P and triose-P intermediates of this cycle,

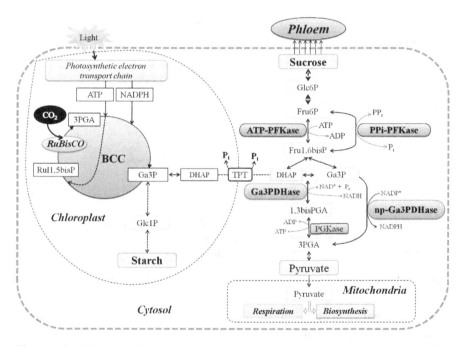

Figure 7.6. Diagram of the carbon flux in a photosynthetic cell of a higher plant. The metabolic network distinguishes carbon fixation in the chloroplast via the BCC, and partitioning initiated from its intermediaries between starch (accumulated within the plastid) and sucrose (produced in the cytosol). Exchange of metabolites between chloroplast and cytosol is detailed with the involvement of the specific translocator of the plastid envelope TPT, being the main exchanger of Pi and triose-P. Abbreviations: PGKase, P-glycerate kinase; Ga3PDHase, phosphorylating glyceraldehyde-3P dehydrogenase (EC 1.2.1.12); np-Ga3PDHase, non-phosphorylating glyceraldehydes-3P dehydrogenase (EC 1.2.1.9); DHAP, dihidroxyacetone-P; 3PGA, 3P-glycerate; 1,3bisPGA, 1,3-bisP-glycerate.

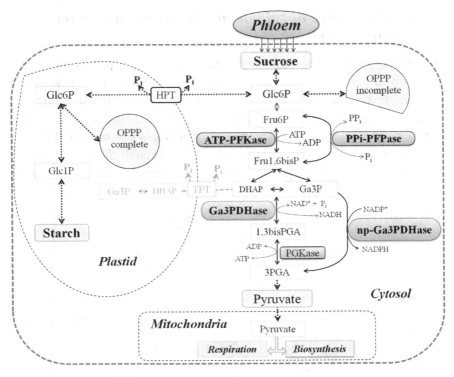

Figure 7.7.   Carbon flux in a heterotrophic higher plant cell. With sucrose as the carbon source, metabolism allocates photoassimilates to be stored as starch. Metabolite exchange between cytosolic and plastid compartments is mediated by specific carriers: the hexose-P transporter (HPT) and the Pi-triose-P interchanger (TPT). OPPP is the oxidative pentose-phosphate pathway; other abbreviations are as in Fig. 7.6.

the respective routes leading to the two major products of photosynthesis, starch and sucrose, are initiated. In this way, photoassimilates are mainly partitioned between chloroplastidic starch (the polysaccharide synthesized from hexose-P intermediates of the RPPP and serving for the temporary accumulation of carbohydrates in leaves) and cytosolic sucrose (the disaccharide used for delivery of photoassimilates through the phloem to other nonphotosynthetic parts of the plant). Details about steps (and their regulation) leading from hexose-P (namely Glc-1P) to starch are given below in this chapter, in the analysis made under the heading "Regulation and Control: Starch Synthesis, a Case Study." As emphasized in Fig. 7.6, major metabolites transported across the chloroplast envelope are triose-P. This process involves a specific translocator that interchanges Pi and triose-P between intracellular compartments, thus making effective photoassimilate partitioning within the photosynthetic cell (see also Emes *et al.* (1999), Iglesias and Podestá (2005), Iglesias and Podestá (2008)). In the cytoplasm, transported triose-P can be the initial substrate for both sucrose synthesis and other metabolic routes.

Photosynthetic (source) tissues export sucrose, which travels through the phloem to the different parts of a plant (sink tissues) (Iglesias and Podestá, 2005). Once sucrose reaches heterotrophic plant cells, it is metabolized in the cytoplasm and specific intermediates are directed to different pathways occurring in their respective intracellular compartments. Figure 7.7 shows the pathway followed in endosperm, a sink tissue where carbon is accumulated as starch inside the amyloplast. As detailed in Fig. 7.7, the major flux of carbohydrates occurs in the opposite direction with respect to that observed in the photosynthetic cell. In the endosperm, carbohydrates move from the cytoplasm to the plastid as sucrose is metabolized, and the intermediate assimilates enter the amyloplast, where starch (the final product) is synthesized and accumulated. In the reserve tissue, starch is the long-term form of glucose storage. As detailed in Fig. 7.7, the entry of carbohydrates into the plastid mainly occurs in the form of hexose-P through a specific translocator (HPT); however, the amyloplast envelope also contains the TPT carrier.

The value of sucrose and starch as biotechnological products is based on their numerous uses. Starch is the quantitatively more important basis of animal food (including man), and it has a broad range of industrial applications (e.g. sizing of paper and board, adhesive in packaging, thickener and gel-based processes, and nano derivatives) (Ballicora *et al.*, 2004; Le Corre *et al.*, 2010; Morell and Myers, 2005; Sivak and Preiss, 1998; Tharanathan, 2005). Starch is the starting material in fermentative methods to produce cyclodextrins, polyols, acids, amino acids, and fructose. Moreover, a current biotechnological application of sucrose and starch is in the field of biorefineries, mainly to generate biofuels (as is the case of ethanol production from sugar cane and maize) and biodegradable plastic-like polymers (Goldemberg, 2007).

A variation on the theme of sucrose and starch as major products of photosynthesis happens in the plant family of Rosaceae (includes many

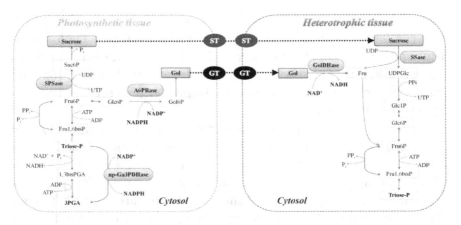

Figure 7.8.   Production of sugar-alcohols as alternative to the sucrose pathway. In the figure, the sucrose pathway has a branch leading to Gol (sorbitol) synthesis in the cytosol of the photosynthetic cell. Gol is a major product for transport of carbon through the phloem to heterotrophic cells where it is stored or utilized to produce fructose. ST and GT are sucrose and glucitol translocator, respectively, in the external membrane of each cell type.

fruit trees such as apple, pear, quince, loquat, peach) that synthesizes sugar-alcohols, glucitol (Gol, or sorbitol), and mannitol. Interestingly, in such distinctive plants, photoassimilates partitioning is more complex because the sugar-alcohol is also a major component that is produced and accumulated at similar levels as sucrose or starch. Figure 7.8 depicts an alternative pathway leading to reduction of carbohydrates to a sugar-alcohol (Gol in the example) in the cytoplasm of photosynthetic and heterotrophic plant cells. Gol metabolism starts in the leaf cell with reduction of Glc-6P to Gol-6P by aldose-6P reductase (A6PRase, also named sorbitol dehy-drogenase) at the expense of NADPH, then followed by hydrolysis of the phosphate group by a specific phosphatase (Figueroa and Iglesias, 2010; Figueroa *et al.*, 2011). The sugar-alcohol is the main metabolite trans-ported via phloem to sink tissues (e.g. fruits) where it is stored in the cytoplasm. Reutilization of Gol involves its conversion to fructose by an NAD-dependent Gol dehydrogenase (see Fig. 7.8).

Beyond the fact that a number of plants accumulating sugar-alcohols have a commercial value (e.g. fruit trees accumulating Gol, or the flavoring species of celery that synthesize mannitol), reduced carbohydrates have relevance in biotechnology because of their role as compatible solutes. It has been demonstrated that sugar-alcohols are beneficial to organisms because they help to maintain osmotic balance, increasing tolerance to oxidative stress and micronutrient deficiency. This opens the potential of performing ME to increase or accumulate these polyols in certain plants to enhance their abilities to cope with adverse environmental conditions (Brown *et al.*, 1999; Shen *et al.*, 1997). Importantly, sugar-alcohols are good low calories food substitutes, and they are very suitable molecules for upgrading biofuel quality and biorefinery compounds (Akinterinwa *et al.*, 2008).

Plants as producers of lignocellulosic biomass

A main final sink of carbon skeletons in plants is represented by the synthesis of cell wall polysaccharides (Delmer and Haigler, 2002). This complex and yet not well understood biochemical process includes cellulose, hemicellulose, and pectin production (Lerouxel *et al.*, 2006). Cellulose is the major component of the plant wall and is the most abundant biopolymer on Earth. The structure of the β-1,4-glucan nonbranched saccharide is linear and organizes in microfibrils after parallel arrangement of the polymer, whose chains tightly pack mediated by hydrogen-bond interactions. Cellulose biosynthesis occurs from UDP-Glc, which is substrate of a membrane-bound cellulose synthase. The enzymatic machinery involved in polymerization integrates protein complexes that arrange as hexameric rosettes of around 30 nm that are visible by freeze-fracture electron microscopy (Doblin *et al.*, 2002; Lerouxel *et al.*, 2006). The synthetic process occurs at the plasma membrane and the polymer is directly deposited into the cell wall.

In the cell wall, cellulose microfibrils are embedded in a matrix that contains other polysaccharides and proteins, some of them glycoproteins. Most of the matrix components are synthesized in the Golgi and vesicles-secreted to the wall. Hemicelluloses are complex molecules providing a cross-linked matrix after association with cellulose microfibrils. Four major classes of polysaccharides can be distinguished in hemicellulose: xyloglucans, (gluco)mannans, glucuronoarabinoxylans, and mixed-linkage glucans. On the other hand, pectins are complex polymeric saccharides with major domains of homogalacturonan and rhamnogalacturonan I, and minor contents of rhamnogalacturonam II. In addition to these components, cell wall structure contains lignin, a complex polymer of lignols, which fills spaces between the polysaccharide components and covalently link hemicellulose. Lignin contributes to confer mechanical strength to the cell wall, and its hydrophobic character is crucial for water management in plants. Lignin is a main component of wood, and it has also been found to play a structural function in red algae (Martone *et al.*, 2009).

Cellulose represents a highly attractive raw material for biofuel production and applications in biorefineries because of its abundance as a major product of CO_2 assimilation by plants and its limited use for human food as well (Carroll and Somerville, 2009). A major problem of cellulosic and lignocellulosic biomass is the difficulty caused by compact structures recalcitrant to deconstruction (Himmel *et al.*, 2007). In fact, crystalline cellulose is highly resistant to chemical or enzymatic hydrolysis, consequently demanding the use of chemicals at relatively high temperature and pressure that increases costs and pollution. Solving this problem is a main challenge in biotechnology. Currently, better biodegrading systems (e.g. by improving catalytic efficiency of cellulases and xylanases), and plant modification by ME to produce "softer" lignocellulosic biomass are being investigated (Carroll and Somerville, 2009; Himmel *et al.*, 2007; Weng *et al.*, 2008). The ME alternative is very important, as specific cultivars could be cultured in low productivity areas for conventional crops in order to obtain biomass highly suitable for biofuel production.

Plants as a source for oils

Figure 7.9 depicts metabolic networks operating in a heterotrophic plant cell specialized in the accumulation of oils (i.e. cell from a seed). Carbon feeding starts with the entry of sucrose from the phloem. Catabolism of the disaccharide in the cytosol renders hexose-P and other metabolites (mainly PEP) that are transported to the plastid for fatty acids synthesis. For the occurrence of this anabolic process, the supply of reducing equivalents is required in the form of NADPH and ATP; both can be provided by the OPPP operating within the plastid from glucose via starch degradation. Fatty acids are exported as acyl-CoA derivatives to the cytoplasm and they serve as substrates for acyl-glycerides production in the endoplasmic reticulum. Triacylglycerols form compact aggregates or oil bodies that differentiate from the soluble cytosol within the heterotrophic cell (see Fig. 7.9).

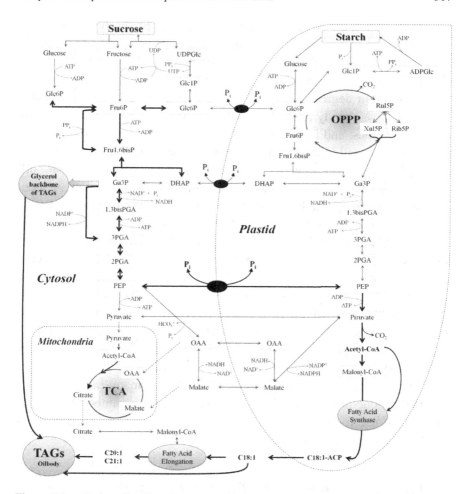

Figure 7.9. Carbon flux in a heterotrophic cell of a higher plant which accumulates triacylglycerides. Carbon is provided as sucrose, whose metabolism in the cytosol produces intermediates (mainly hexose-P, triose-P, and PEP) that are taken up by the plastid for synthesis of starch and fatty acids. Starch in the plastid represents a storage of glucose that is utilized for production, via the OPPP and glycolysis, of NADPH, ATP, and acetyl-CoA that are further metabolized for synthesis of fatty acids. The latter are transported to the cytosol and serve for acylglycerols synthesis in the endoplasmic reticulum. TAG is the abbreviation for triacylglycerides.

The metabolism of oil production is relevant for biofuel production and other biomaterials, such as biodiesel and polymeric polyols (Lligadas *et al.*, 2010; Scott *et al.*, 2010). Important is to not only understand how to maximize TAG production in plants but also manipulate the size and the insaturation degree of the fatty acids present in oil bodies. Valuable strategies have already been implemented to produce biodiesel from plant triacylglycerides, and important amounts of them are being utilized as an alternative to fossil fuels. The process involves extraction and hydrolysis of triacylglycerides from a plant (e.g. rapeseed, soybean, sunflower, castor oil) or green alga after which fatty acids are trans-esterified with low molecular weight alcohols (methanol or ethanol) to obtain the biofuel. Although biodiesel is more environment friendly than fossil fuel, its production generates some problems that need to be solved. For example, production of biodiesel generates glycerol as a main by-product (about 10% w/w), converting the polyol in a low-value (and even waste) chemical (da Silva and Bicego, 2010; Lligadas *et al.*, 2010). Solutions to this problem need the development of biotechnological strategies: (i) use of glycerol as a substrate for growing microorganisms with different purposes (e.g. single cell additives, production of recombinant proteins, chemicals, biofuels) (Giordano *et al.*, 2010; O'Grady and Morgan, 2010); (ii) transformation of glycerol into dihydroxyacetone, a high-value chemical with many industrial uses, and (iii) production of 1,3-propanediol, an emerging commodity chemical of relevant use for polymers synthesis in biorefinery-type of approaches (da Silva and Bicego, 2010).

Distinctive characteristics of metabolic reactions in the cytoplasm of plant cells with relevance for carbon and energy partitioning

Analyzing the assimilates distribution in metabolic networks spread throughout different intracellular compartments and tissues shows that hexose-P, triose-P, PEP, and pyruvate constitute key nodes in plants (Figs. 8.6, 8.7, and 8.9). The metabolic processing of these metabolites determines the existence of branched central catabolic pathways (e.g. glycolysis). The latter implies that precursors channeling into different alternative routes have critical consequences for the flux distribution of energetic and reducing equivalents in the cell under physiological or stressful conditions. Another distinctive feature includes the presence of compounds related with signaling and regulation, as it is the case for trehalose and its modified intermediates. A complete knowledge of the full picture and its pecularities is relevant for a comprehensive understanding of plant metabolic networks, and for its rational manipulation.

One specific example is given by the central role played by triose-P in plant metabolism. Plant cells possess three different glyceraldehyde-3P dehydrogenases (Iglesias, 1990); one is the classical cytoplasmic phosphorylating enzyme (Ga3PDHase, EC 1.2.1.12, GAPC) involved in glycolysis, and catalyzing the reaction where oxidation of Ga3P is coupled to phosphorolysis to render the "high energy" compound 1,3-bisP-glycerate:

$$\text{Ga3P} + \text{NAD}^+ + \text{Pi} \leftrightarrow 1,3\text{-bisP-glycerate} + \text{NADH} + \text{H}^+$$

The chloroplastic form of this enzyme (GAPAB, EC 1.2.1.13) catalyzes a similar reaction, except that it can use either NAD^+ or NADP^+ as cofactor. In fact, physiologically this enzyme is involved in the only reducing step of the BCC (see also Fig. 7.3):

$$1,3\text{-bisP-glycerate} + \text{NADPH} + \text{H}^+ \leftrightarrow \text{Ga3P} + \text{NADP}^+ + \text{Pi}$$

The third form is the cytoplasmic enzyme, found only in plants and some bacteria (Iglesias, 1990), which has the characteristic of being an NADP^+-specific nonphosphorylating Ga3PDHase (GAPN, np-Ga3PDHase, EC 1.2.1.9). In this reaction, oxidation is not coupled to phosphorolysis but to hydrolysis, resulting in the practically irreversible production of 3PGA:

$$\text{Ga3P} + \text{NADP}^+ + \text{H}_2\text{O} \rightarrow 3\text{PGA} + \text{NADPH} + 2\text{H}^+$$

Consequently, triose-P is a critical intermediary metabolite in the distribution of carbon and energy fluxes in plants. The translocator-mediated exchange of Pi and triose-P between chloroplast and cytosol is associated with the occurrence of multiple Ga3PDHases. The scheme in Fig. 7.10 is a detailed picture of the metabolic reactions in which Ga3P is involved. It can be observed that the partition of triose-P between chloroplast and the cytosol can function as a shuttle mechanism to export carbon assimilates and photosynthetically generated NADPH and ATP between the two compartments.

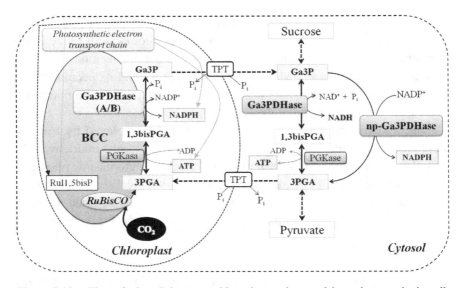

Figure 7.10.   Flux of triose-P between chloroplast and cytosol in a photosynthetic cell. The specifically mediated exchange through TPT of Pi and triose-P generated by carbon fixation in the BCC is shown. Metabolism of triose-P in each compartment by the respective Ga3PDHases (phosphorylating and nonphosphorylating) constitutes a mechanism to export ATP and NADPH photosynthetically generated in the chloroplast to the cytosol.

The presence of cytoplasmic NAD^+-dependent Ga3PDHase and $NADP^+$-dependent np-Ga3PDHase provides glycolysis with alternative routes to convert Ga3P into 3PGA (see Fig. 7.10). The pathway involving the NAD^+-dependent enzyme will produce NADH and ATP, since this reaction couples oxidation to production of 1,3-bisP-glycerate, whose high energy content allows substrate-level phosphorylation in the following step of glycolysis catalyzed by PGKase (see Figs. 7.6, 7.7, 7.9, and 7.10; see also Iglesias (1990)). If oxidation of Ga3P takes place via np-Ga3PDHase, the pathway is energetically uncoupled and irreversible. Under these conditions, NADPH will be generated which could be relevant since OPPP is fully operative in the plastid but incomplete in the cytosol (see Fig. 7.9). Bioenergetically, such a shortcut taking place in cytoplasmic glycolysis of plant cells is nontrivial. Supposedly, the alternative to produce ATP or NADPH should be regulated, but this is not fully understood, although progress has happened during the last decade (Bustos and Iglesias, 2002; Bustos and Iglesias, 2003).

In agreement with the regulatory hypothesis of the two Ga3PDHases operating in the cytosol of plant cells, it was established that in nonphotosynthetic tissues (endosperm, shoots) np-Ga3PDHase is subjected to post-translational modification by phosphorylation (Bustos and Iglesias, 2002). Phosphorylation, by itself, has no effect on the enzyme kinetics, but it is a requisite for its interaction with 14-3-3 regulatory proteins (Bustos and Iglesias, 2003). When forming the latter complex, np-Ga3PDHase exhibits reduced activity (by about 3-fold) and enhanced sensitivity to regulation by adenylates and PPi (a metabolite of quantitative relevance in plant cells, see below in this section). Because the interaction between the phosphorylated enzyme and the regulatory protein is released by Mg^{2+}, the regulatory scenario supposes that activity levels of np-Ga3PDHase can vary within near one order of magnitude scale, depending on the relative content of divalent cations and intermediates of energetic metabolism in the cytosol (Bustos and Iglesias, 2002; Bustos and Iglesias, 2003). Furthermore, proteomic and interactomic (14-3-3-partners) approaches have also identified the phosphorylating NAD^+-dependent enzyme (GAPC) in a post-translational modified state (Alexander and Morris, 2006; Cotelle *et al.*, 2000), although no study relating to it with possible changes in activity and/or regulation has been performed. That posttranslational modification of np-Ga3PDHase occurs in heterotrophic but is absent in photosynthetic cells can be understood as due to plentiful supply of ATP and NADPH in photoautotrophy. Conversely, the limited availability of photoassimilates in heterotrophic tissue would render the generation of energy and reducing equivalents critical.

The key role played by both cytosolic Ga3PDHases in plant metabolism has been put in evidence by studies performed with Arabidopsis mutants deficient in each one of the enzymes (Rius *et al.*, 2006; Rius *et al.*, 2008). Results showed that phosphorylating Ga3PDHase function is crucial for carbon flux and mitochondrial function. Plants deficient in this enzyme exhibited decreased ATP levels and respiratory rate, delayed growth, morphological alterations in siliques and in embryo development, and a low number of seeds (Rius *et al.*, 2008). The null-mutants in np-Ga3PDHase, were characterized by increased levels of Glc-6P dehydrogenase and other cytoplasmic enzymes, and higher levels of reactive oxygen species (ROS) (Rius *et al.*, 2006). Moreover, the nonphosphorylating enzyme could be critically involved in the generation of NADPH which by supplying electrons to the antioxidant defenses will help cope with excess ROS (Bustos *et al.*, 2008). Thus, wheat leaves exposed to high levels of ROS contain relatively increased amounts of active np-Ga3PDHase, which is in agreement with the modulatory role of the intracellular redox environment exerted by this enzyme.

The absence of inorganic pyrophosphatase and the consequent cytoplasmic accumulation of PPi (up to 0.5 mM) is another feature of plant cells (Figueroa *et al.*, 2011; Iglesias and Podestá, 2005; Plaxton and Podestá, 2006). PPi is significant for plant metabolism, being an alternative phosphoryl donor operating preferentially under stressful conditions. In fact, transgenic plants expressing inorganic pyrophosphatase from *E. coli* exhibited highly reduced levels of PPi that caused a dramatic inhibition of plant growth (Plaxton and Podestá, 2006). PPi is also directly involved in the conversion of Fru-6P to Fru-1,6-bisP, which in plants can occur by alternative routes (see Figs. 7.6, 7.7, and 7.9). The latter reaction constitutes a regulatory point of glycolysis in different organisms, but in a different manner as compared with plants. The classical pathway involves the reaction catalyzed by ATP-dependent phosphofructokinase (ATP-PFKase, EC 2.7.1.11):

$$Fru\text{-}6P + ATP \rightarrow Fru\text{-}1,6\text{-}bisP + ADP$$

Plants also have a PPi-dependent enzyme which uses PPi as a phosphoryl donor (PPi-PFKase, EC 2.7.1.90):

$$Fru\text{-}6P + PPi \leftrightarrow Fru\text{-}1,6\text{-}bisP + Pi$$

The regulation of this alternative glycolytic reaction is exerted by specific allosteric effectors that modulate the activity of each PFKase. The PPi-dependent enzyme is significantly activated at low concentrations of Fru-2,6-bisP and is inhibited by PEP (a metabolite that also affect levels of Fru-2,6-bisP by inhibiting its synthesis). On the other hand, ATP-PFKase is insensitive to Fru-2,6-bisP, but it is inhibited by PEP. The differential functioning of these enzymes is of relevance for plant glycolysis, particularly under stress. For instance, under Pi starvation, the pathway using PPi-PFKase is dominant due to the fact that Pi inhibits this enzyme and activates the ATP-dependent enzyme (Iglesias and Podestá, 2005; Plaxton, 1996; Plaxton and Podestá, 2006).

Pyruvate is not necessarily the final product of glycolysis in plants, but PEP constitutes a key metabolite in that pathway (Iglesias and Podestá, 2005; Plaxton and Podestá, 2006; see also Fig. 7.2). In fact, PEP is an important allosteric regulator of many enzymes (as the above stated effects on ATP- and PPi-PFKase and on the synthesis of Fru-2,6-bisP), and it also is a branch-point metabolite. PEP is a substrate for: carboxylases (PEP carboxylase and PEP carboxykinase), a specific phosphatase, and pyruvate kinase, thus representing in plants a metabolic node which connects glycolysis, gluconeogenesis, replenishment of intermediaries of the tricarboxylic acids cycle, and nitrogen assimilation (Iglesias and Podestá, 2005; Plaxton, 1996; Plaxton and Podestá, 2006).

Another unique metabolic feature exhibited by plants concerns the signaling role of the nonreducing disaccharide trehalose [α-D-glucopyranosyl-$(1\rightarrow1)$-α-D-glucopyranoside] on carbon metabolism. The occurrence and functionality of trehalose in plants is a new emerging research subject, whose significance is gaining momentum (Paul, 2008; Paul *et al.*, 2008). For many years it was thought that trehalose had minimal (if any) functional relevance in plants, and that its putative role was replaced by sucrose. A rationale for the latter was that detectable levels of trehalose were restricted to a few plants called resurrection plants that are resistant to extremely dry conditions. This view changed after the advent of genomics of *Arabidopsis thaliana* and other species, showing the presence of several genes related with trehalose metabolism. Studies on ME of the trehalose pathway in plants reinforced the importance of this dissacharide. By 2002, it was demonstrated that the gene related with trehalose-6P (Tre-6P) was indispensable. Further research showed that even trace amounts of trehalose (less than 10 μM) play a key role in coordinating metabolism with plant development (Paul *et al.*, 2008). Relevant effects of Tre-6P are exerted in embryo development, normal vegetative growth in leaves, and the transition to flowering. It regulates starch accumulation through modulation of ADPGlcPPase mediated by posttranslational redox modifications via thioredoxin. Besides, Tre-6P is also an allosteric effector of protein kinases, acting as a sensor of sucrose status (Paul, 2008). Synthesis of trehalose in plants involves the reactions catalyzed by Tre-6P synthase and Tre-6P phosphatase:

$$Glc\text{-}6P + UDPGlc \leftrightarrow Ter\text{-}6P + UDP$$

$$Tre\text{-}6P + H_2O \rightarrow Tre + Pi$$

whereas its degradation occurs mainly via trehalase-catalyzed hydrolysis:

$$Tre + H_2O \rightarrow 2Glc$$

Trehalose is an osmolyte with protective ability due to its stabilizing properties of the structure of proteins, other macromolecules, and lipid bilayers. The presence of trehalose in living organisms can help under unfavorable conditions and be used as a biotechnological tool for introducing resistance to stress.

MCA and MFA Studies in Plants

Studies carried out with reversed genetics helped to understand carbon flux and its regulation in higher plants (Stitt, 1999). Reversed genetics combines metabolic biochemistry and molecular genetics to obtain transformed plants, where the expression of one specific enzyme is progressively decreased and/or increased. This can be done through classical genetics, by varying the number of functional copies of a gene; or through plant transformation utilizing antisense or sense constructs of specific genes. After reverse engineering, quantitative analysis of metabolic fluxes is performed. Using this experimental approach, the control exerted by different enzymes on the photoassimilation flux was determined. The analysis comprised the study of enzymes catalyzing irreversible (fine-regulated) as well as reversible (nonregulated) steps from RPPP, and the pathways for starch and sucrose synthesis in plant cells.

Insightful results were obtained on the effects observed on metabolic fluxes and their control in photosynthetic cells of higher plants (Stitt, 1999). It was determined that control of flux is usually distributed between several enzymes catalyzing both regulatory and nonregulatory steps. Additionally, it was observed that because the relative levels of regulated (generally in excess) and nonregulated enzymes are different, it could be possible to compensate for metabolic changes and also avoid severe flux limitation under certain ambient conditions. It was further shown that the contribution of certain enzymes to the control of flux depends on the short-term conditions under which the measurement is performed, as well as on the long-term environmental conditions for plant growth (Stitt, 1999).

These principles were demonstrated by analyzing the pattern of flux control of carbon assimilation exerted by enzymes from RPPP. Under certain conditions, a decrease of 40% or more in the activity of finely regulated enzymes of the BCC led to slight or no inhibition of photosynthesis. Among these enzymes are those catalyzing irreversible reactions, namely Rubisco, Fru-1,6-bisP phosphatase, sedoheptulose-1,7-bisP phosphatase, and P-ribulokinase (see Fig. 7.3). On the other hand, changes in the levels of nonregulated enzymes, such as aldolase and transketolase, considerably affected carbon photoassimilation under the same conditions (Stitt, 1999). It has also been reported that variations in Rubisco levels exert low control (control coefficient of 0.1) or high control (control coefficient of 0.9) of the photosynthetic flux depending on whether light conditions are constantly moderate or suddenly increased from a low steady level, respectively.

Of interest are also studies showing how compartmentalization and the existence of branch points are important factors to be considered for MCA or MFA in higher plants. As an example, it has been pointed out that the different contribution of chloroplastic and cytosolic P-glucoisomerase (PGI, catalyzing the reaction: Glc-6P $\Leftrightarrow$ Fru-6P) to photoassimilates partitioning between starch and sucrose in a photosynthetic cell (Stitt, 1999). Figures 7.2, 7.8, and 7.9 show that in photosynthetic cells PGI occurs in two different subcellular compartments. In the chloroplast, starch is synthesized from Fru-6P (an intermediate metabolite of RPPP) by reversible reactions catalyzed by PGI and P-glucomutase (PGM, catalyzes the reaction: Glc-1P $\Leftrightarrow$ Glc-6P), followed by the irreversible step mediated by ADPGlcPPase and finally by reactions of SS and BE (Ballicora *et al.*, 2003; Ballicora *et al.*, 2004; Iglesias and Podestá, 2005). Also from intermediates of the RPPP, photoassimilates partition to the cytosol for sucrose synthesis via the Pi-triose-P counter-exchange. From triose-P, Fru-6P is produced and then converted to Glc-6P by cytosolic PGI; the route is then continued to sucrose synthesis (Iglesias and Podestá, 2005). Experiments carried out in *Clarkia xanthiana* mutants, which exhibit decreased expression of cytosolic or chloroplastic PGI, showed different responses in starch and sucrose fluxes. This demonstrated that the contribution of an enzyme is related to the pathway in which it is involved, rather than with the reaction the enzyme catalyzes. In these studies, it was found that a decrease of plastidic PGI inhibits (in high light but not in low light conditions) rates of sucrose build up as well as of photosynthesis. A decrease in cytosolic PGI instead reduces the rate of sucrose production (in low light but not in high light conditions), this leading to a compensating increase of starch synthesis with no effect observed on the rate of photosynthesis (Stitt, 1999).

The control of the flux of assimilates also needs to take into account the high degree of compartmentalization and different cell types (photosynthetic and heterotrophic) existing in plants. This was clearly exemplified in studies performed on the genetic manipulation of starch synthesis (Stark *et al.*, 1992). It was shown in the increase in ADPGlcPPase activity within potato amyloplasts. Expression of a bacterial gene coding for a mutant, unregulated, ADPGlcPPase produces plants that accumulate significantly higher amounts of starch in their tubers (see below). In order to obtain transgenic plants that provide a product of higher quality (Collins and Shepherd, 1996), it was necessary to direct the transformer gene only to amyloplasts; thus, the gene was introduced under the control of a tuber-specific patatin (Stark *et al.*, 1992). Otherwise, expression of the gene in a constitutive manner (i.e. under the control of CaMV 35 S promoter) is detrimental for plant growth and development. The reason for the latter is that if ADPGlcPPase activity is increased in the chloroplast, an excess of starch is accumulated in the leaf, which then reduces sucrose synthesis and availability for export to actively growing tissues (Stark *et al.*, 1992). This demonstrates that ADPGlcPPase is a rate-controlling step of starch synthesis provided is fully activated. Moreover, the different pathways followed by photoassimilates in source or sink tissues determine a distinctive contribution of ADPGlcPPase to carbon flux and partitioning in different plant cells.

Regulation and Control: Starch Synthesis: A Case Study

Starch is the major storage product of photosynthesis, as up to 30% of the carbon photoassimilated by plants is channeled to this polysaccharide under optimal conditions (Ballicora *et al.*, 2004). Thus, the regulation of starch levels in plants has a tremendous potential impact on the primary productivity on earth. We will analyze the manipulation of the starch biosynthetic pathway in order to produce a significant increase of polysaccharide accumulation in crops.

Starch synthesis from Glc-1P in plants occurs through three reactions successively catalyzed by ADPGlcPPase, SS, and BE (Ballicora *et al.*, 2003, 2004; Iglesias and Podestá, 2005). In the first step, ADPGlc is produced from Glc-1P and ATP, mediated by ADPGlcPPase. Then, different isoforms of SS utilize the sugar-nucleotide as the glucosyl donor to elongate an α-1,4-glucan chain. Specific branching isoenzymes introduce α-1,6-branch points in the polysaccharide.

The reaction catalyzed by ADPGlcPPase is irreversible *in vivo* and constitutes a key regulatory step in the pathway. Plant ADPGlcPPase is a highly regulated enzyme. One level of regulation involves posttranslational reduction via thioredoxin (Ballicora *et al.*, 2003, 2004), in a mechanism ultimately modulated by Tre-6P (Paul *et al.*, 2008). On the other hand, the enzyme is allosterically regulated by 3PGA (activator) and Pi (inhibitor). The well-known interplay between both regulatory metabolites of the enzyme has been highlighted after showing that ADPGlcPPase regulatory properties exhibit ultrasensitive behavior in molecularly crowded environments (Aon *et al.*, 2000, 2001; Casati *et al.*, 2000; Gomez *et al.*, 1999; Gomez-Casati *et al.*, 2001, 2003). Ultrasensitive systems exhibit sensitivity amplification (i.e. the percentage change in a response, compared with the percentage change in the stimulus). Amplification factors for ADPGlc synthetic rates ranging from 11- to 19-fold were described as a function of 3PGA levels. An advantage of this mechanism could be given by the fact that in the light–dark transition the system may be quickly deactivated (i.e. a sharp decrease of starch levels), thereby hindering wastage of ATP. In addition, starch synthesis during the dark–light transition would operate as an ultrasensitive switch-like device that can amplify 14- to 15-fold the levels of starch for a narrow 3PGA/Pi ratio.

Ultrasensitivity has been defined as the response of a system that is more sensitive to changes in the concentration of the ligand than is the normal hyperbolic response given by the Michaelis–Menten equation (Goldbeter and Koshland, 1984; Koshland, 1998). As a device in biochemical networks, one main advantage of ultrasensitivity is that it allows a several-fold increase in flux through a metabolic step over a narrow range of change in substrate (Koshland, 1998) or effector concentrations (Aon *et al.*, 2000, 2001; Casati *et al.*, 2000; Gomez *et al.*, 1999; Gomez-Casati *et al.*, 2001, 2003). In this sense, ultrasensitivity functions as a switch-like device that filters out small stimuli by restricting the range and threshold of stimulation (Ferrell and Machleder, 1998).

Indeed, ADPGlcPPase is an enzyme that functions as an information-processing biochemical device (i.e. sensing the levels of its allosteric effectors: 3PGA and Pi) besides its well-known role in synthesizing ADPGlc, the glucosyl donor for starch synthesis in plants, or glycogen in cyanobacteria. Thus, it becomes relevant to understand its dual information-mass transforming abilities quantitatively, along with the impact on polysaccharide synthesis regulation. Kinetic data on cyanobacterial ADPGlcPPase were obtained *in vitro* under aqueous or polyethyleneglycol (PEG)-induced crowded environments (Casati *et al.*, 2000; Gomez *et al.*, 1999), and *in situ* (Gomez-Casati *et al.*, 2001, 2003). A mathematical model was formulated that describes starch metabolism in cells performing oxygenic photosynthesis (Aon *et al.*, 2001; Gomez-Casati *et al.*, 2003). Ultrasensitive behavior was also observed *in vitro* in ADPGlcPPase from spinach leaves (Gomez-Casati, Aon and Iglesias, unpublished results). The mathematical model takes into account the ADPGlc production by ADPGlcPPase as well as starch synthesis and degradation. Simulations performed with this model allowed to reproduce the ultrasensitive behavior of ADPGlcPPase and its corresponding amplification factors under crowding conditions. Results from this model allowed us to establish that polysaccharide synthesis is also ultrasensitive (Aon *et al.*, 2001; Gomez-Casati *et al.*, 2003).

The results of the model explained above, and shown in Fig. 7.11, clearly depict the links existing between Regulation and Control. There is a phase relationship between the maximum of amplification attained by ADPGlcPPase rate and the accumulation of starch, as a function of the 3PGA/Pi ratio. At arrow 1 (Fig. 7.11), the enzyme is maximally sensitive toward the stimulus given by the ratio between the two allosteric effectors. The maximal amplification implies a maximal relative increase in velocity in response to the stimulus. This is the Regulatory Phase; using the terminology of MCA the elasticity coefficients are maximal, whereas the control coefficients are minimal (see Chap. 5). Arrow 2 indicates the phase of behavior where the system is only minimally sensitive to the stimulus (Fig. 7.11).

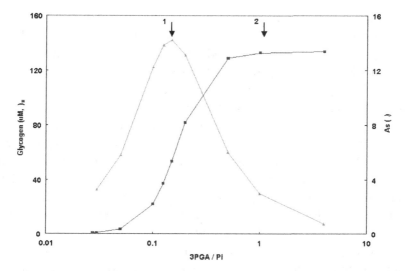

Figure 7.11. Steady-state levels of storage polysaccharides and amplification factors exhibited by ADPGlucose pyrophosphorylase (ADPGlcPPase) as a function of the ratio 3PGA/Pi. Carbon metabolism from G1P toward storage polysaccharides (cyanobacterial glycogen or starch) synthesis and subsequent degradation was modelled with a system of two ordinary differential equations (ODEs). The two equations represented the dynamics of ADPGlc and glycogen (GLY), i.e. the state variables, according to the following kinetic scheme:

$$ATP + G1P \xrightarrow{ADPGlcPPase} ADPGlc + PPi$$

$$ADPGlc + GLY_n \xrightarrow{GLYsy} GLY_{n+1} + ADP$$

$$GLY \xrightarrow{GLYdeg}$$

V_{AGPase} represents the reaction rate catalyzed by *ADPGlcPPase*, and V_{GLYsy}, V_{GLYdeg} the glycogen synthesis and degradation, respectively. The ultrasensitive response of V_{AGPase} toward its allosteric activator 3PGA, elicited by orthophosphate, Pi, and polyethyleneglycol (PEG)-induced molecular crowding, operating under zero- or first-order conditions with respect to its substrates, G1P (Glc-1P) and ATP (Casati *et al.*, 2000; Gomez *et al.*, 1999; Gomez-Casati *et al.*, 2001), was taken into account in the rate expression of the enzyme (Gomez-Casati *et al.*, 2003).

Otherwise stated, the enzyme is saturated thus becoming a rate-controlling step. This is the Controlling Phase.

A biochemical switch-like ADPGlcPPase can then function as a regulatory or control device depending upon the level of stimulus, entraining different polysaccharide levels. Indeed, at low 3PGA/Pi ratios increasing the cellular levels of ADPGlcPPase by the techniques of molecular biology will not translate into higher levels of starch, despite the potentiality of the enzyme to regulate the flux under those conditions. Higher 3PGA/Pi ratios should be necessary to allow the enzyme express its maximal potential. In good agreement with the latter, the successful transformation of plants to increase the amounts of starch accumulated in storage tissues, required the expression of a gene (*glgC*) coding for a mutant ADPGlcP-Pase from *E. coli*, insensitive to regulation and exhibiting high specific activity (Collins and Shepherd, 1996; Stark *et al.*, 1992). Thus, in transgenic plants, the pathway of starch synthesis operates under nonregulated conditions and continuously in the phase of behavior minimally sensitive to the stimulus, with ADPGlcPPase activity maximized and steadily controlling the flux (Fig. 7.11, arrow 2). The result was that transgenic plants accumulated over 30% higher amounts of starch. Interestingly, the expression of an ADPGlcPPase sensitive to regulation showed no modification in the levels of polysaccharide (Stark *et al.*, 1992). In other words, enhancing the amount of enzyme in the Controlling Phase was effective for increasing the amount of final product, as expected according to the analysis presented in Fig. 7.11. All these results demonstrate that, when fully activated, ADPGlcPPase may be a main rate-controlling step of starch synthesis.

The Relevance of Unicellular Algae: Auto-, Hetero-, and Mixo-Trophy

The use of unicellular photosynthetic microorganisms in productive processes constitutes the basis of applied phycology. Phycologists call microalgae to such kind of phototrophic microorganisms, including true algae (eukaryotes) as well as cyanobacteria (Richmond, 2004). Some species of cyanobacteria have the advantage of being able to fix CO_2 and N_2 phototrophically. Metabolically, cyanobacteria resemble a chloroplast from a higher plant. Autotrophy in cyanobacteria is supported by oxygenic photosynthesis, a process that was originated on Earth by these microorganisms some 3.6 billion years ago, transforming the planet and creating the current biosphere (Archibald, 2009; Gould *et al.*, 2008). The endosymbiotic acquisition of a cyanobacterium by a heterotrophic eukaryote generated plastids that bestowed the autotrophic character (Archibald, 2009; Gould *et al.*, 2008). From this process, primary endosymbionts originated, which include glaucophyta, green algae (mainly represented by *Chlamydomonas* spp. and from where plants descent) and red algae, that are organisms with plastids having double membrane. Subsequent secondary and tertiary endosymbiotic events involved the engulfment of the primary endosymbiont by another heterotrophic eukaryote, generating photosynthetic organisms with plastids having multiple membranes. Examples of secondary endosymbiosis are *Euglena* spp. algae and diatoms (e.g. *Phaeodactylum* spp.), photosynthetic microorganisms characterized by possessing chloroplasts surrounded by three or four membranes (Gould *et al.*, 2008). Some of the secondary endosymbionts have subsequently reverted to heterotrophic organisms, becoming important pathogenic species such as *Plasmodium* and *Trypanosoma* that still possess remnant plastid structures (Archibald, 2009).

All these evolutionary events created a great variety of lifestyles with a wealth of metabolic capacities. This metabolic versatility composes a solid basis for developing applied phycology strategies. Some microalgae display the ability for growing photosynthetically, heterotrophically, or mixotrophically (Perez-Garcia *et al.*, 2011; Richmond, 2004). The use of these microorganisms in productive processes has obvious advantages based on the feasibility of manipulating growth conditions and development. The photosynthetic characteristic of microalgae is advantageous because of their low nutrient requirements and low costs in their production. However, under certain circumstances photoautotrophy can create limitations, thus the option of heterotrophic growth is relevant to overcome the problem (Perez-Garcia *et al.*, 2011). Advantages of heterotrophy are: (i) photo-dependency can be tied to reduced production of a desired product; (ii) requirements of light energy for feeding the process can result more expensive than the supply of an organic carbon source; (iii) it could be convenient to use a carbon substrate that is a waste or by-product (as in the above stated case of glycerol as a by-product in biodiesel production), but useful for feeding microalgae (O'Grady and Morgan, 2010); (iv) heterotrophy can offer more choices for adjusting culture conditions. Main microorganisms of interest for the possibility of doing photoauto-, hetero-, or mixo-trophy, are the green alga

Chlamydomonas reinhardtii and the diatom *Phaeodactylum tricornutum* (Ginger *et al.*, 2010; Richmond, 2004).

Applied phycology is being extended to a wide range of biotechnological processes, including food production (for humans and animals, e.g. for aquaculture), fertilizers, chemicals, and recombinant proteins, as well as in bioremediation (Richmond, 2004). Lately, microalgae proved their usefulness in the fields of biofuels and biorefineries (Pittman *et al.*, 2011; Richmond, 2004). A good potential is found in microalgae as a source of biomass (for ethanol production) or lipids (for biodiesel). Microalgae also produces methane and hydrogen (Pittman *et al.*, 2011; Richmond, 2004; Rupprecht *et al.*, 2006). Generation of methane and hydrogen by photosynthesis represents a very effective process in terms of reduction of CO_2 in the atmosphere, especially for hydrogen whose combustion produces H_2O. Hydrogen bioproduction is mediated by hydrogenase, an enzyme of complex structure that catalyzes the reversible two-electron oxidation of H_2, paradoxically one of the simplest redox reactions (Ghirardi *et al.*, 2007; Rupprecht *et al.*, 2006). Biological production of hydrogen is solar-powered in oxygenic photosynthetic organisms but also occurs when using organic carbon as a source of necessary H^+ and e^-. For instance, certain algae can use starch to produce hydrogen under anaerobic conditions (Rupprecht *et al.*, 2006). Since the reaction of hydrogen production by hydrogenase is inhibited by oxygen, the ability of some microalgae to exhibit variable trophic growth is advantageous, since limitations caused by oxygenic photosynthesis can be overcome (see Chap. 4, for a specific example).

Concluding Remarks

MCE as applied to plants needs more development although at present it is receiving considerable interest. Biofuel production is one main research field where MCE is being actively utilized. Photosynthetic organisms (cyanobacteria, microalgae, and higher plants) are complex biological systems mainly due to the occurrence of multiple metabolic routes involving pathway duplication, with each path operating in different intracellular compartments. In fact, although much relevant information on plant metabolism (and its regulation) has been obtained, a full understanding of the complex scenario is lacking. Beyond its complexity, plant metabolism provides a rich source of metabolic strategies and molecular tools to be applied in biotechnology.

Plants are highly appropriate for MCE because of their autotrophic character that allows them to be considered as self-sufficient biological factories. Techniques for growing plants *in vitro* and for their genetic transformation have been developed, and many important biotechnological goals have utilized plants for obtaining novel or higher quality products. With the improvement of "omics" technologies, special efforts are being focused on further characterizing carbon and energy fluxes as well as their control and regulation. This is a very necessary step for optimizing conditions and tools for rationally improving plant performance in different bioprocesses. The whole scenario looks exciting, both for the possibility of accomplishing a better understanding of plant metabolism and physiology, as well as the many relevant applications deriving from such knowledge.

Chapter 8

Animal Cell Culture Techniques

General Remarks: Cells Cultured as Organisms

Considerable advances in animal cell culture techniques enable the production of cell lines, not only from human tissues and organs but also from invertebrates, (especially insects, amphibia, and fish). The nutritional principles and growth conditions employed are broadly based on those established for microorganisms, although "complex" i.e. chemically undefined growth media are routinely employed on account of the greater diversity of requirements. Thereby, an element of variability of growth rates and yields is unavoidable: even for bacterial growth serious problems of reproducibility have been documented (e.g. for *Salmonella enterica serovar Typhimurium* where the use of "peptones" deficient in a single amino acid (tyrosine) leads to the development of phenotypically aberrant organisms that lack the flagella and immunological determinants, normally regarded as key diagnostic characteristics (Gray *et al.*, 2006; Gray *et al.*, 2008). This example emphasizes the need for rigorous quality control of media composition, and (as is almost invariably the case) where sera are employed, such considerations become paramount.

Yeast extracts provide many essential nutrients especially the B-group vitamins (other than B_{12}). Nutrients required for growth may be classified as: (a) sources of the 'major' elements C, H, and N that account for the bulk of the biomass produced; (b) sources of the minor elements P, K, S, and Mg; (c) Vitamins and growth factors; (d) sources of "trace" elements. Carbon sources serve a dual function: i.e. as organic chemical structures acting as building blocks for the assembly of all the carbon skeletons of cellular components, and also as sources of catabolic energy.

Oxygen as the terminal electron acceptor provides the other essential component for energy generation in aerobic growth. When O_2 is in plentiful supply, compounds serving as carbon and energy sources are limiting for carbon. Under O_2 limitation such compounds are almost always limiting for energy (Bauchop and Elsden, 1960). Other electron acceptors (e.g. NO_3^-, SO_4^{2-}, CO_2) can act as alternatives to O_2. An historical perspective on these principles of microbe and cell cultivation (Panikov, 1995; Pirt, 1975) is especially important where heterologous protein expression requires high cell density (Shiloach and Fass, 2005) and for this purpose bacterial and yeast cultures are used extensively. Recipes and methods for axenization of protist cultivation pioneered by Hutner continue to be developed (Goldberg, 2004), but many species elude culture even now. The cultivation of some parasitic species, e.g. the erythrocytic stage of plasmodia, represents achievements of great significance (Trager, 1964).

The design and operation of fermentation systems has been a major endeavor from the earlier phase (Atkinson and Mavituna, 1983) to the present, where scale-up from automated growth monitoring in the wells of microtiter plates to stirred tank fermenter scale (1.4 L) (Kensy *et al.*, 2009) can be successfully performed. Oxygen transfer rates are of the utmost importance in the scale-up of aerobic fermentations (Garcia-Ochoa and Gomez, 2009). Anaerobic cultivation methods on plates of solid medium in anaerobic jars and in liquid media in anaerobic cabinets require many precautions for exclusion of atmospheric O_2. Supplies of gases frequently contain sufficient O_2 to inhibit growth of strict anaerobes, and pre-purification using catalyst-containing columns may be essential. Butyl-rubber or stainless steel tubing is also recommended, and it must be remembered that some plastics (e.g. silicone rubber) are highly permeable to gases. Thus scavengers, such as ascorbate and cysteine, and use of indicators for trace contamination, e.g. resazurin solution, are commonly employed (Willis, 1977). Methods for estimating microbial numbers and biomass are also generally applicable to cell lines. Coulter Counter and flow cytometry (Lloyd, 1993; Lloyd *et al.*, 1982) as automated methods for cells in suspension supplement traditional methods (Jones, 1979).

Mammalian Stem Cell Culture

The burgeoning and highly specialized field of stem cell biology is beyond the scope of this book with enormous potential clinical applications (Hemmat *et al.*, 2010). Pluripotent cells are derived from adult, embryonic, and perinatal sources and their developmental paths and proliferative and expansive capacities are critically dependent on culture conditions, e.g. for adipose tissue engineering (Hillel and Elisseeff, 2010) and a host of other organ repair programs.

Gene expression profiles during early differentiation of mouse embryonic stem cells provide new insights into the mechanisms controlling stem cell differentiation as a key to future advances in tissue and organ regeneration and repair (Mansergh *et al.*, 2009).

Fish Cell Lines

Cultured fish cells have many applications in cell physiology, biology, carcinogenesis, transgenics, and toxicology. An earlier catalog of cell lines derived from 74 species or hybrids representing 34 families of teleost fishes (Fryer and Lannan, 1994) has recently been updated (Lakra *et al.*, 2010). This new information includes culture of cells from brain, eye muscle, fin, heart, liver, ovary, skin, spleen, swim bladder, and vertebrae, totaling 283 examples worldwide. The majority of these investigations are carried out in Asia, and freshwater, marine, and brackish water fishes are included.

This review stresses the authentication, applications, and cultivation practicalities of research with the cell lines, and stresses the problems inherent in overpassage and cross-contamination. Examples of the usefulness of fish cells are the cultivation of important parasitic protists that have resisted other attempts at *in vitro* methods. Even successes with species of mammalian origin (e.g. *Enterocytozoon bienusi*) have been achieved (Monaghan *et al.*, 2009). A fish gill cell line RTgill-WI, obtainable from the American Type Culture Collection, is ideal as a model for testing epithelial resistance, aquatic toxicology, and gill diseases caused by viruses and protists (Lee *et al.*, 2009).

Amphibian Cell Lines

Cultivation of tissues and cells from amphibians has been described by Kloas *et al.* (1999), who used hepatocytes (Zhengan, 1978), and Laskey (1970), who used antibiotic cocktails to minimize bacterial contamination. As an example, a permanent cell line from the Bullfrog (*Rana catesbeiana*) (Wolf and Quimby, 1964) was taken through 57 subcultures over a period of almost 3 years.

Insect Cell Lines

Insect cell cultures are known to be extremely useful for heterologous protein expression with complex pest-translational modifications using the Baculovirus Expression Vector System (BEVS) or other means of transfection (Goosen, 1991; Pfeifer, 1998; Taticek *et al.*, 2001), e.g. of lepidopteran or *Drosophila* cells. Initially a single cell line from a pupal ovarian tissue served as a stable model. Invitrogen and Gibco provide detailed recipes and procedures and unlike most other cell lines those from insects require no CO_2 buffering systems and can be grown in Wave Bioreactors (Weber *et al.*, 2002). Smagghe *et al.* (2009) point out that more than 500 insect cell lines have been utilized from many different orders and species and are useful in testing insecticides, and as research tools in virology and insect immunity (Sudeep *et al.*, 2005), as well as for the production of high value recombinant proteins for medical uses using serum-free media (Ikonomon *et al.*, 2003).

Mammalian Cells

Detailed manuals on the laboratory facilities, growth requirements, culture conditions, and cryopreservation protocols provide comprehensive coverage (Davies, 2009; Feder and Tolbert, 1985; Lubiniecki, 1990). Some examples of commonly used cell lines are given in Table 5.1. Many biotechnological applications for the production of high value pharmaceutical products (vaccines, growth factors, and hormones) can now be carried out using transgenic strains of bacteria (often *Escherichia coli*) or

Table 8.1   Some commonly used cell lines

Cell line	Cell type and origin
3T3	Fibroblast mouse
BHK21	Fibroblast (Syrian hamster)
MDCK	Epithelial cell (dog)
HeLa	Epithelial cell (human)
PtK1	Epithelial cell (rat kangaroo)
L6	Myoblast (rat)
PC12	Chromaffin cell (rat)
SP2	Plasma cell (mouse)
COS	Kidney (monkey)
293	Kidney (human); transformed with adenovirus
CHO	Ovary (Chinese hamster)
DT40	Lymphoma cell for efficient targeted recombination (chick)
R1	Embryonic stem cells (mouse)
E14.1	Embryonic stem cells (mouse)
H1, H9	Embryonic stem cells (human)
S2	Macrophage-like cells (Drosophila)
BY2	Undifferentiated meristematic cells (tobacco)

Many of these cell lines were derived from tumors. All of them are capable of indefinite replication in culture and express at least some of the special characteristics of their cell of origin. BHK21 cells, HeLa cells, and SP2 cells are capable of efficient growth in suspension; most of the other cell lines require a solid culture substratum in order to multiply.
Source: Copyright © 2002, B. Alberts, A. Johnson, J. Lewis, M. Raff, K. Roberts, and P. Walter; Copyright © 1983, 1989, 1994, B. Alberts, D. Bray, J. Lewis, M. Raff, K. Roberts, and J. D. Watson.

yeasts (e.g. *Saccharomyces cerevisiae, Yarrowia lipolytica*, or high-level secretory strains of *Pichia pastoris* (Gurramkonda *et al.*, 2010) thereby removing the need for largescale mammalian cell technology and for growth media containing animal products (with the attendant possibility of prior contaminants). The prime example of this is insulin, in which case almost two decades of engineering of producer strains of *S. cerevisiae* in the Novo Nordisk A/S laboratories has resulted in almost the whole of the global supply of this hormone (>7 t/annum: the largest output of any pharmaceutical) (Kjeldsen and Pettersson, 2003). Another rapidly developing technology involves the heterologous expression in yeasts of mammalian steroidogenic P4505 and the corresponding electron

Table 8.2   Human and other cell lines

Cell line	Cell type and origin
Human	
National Cancer Institute's 60 cancer cell lines	ESTDAB database http://www.ebi.ac.uk/ipd/estdab/directory.html
DU145	Prostate cancer
Lncap	Prostate cancer
MCF-7	Breast cancer
MDA-MB-438	Breast cancer
PC3	Prostate cancer
T47D	Breast cancer
THP-1	Acute myeloid leukemia
U87	Glioblastoma
SHS5Y	Human neuroblastoma cells, cloned from a myeloma
Saos-2 cells	Bone cancer
Primate	
Vero (African green monkey Chlorocebus)	Kidney epithelial cell line initiated 1962
Rat tumor	
GH3	Pituitary tumor
PC12	Pheochromocytoma
Mouse	
MC3T3	Embryonic calvarial
Plant	
BY-2 cells	Tobacco cells kept as cell suspension culture, they are model system of plant cell
Other species	
ZF4 and AB9 cells	Zebrafish
Madin-Darby Canine Kidney (MDCK)	Epithelial cell line
A6 ki	Xenopus

transport proteins required for hormone production (Novikova *et al.*, 2009).

It is necessary to mimic *in vitro* physiological oxygen tensions when culturing cells *in vitro* and rather little attention has been paid to this (Toussaint *et al.*, 2011). Methods for producing and controlling dissolved O_2 have been reviewed (Lloyd, 2002): as levels *in vivo* in tissues are often $<10\,\mu M$, there is a common tendency to provide too much O_2. Where inappropriately elevated oxygenation conditions have been employed, oxidative stress deranges differentiation of stem cells, so that it may not proceed along anticipated routes. Premature arrests of cell division, senescence, or even switched phenotypes are possible consequences.

Novel experimental *in vitro* and *in vivo* (Khan, 2009) models of mammalian blood-brain barrier have been elaborated using cultures of primary brain microvascular endothelial cells (Alsam *et al.*, 2003), which exhibit high trans-endothelial electrical resistance ($\sim$2000 Ohm/cm^2). These confluent cell layers with their tight junctions present barriers even for small molecules (e.g. dyes and antibodies) and are thus of tremendous value in studies of host–parasite interactions.

Chapter 9

Metabolic and Cellular Engineering and Bioprocess Engineering in the Industrial Production of Therapeutic Proteins

Metabolic and Cellular Engineering (MCE) is at an interphase between fundamental and biotechnological research. The current stage of MCE is focused on identification of the overall regulation and control of networks. Systems Bioengineering approaches directed to identifying the overall control and regulation of networks characterize the present developmental stage of MCE. Network components can be metabolic and regulatory as in cells and also as steps and components of an industrial bioprocess. Physiological Engineering, Industrial Systems Biology or TDA (as in this book), are synonymous terms used in Systems Bioengineering, all with the common goal of understanding the function of whole networks. When applied to cells, whole system approaches seek to design metabolic or regulatory pathways. Thus an integrated scheme includes many experimental disciplines (microbial physiology, genetic engineering, molecular biology, fermentation technology, biochemistry, and computational modeling) and theoretical tools (Metabolic Flux Analysis, Metabolic Control Analysis, and kinetic and networks modeling).

Bioprocesses underpin the food and pharmaceutical industries. In this context, Systems Bioengineering strives to perform Bioprocess Engineering as a thorough transdisciplinary effort aimed at manufacturing a product with standards of quality, safety, and efficacy. Its success is the result of a deep understanding of correlations existing among major process parameters, input materials, equipment, operator capabilities, and cell physiology.

MCE, as the major theme of this book, is now considered within the transdisciplinary effort demanded by the design of a bioprocess at an industrial level. In this context, MCE plays a central role in the overall Bioprocess Engineering endeavor, focusing on the rational design of cells directed to a specific biotransformation. By paying attention to the organism, MCE also demands a rigorous understanding of the key variables/parameters of a cell's physiology and metabolism, and their interactions. Assessing the organism's potential for redirecting metabolic fluxes to the product of interest, as a function of environmental factors or genetic interventions, is critical for MCE. At the meeting point between cell physiology and factors determining the success bioprocess intended for industrial production is where we now direct the focus of our analysis.

Bioprocess Engineering in the Industrial Production of Therapeutic Proteins

Biopharmaceutical manufacturers have always faced the challenge of finding ways to make living organisms produce proteins with desired characteristics, purifying them from complex mixtures with economically feasible yields, and formulating them into stable, medically useful products. These challenges are compounded by the variability in: (i) raw material quality, (ii) equipment components, (iii) environment within the manufacturing facility, and (iv) operator skills. As those who have struggled with these issues know so well, the quality of the biological product depends to a large extent on the design and control of the manufacturing process (Rathore and Winkle, 2009).

In the pharmaceutical industry, the bioprocess design for effective production of a therapeutic molecule follows a systems approach. Its development and optimization as well as its effective and safe use involve the consecutive use of recombinant DNA and fermentation technologies as well as mathematical modeling.

After selection of the appropriate structure/function molecular properties, the process development, process scale-up, and final delivery to the patient, a transdisciplinary approach is essential. At a later stage, molecular biology methods, pharmacokinetics, and toxicological design determine safe and effective dosage.

As in the case of the TDA approach adopted in this book, a special emphasis is placed on the iterative nature of the whole Bioprocess Engineering. At different stages, the manufacturing of a biomolecule requires the understanding of the interaction between the organism and its environment (e.g. media composition, incubation parameters) and how that interaction impacts productivity and product quality. This topic can be systematically addressed through the *Design of Experiments* (DoE). DoE comprises different statistical tools directed to the quantitative assessment of existing correlations between process parameters (e.g. pH, dissolved O_2, temperature), control capabilities, and input materials. Multivariate analysis is utilized to obtain a thorough understanding of all the interrelationships affecting the quality of the therapeutic product.

DoE pursues the identification of the operating parameters and components of the bioprocess which have a significant negative impact on product quality, eventually compromising its efficacy and safety for the patient. Importantly, DoE is applied iteratively, i.e. the result of each parameter combination will be tested on product quality and quantity until an optimum is achieved. The surface described by the optimal combination of the most successful significant parameters of a process. The optimal surface determines the operational limits of the process, which is given by the tolerated range of variation permitted for each of the significant bioprocess parameters.

The correct application of DoE constitutes the sound scientific basis of the so-called Quality-by-Design (QbD) paradigm. By following QbD industrial manufacturing seeks to incorporate quality standards into Bioprocess Engineering. This practice is oriented by Regulatory Agencies, e.g. Food and Drug Administration (FDA) through different Good Manufacturing Practices (GMP) guidelines aimed at reassuring product quality/safety/efficacy as a result of a deep scientific understanding of the whole bioprocess. Three important documents were published as part of the International Conference on Harmonization (ICH) guidelines: (i) Q8 Pharmaceutical Development (FDA, 2006a) describes the expectations for the Drug Product Pharmaceutical Development, (ii) Q9 Quality Risk Management (FDA, 2006b), presents approaches to producing quality pharmaceutical products using current scientific and risk-based principles, and (iii) Q10 Quality Systems Approach to Pharmaceutical GMP Regulations (FDA, 2006c) provide a model for an effective quality management system for the pharmaceutical and biotech industry.

QbD is defined as a more scientific, risk-based, holistic, and proactive approach to pharmaceutical development. The QbD concept promotes the understanding of the product and manufacturing process at an industrial level: building in quality, not testing it. Its implementation for biopharmaceutical products is occurring, while we write. It represents a very welcome challenge for the Industry and the US Regulatory Agency because it requires the establishment of a common nomenclature. The underlying science supports the introduction of new technology to monitor and control the production of biomolecules while providing a solid scientific basis for new drug presentations at the FDA or any regulatory agency in the world.

It is crucial to public health that the drugs upon which we depend are safe, efficacious, and of consistent quality. Safety is initially evaluated based on toxicological studies using animal models, and efficacy based on the first-time-in humans (FTIH) studies. Safety and efficacy are continuously evaluated during more advanced stages, e.g. clinical trials with more representative groups of patients.

Another aspect of product quality is batch consistency, i.e. the reproducibility of the process performance for producing the biomolecule of interest with the aim of minimizing all possible sources of variability. The specifications are defined as a list of tests and appropriate acceptance criteria which are numerical limits; specifically they constitute critical quality standards proposed and justified by the manufacturer and approved by regulatory authorities. Lot and lot-to-lot variability depend, to a large extent, on the quality of raw materials, the design of the manufacturing process, and its control systems. Thus, the essence of QbD is the incorporation of safety and consistency into product quality while keeping the biomolecule critical quality attributes during bioprocess development.

To realize the full benefits of quality, one must develop a thorough understanding of the interrelationship between the attributes of the input materials, the process parameters, the physiology of the living organism, and the characteristics of the final product. With this information at hand, it is possible to manufacture each unit of product with the desired quality standard and a very high degree of confidence. Of particular note in this regard is the quality control system known as Process Analytical Technology (PAT) that has been applied with great success to manufacturing operations outside the pharmaceutical industry. In 2004, the FDA published the guidance for industry on PAT to promote the development and implementation of the agency's "Product Quality for the 21st Century Initiative" (FDA, 2004). The PAT initiative stimulates the use of innovative technologies whose implementation can give the quality assurance, more specifically rapid and real time analytical technologies to have direct control of product quality during manufacturing. Included are near-infrared spectroscopy used to monitor metabolite concentrations, online biomass, and product accumulation during high-cell density microbial fermentations. All these analytical tools are readily available, opening the door to continuous processing and real time release. These changes in product development and expression system technology have driven, and relied upon, parallel advances in manufacturing sciences.

Bioprocess Design and Product Development

Empirical development of production processes and losses during manufacturing constitute a sensitive topic for the FDA. The limited understanding of manufacturing problems has been highlighted by the regulatory review agency panel in the application supplements stemming from new drug applications (NDAs), biological license applications (BLAs), and abbreviated new drug applications.

We now turn our attention into bioprocess design. Figure 9.1 depicts the whole design of the bioprocess and product development. In the following we will treat each step of the process in detail at the laboratory scale. Laboratory-based bioprocess design and optimization will have to be transferred to the scale-up phase of the industrial process (Stage 6, Fig. 9.1).

Stage 1: Molecule design

The first stage of the product and process development is represented by the molecule design, which focuses on molecular characteristics/attributes. Important traits are molecular weight, types of nonspecific posttranslational modifications (PTMs, e.g. glycation, acetylation, phosphogluconoylation/gluconoylation), and expected posttranslational modification patterns

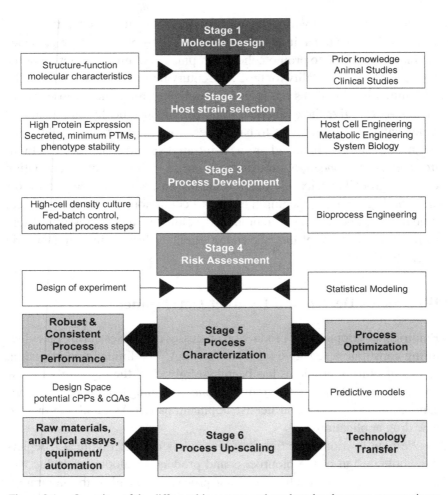

Figure 9.1.   Overview of the different bioprocess and product development stages prior to transfer to commercial scale.

(e.g. glycosylation, S-S bonds). Building-in quality into biomolecules requires understanding of the molecular structure-function characteristics, and provides the basis for rational protein engineering. Knowledge about structural-functional relationships helps define the molecular attributes that are important for patient safety, therapeutic efficacy, and *in vivo* pharmacokinetics (Table 9.1).

Table 9.1   Molecule design strategies to maximize clinical performance (safety/efficacy) while minimizing potentially negative impacts on quality

Molecule modification	Description of modification	Molecule engineering/ methods	References
Amidation/ Deamidation	Damages the amide-containing side chains of the amino acids asparagine and glutamine	Replace all Asn and Gln residues of heterologous protein	(Stephenson and Clarke, 1989)
Acetylation	Reaction involving the replacement of the hydrogen atom of a hydroxyl group with an acetyl group	In living cells acetylation occurs as a co- and post-translational modification, e.g. histones, p53, and tubulins. Heterologous proteins exposed to host acetyltransferases	(Polevoda and Sherman, 2003)
Aggregation	Formation of chemical bond between 2 or more monomers. Disulfide bond from previously unpaired free thiols is a common mechanism for covalent aggregation. Tyrosine oxidation may also result in covalent aggregation	Protein engineered to block sulfhydryl groups, change hydrophobic clusters or glycine repeats by altering the amino acid sequence	(Cromwell *et al.*, 2006; Zhang and Czupryn, 2002)
Glycosylation	Addition of a glycosyl group by enzyme activity to either asparagine, hydroxylysine, serine, or threonine, resulting in a glycoprotein	Engineer desired glycosylation sites, if glycosylation is required for potency/ efficacy	(CMC, 2009)
Phosphorylation	Enzyme addition of a phosphate molecule to a polar R group of an amino acid residue, induce protein conformational change. Phosphorylation on serine is the most common, followed by threonine	Understanding the cell "state" at production requires knowing the phosphorylation state of its proteins, and impact on heterologous proteins	(Chang and Stewart, 1998)

(Continued)

Table 9.1   (*Continued*)

Molecule modification	Description of modification	Molecule engineering/ methods	References
Phospho-gluconoylation	Spontaneous modification by phosphogluconolactone, a reactive cell intermediate, mass shift of molecule by covalent bond with gluconolactone or its phosphorylated form	Co-expression of enzyme degrading phosphogluconolac-tone, so removal of intracellular accumulation	(Aon *et al.*, 2008)
S-S bonds	A covalent bond, usually derived by the coupling of two thiol groups. Disulphide bonds play an important role in the folding and stability of some proteins, usually those secreted to the extracellular medium	Removal of unpaired cysteine residues	(Ruoppolo *et al.*, 2000)
Degradation products/ proteolysis	Directed degradation (*digestion*) of proteins by proteases	Protease identification, engineering cleavage sites, or protease inhibitors, proteomic technology for identifying proteolytic substrates	(Barrett *et al.*, 2003)
Glycation	Nonenzymatic addition of sugars to different amino acids, a haphazard process that impairs the functioning of biomolecules	Design on host engineering to reduce intracellular pools of sugars, reactive intermediates during production process	(Vlassara, 2005)
Carbamylation	Protein amino groups reacts with isocyanic acid (cysteine, lysine, arginine). Ample evidence of protein carbamylation *in vivo* or by use of urea	Avoid the use of complex nitrogen sources during development/media optimization and protein characterization	(Barrett *et al.*, 2003)
Intentional modifications	By increasing half-life in systemic circulation so different administration route and/or frequency, higher bioavailability or better efficacy	Design of pegylated proteins, glycosylation, or fusion proteins	(Shayne, 2007)

Product attributes that are linked to safety and efficacy are considered "critical" and must be identified and controlled to ensure that the product meets the required quality profile. An important example is given by the A-Mab, a humanized IgG1 monoclonal antibody that was intended to treat the non-Hodgkin's Lymphoma (CMC, 2009). Aggregation, glycosylation, and host cell protein contamination were assessed as high critical quality attributes (cQAs) for their impact on efficacy and safety. Subsequently, acceptable ranges for the cQAs are used to establishing design boundaries of the production process (Fig. 9.1). At later stages, the cQA ranges become specification requirements to be met in order for therapeutic use.

Stages 2 and 3: Host strain selection and process development

Selection of the microorganism as the preliminary most convenient expression system for the selected molecular product is the next step after the molecule design. At this stage, process developers participate to determine protocols, e.g. (i) high cell density fed-batch cultures, (ii) a completely defined composition of the medium culture, (iii) secretion of the molecule to the extracellular medium rather than intracellular accumulation either as soluble product or as inclusion bodies, (iv) minimum level of undesired modifications of the molecular product throughout the culture, (v) minimal addition/feed, and (vi) few interventions during culture incubation, and if it is possible an automated control system with feedback loop controls for online monitoring of critical process parameters (cPPs, e.g. dissolved oxygen). cPPS are defined as the operational parameters that have a direct impact on the cQAs.

Clearly, recombinant DNA technology is involved throughout the design process aimed at bestowing the protein with beneficial therapeutic traits. The improvement of productivity in the host cell involves strong promoters, best signal peptides for correct protein sorting out, and decreasing PTMs. Improving robustness from the early stages of development until commercial late phase represents a very important goal (Fig. 9.1).

Robustness is a trait not genetically encoded thus inherent to the phenotype of the cell line. As such robustness depends upon culture conditions and the control system. The expertise relating to process robustness for consistent selected molecule production is the next phase in biopharmaceutical development. Informed evidence is necessary for a procedure that accounts for critical parameters and is sufficiently robust to give reliable and consistent product from batch to batch.

Stage 4: Risk assessment

A preliminary evaluation of the potential cPPs and cQAs, using a risk-based approach is performed to assess the safety and efficacy of the product. (FDA, 2006b).

In an ideal world, data from toxicological and early clinical studies should provide insights into cQAs related with drug safety and efficacy. Sometimes the information is scarce and decisions have to be made based on the best case scenario. Therefore, a risk-based decision is conducted before defining the experimental design for proving process robustness and consistency, a tool named Failure Mode and Effects Analysis (FMEA) is utilized for evaluating risk. FMEA results will assess the degree of risk for every process parameter in a systematic way, prioritizing the experiments necessary to understand the impact of key operational parameters affecting the overall process performance and product quality attributes (Seely and Haury, 2005).

FMEA gives a quantitative risk score to parameters that need to be taken into account for evaluating the process robustness. This estimation is based on three aspects: severity, occurrence, and detection. *Severity* is defined as the seriousness of potential failure effects and their impact on the final product quality and clinical application. The scale goes from 1 (not severe) to 10 (extremely severe). *Occurrence* measures the failure frequency. In this case, the scale runs from 1 (for infrequently, 1 failure per 1,000 production batches) to frequently (greater than 1 failure per 5 production batches). *Detection* scores the probability of timely failure recognition and its correction before use of the end product. The detection scale goes from 1 (almost certain) to 10 (no system is in place to prevent the cause of failure).

All three scores are multiplied to provide the risk priority number (RPN). The RPNs are then ranked to identify parameters that represent a risk high enough to be included in the process robustness study (Table 9.2). The RPN scores are typically represented in a Pareto chart. This plot allows a quick

Table 9.2 FMEA conducted to decide based on RPN scores which parameters of the bioprocess represent the highest risk for critical quality attributes of the molecular product that directly affect its safety/efficacy

Index number	Process step	Process parameter failure	CQAs affected	Potential effect(s) of failure	Severity	Potential cause(s) of failure	Occurrence	Current process controls	Detection	RPN
184	Production Bioreactor	Product variant	Yes	high % of intact product	10	Efficacy decrease	10	No measurement available	10	1000
43	Production Bioreactor	Foam control	None	Slow cell growth	4	Foam level detector fails	7	Foam detector probe, visual check by operator	7	196
1	Production Bioreactor	Temperature control	Yes	High protease activity	4	Control system fails or cool water supply fails	4	Control system in place, backup power supply, cooling unit maintenance	4	64
26	Production Bioreactor	Harvest Cooldown	Yes	Product loss by high protease activity	4	Temperature control system fails	4	Control system in place, backup power supply, cooling unit maintenance	4	64
23	Production Bioreactor	Fed-batch time	None	Product titer	4	Feed pump failure, wrong feed program loaded by operator	1	Pump maintenance, most recent version kept by automation	7	28

Note: The index number (first column) corresponds to the number assigned to the process step that could potentially fail. The process step is related with the phase of production in the bioreactor.

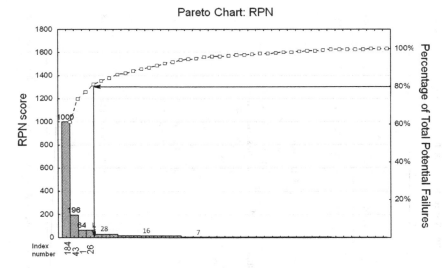

Figure 9.2.   Pareto chart showing the RPN scores of process parameters at the bioreactor production phase. The left and right *y*-axis scales correspond to the RPN score and the percentage of all potential failures, respectively. Process steps with RPNs above 50 are shown in Table 9.2, and whose index numbers are 184, 43, 1, 26 (1st column of Table 9.2). These parameters will potentially cause 80% of the bioprocess failures.

visualization of which parameters are more influential according to the general principle that 20% of parameters cause 80% of the failures. The operational process parameters with scores higher than the cut-off (RPN = 50) are then considered for the DoE study aimed at testing the process robustness (Fig. 9.2). As an example, the molecular product of interest (a protein) exhibits the highest RPN score (Table 9.2) due to the overall potential failure to obtain the proper cleavage of the leader sequence during its cellular processing. In this particular case, all scores are maximal, i.e. severity because of significant amounts of unprocessed protein found; occurrence because it diminishes the molecule's efficacy; and detection because of the absence of a proper assay to detect the product variant while it is being produced. The RPN score of the process step (Product variant, Table 9.2) with index number 184(= 1,000), ranks at the top of the Pareto chart (Fig. 9.2). Essentially the product variant is treated as a process parameter related to the host cell line, which needs to be engineered to preclude or minimize the formation of this PTM.

Overall, the results obtained by FMEA are essential to detect sensitive parameters of the bioprocess, the failure of which will have a considerable impact on cQAs of the molecular product, thus directly affecting product efficacy and patient safety.

Stage 5: Process characterization

The evaluation of process robustness is performed using the DoE approach. DoE is a statistical methodology seeking to establish any correlation between process parameters, process performance (e.g. productivity), and product quality. DoE is also aimed at minimizing the numbers of experiments (e.g. fermentation runs) necessary for collecting the maximal amount of information, according to the objectives of the study. The evaluation of the statistically significant correlations between parameters helps to understand how cell function within a certain environmental condition is affected by all possible combinations of the different process parameters (e.g. pH, temperature, dissolved oxygen (DO), antifoam concentration). Also DoE facilitates to understand how different combinations of environmental conditions result in the production of a certain level of the product of interest, and the potential occurrence of PTMs.

According to the FDA, the *design space* is defined as: "The multidimensional combination and interaction of input variables (e.g. material attributes) and process parameters that have been demonstrated to provide assurance of quality" (FDA, 2006a). The *design space* should inform about how well the process will meet the cQAs (Fig. 9.1). In practice it usually deals with multiple input and output variables. Performing the simultaneous optimization of a multiple response surface to different input variables and parameter combinations of the bioprocess to define the design space can be a challenging task (Stockdale and Cheng, 2009).

A multidimensional design space using a Bayesian predictive approach has been proposed (Peterson, 2004). Recently developed algorithms (implemented in Statistica software) can quantify the design space by computing the multidimensional space, where the probability of all cQAs meeting the specifications simultaneously is higher than some pre-specified criterion. However, this type of calculation ignores the modeling error (which is inevitable) and the correlation among the cQAs. Instead, the Bayesian approach assesses the quality assurance by computing the probability of all the cQAs that meet the expected specified values simultaneously. It also takes into consideration the modeling error and cQA correlation structure (Fu *et al.*, submitted). It is so far the most sophisticated and powerful approach to quantifying the design space. This approach consists of building a response surface model for each cQA, and calculating the probability that the cQAs will meet specification over a multidimensional grid of operational conditions (Peterson, 2004; Stockdale and Cheng, 2009). The design space is then mathematically defined as the region where the probability of meeting the specifications simultaneously is higher than some pre-specified value, e.g. 0.9 (90% probability to achieving every single specified cQA value).

Figure 9.3 shows an example of a stepwise approach to optimize a fermentation process and quantify the process design space. Three potential cPPs (temperature, pH, and DO) were identified through FMEA analysis. In this case, the potential cQA represented by an undesired product derivative,

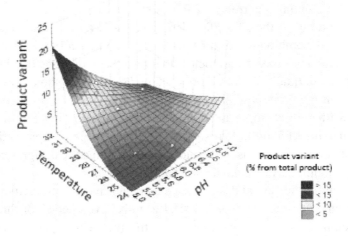

Figure 9.3. Three-dimensional surface response of the product variant (a critical quality attribute, cQA; see Table 9.2), based on experimental pH and temperature variation ranges planned in the DoE for process optimization. The white dots on the 3D surface represent the bioprocess response to all possible cPPs combinations.
Source: Adapted from Fu *et al.* (2011).

or product variant (Table 9.2), is plotted as the bioprocess response and is quantified as the relative amount (percentage) of the total product. A three-factor statistical DoE approach was used to optimize the cPPs in a wide range. All cPPs were simultaneously varied at the high or low level. Figure 9.3 shows all factorial points (i.e. the percentage of product variant obtained under all possible combinations between the three cPPs) as a 3D surface response versus pH and temperature. DO is not shown because it was found that it has no impact on the percentage of product variant after evaluating the interaction effects among the three cPPs. According to the results obtained, the model predicts that high temperatures and low pH induce an undesirably high percentage of product variant (Fig. 9.3). Combining either low or high values of pH and temperature minimize product variant, thus indicating which combinations to adopt for the bioprocess optimization.

Analysis of the individual responses from the cQAs in the bioprocess shows that each one is differently affected by the cPPs included in the model. The process optimization has to consider all the interactions, and find the right balance to achieve the specifications for all the cQAs. Since a particular combination of cPPs set points may be beneficial to one cQA while compromising others, it is important that the overall process cQAs responses is considered in order to ensure reliable and consistent performance.

Predictive quadratic multivariate models were built for each cQA using the entire DoE experimental space. These predictive models were then used to calculate the probability that the proposed cQAs measured from samples will simultaneously meet the respective process specifications. This approach offers benefits over traditional modeling approaches as it provides quantitative information to describe the modeling process and uncertainty (Fu *et al.*, submitted).

Figure 9.4 shows the probabilities calculated for all cQAs according to the Bayesian predictive approach. The areas of high and low probabilities are represented by the big and small squares, respectively, with their corresponding numerical values (Fig. 9.4). The design space of the bioprocess corresponds to the region where the probabilities are greater than or equal to some pre-specified value. The probability decreases as the cPPs combinations approach the extremes (e.g. low values of pH and temperature) (Fig. 9.4). The model shows that there is large design space to meet the cQAs measured from fermentation samples simultaneously.

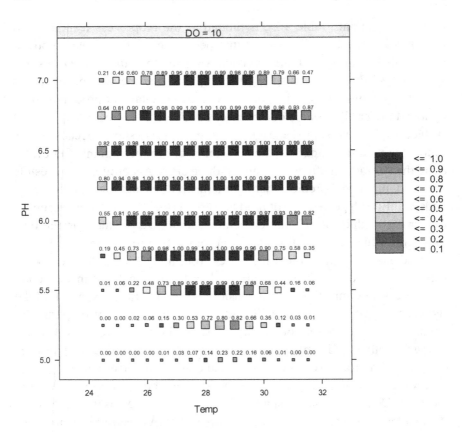

Figure 9.4.   Definition of the design space according to the Bayesian approach.
Source: Adapted from Fu *et al.* (2011).

In summary, first DoE seeks to define the range of potentially cPPs. A systematic evaluation is performed of all the values of a parameter within a certain range in interaction with other parameters. Finally, the impact of selected parameter combinations on defined responses of the production bioreactor with respect to cQAs is evaluated. The set point combination of the cPPs defines the *design space*. This multidimensional design leading to the experimental space specifies the limits within which a process will perform reliably and consistently to give a product quality that meets critical quality attributes.

Chapter 10

Systems Biology of Metabolic
and Cellular Engineering

Biological Complexity and Systems Bioengineering

The understanding of how a cell works is a fundamental step in our pursuit of creating fully artificial living systems that are able to respond to the environment, as well as to evolve and reproduce. One main question underlying this endeavor is how the mass–energy and information networks of the cell interact with each other to produce a certain phenotype or, more specifically, a cellular response leading to an altered phenotype (Fig. 1.10).

The mass–energy transformation networks, comprising metabolic and transport processes (e.g. metabolic pathways, electrochemical gradients), give rise to the metabolome and fluxome, which account for the whole set of metabolites and fluxes, respectively, characterizing the cell. The information-carrying networks include the genome, transcriptome, and proteome, which account for the whole set of genes, transcripts, and proteins, respectively, possessed by the cell.

Signaling networks modulate (activating or repressing) the interactions between information and mass–energy transducing networks, thus mediating between the genome–transcriptome–proteome and metabolome–fluxome (Fig. 1.10). As such, signaling networks pervade the whole cellular network playing the crucial role of influencing the unfolding of its function in space and time.

The output of signaling networks consists of concentration levels of intracellular metabolites (e.g. second messengers such as cAMP, AMP, phosphoinositides, reactive oxygen or nitrogen species), ions, proteins or small peptides, growth factors, and transcriptional factors. Through signaling networks, cells modulate, suppress, or activate gene expression (transcription, translation), whole metabolic pathways (e.g. respiration and gluconeogenesis in carbon catabolite repression) or certain enzymatic reactions within them. According to their level, intracellular messengers may act as allosteric effectors (positive or negative) on enzymes whose action reverberate on whole metabolic pathways that take part in crucial cellular mechanisms in response to environmental challenges (e.g. oxygen or substrate shortage) or cues (e.g. light, temperature). Thus, the complexity of the question on how the mass–energy and information networks of the cell interact with each other to produce a certain phenotype, stems from the dual role of, e.g. metabolites or transcriptional factors, since they are at the same time a result of the mass–energy or information networks while being active components of the signaling networks that will activate or repress the networks that produced them. The presence of these loops, in which the components are both cause and effect, together with their self-organizing properties, sustained by a continuous exchange of energy and matter with the environment, are the most consistent and defining traits of living systems (Aon *et al.*, 2007a; Capra, 1996; Luisi, 2006; Varela *et al.*, 1974). A successful approach to the engineering of cellular systems needs to account for this complexity. The emerging trend of approaches to engineering at whole system level represents a step forward but will not be enough unless we acknowledge the true complexity of the spatiotemporal organization of living cells.

Cellular Systems Biology

We are challenged to address the question of how a cell works during its main life stages, i.e. growth, division, differentiation, and death. In order to achieve this knowledge, a more fundamental understanding is needed about the dynamics and role of signaling networks as well as cytoplasmic organization on cell physiology. Considering cells as integrated multilayer mass–energy–information networks (Figs. 1.10 and 5.1) the following distinct and general properties characterize their function:

(1) Multiple temporal and spatial scales of interacting dynamic systems at all levels but with different relaxation times.

(2) The overall control and regulation of the cellular network is distributed, i.e. all nodes (e.g. metabolites, ions) are controlled by all edges, and all edges (e.g. enzymes, channels) control and are controlled, although to a different extent, by all other edges.
(3) Signaling networks connect all layers from genome to fluxome.

For this formidable task, we have at hand the following resources:

(1) the ability and the computational power to mathematically model very complicated systems, and analyze their control and regulation, as well as predict changes in qualitative behavior;
(2) an arsenal of theoretical tools (each with its own plethora of methods);
(3) high throughput technologies that allow simultaneous monitoring of an enormous number of variables;
(4) powerful imaging methods and online monitoring systems that provide the means of studying living systems at high spatial and temporal resolution for several variables simultaneously;
(5) the possibility of employing detailed enough bottom-up mathematical models that may help rationalize the use of key integrative variables, such as the membrane potential of cardiomyocytes or neurons, in top-down, conceptual, models with a few state variables.

In the following sections we describe through specific examples the three main traits characterizing complexity in cell function, as referred in the present section.

An *In Silico* Cell: Multiple Temporal and Spatial Scales of Interacting Dynamic Systems in the Heart

In order to visualize and analyze in a direct way the multiple temporal and spatial scales involved in a complex biological system, we will utilize a mathematical model of the cardiomyocyte that integrates both excitation-contraction (EC) coupling and mitochondrial energetics (ME) (Cortassa *et al.*, 2006). The ECME model represents a leap forward within a long chain of previous achievements in which many authors were involved (Cortassa *et al.*, 2003; Cortassa *et al.*, 2006; DiFrancesco and Noble, 1985; Hodgkin and Huxley, 1952; Jafri *et al.*, 1998; Luo and Rudy, 1991; Luo and Rudy,

1994; Rice *et al.*, 1999; Rice *et al.*, 2000; Winslow *et al.*, 1999). The dynamics of the ECME model is described by a system of 50 ordinary differential equations accounting for essential electro–mechanical–energetic functions of the cardiomyocyte.

This computational model has been validated through its ability to: (i) recapitulate the linearity between cardiac work and respiration in the heart (Cortassa *et al.*, 2006), (ii) reproduce the rapid time-dependent changes in mitochondrial NADH and Ca^{2+} in response to abrupt changes in workload (Cortassa *et al.*, 2003; Cortassa *et al.*, 2006), (iii) simulate experimentally observed oscillations in mitochondrial membrane potential, NADH, glutathione, ROS (Aon *et al.*, 2003; Cortassa *et al.*, 2004), and (iv) in action potential duration during mitochondrial oscillations (Aon *et al.*, 2003) with an ECME model further accounting both for mitochondrial ROS-induced ROS release (RIRR), and the link between the mitochondrial energy state and electrical excitability mediated by the sarcolemmal K_{ATP} current (ECME–RIRR model) (Zhou *et al.*, 2009).

Using the ECME model, we calculated the time course of the state of electrical, mechanical, and energetic processes during a single beat of a cardiomyocyte. This calculation reveals an essential feature of a complex system (e.g. heart cell): the multiple temporal and spatial scales involved; these appear as nested networks (networks within networks) of interacting dynamic systems (Fig. 5.1). Spatially, several compartments (sarcoplasmic reticulum, mitochondria, dyad, sarcolemma, bulk cytoplasm) are involved (Fig. 10.1). Temporally, the different processes involved exhibit distinct relaxation times: mitochondrial energy fuels the electrical and contractile machineries of the heart cell on a slower time scale (few seconds) compared with the electrical processes (milliseconds), which are followed by mechanical events associated with the force of contraction exerted during systole and subsequent relaxation.

The action potential (AP) constitutes a crucial integrative variable of the heart cell. A complex array of channels underlie the AP, and their dynamics results from sequential as well as simultaneous interplay of Na^+, Ca^{2+}, and K^+ currents through sarcolemmal channels. The AP, ionic currents, ATP-consuming pumps, and mitochondrial energetic variables, are shown in Fig. 10.2 during a single beat, when Ca^{2+}-induced Ca^{2+}-release (CICR) is turned on (see figure legend for further explanation). After voltage-gated Na^+-channels in the sarcolemma are activated, the inward Na^+-current

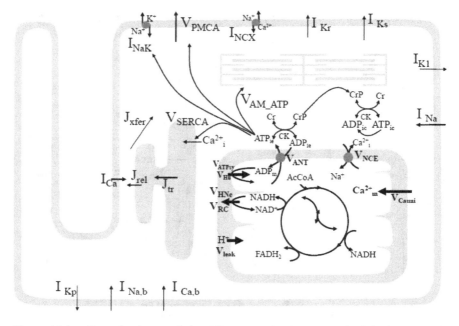

Figure 10.1. General scheme of the ECME model. The electrophysiological module includes the main ion transport processes involved in EC coupling, accounting for the transport of Ca^{2+}, Na^+, and K^+ across the sarcolemma, Ca^{2+} transport inside and across the sarcoplasmic reticulum (SR) membrane, and Ca^{2+} handling by mitochondria. Five different Ca^{2+} compartments are defined including the mitochondrial matrix, the dyadic subspace (extending between the membranes of the T-tubule and of the junctional SR), the junctional and network SR, and the myoplasmic compartments. Extracellular Ca^{2+} is considered an adjustable parameter. The mitochondrial module describes the production (F1, F0 ATPase) and transport (ANT) of ATP, Ca^{2+} transport, and Ca^{2+} activation of the TCA cycle dehydrogenases. The creatine kinase (CK) reaction occurs near the mitochondria (but does not include potential limitations imposed by the outer membrane) and the creatine phosphate (CrP) diffuses to the cytoplasmic compartment where there is another pool of CK that catalyzes the regeneration of cytoplasmic ATP_{ic} to fuel constitutive cytoplasmic ATPases (labeled "cyto"). Also the total pool of adenine (CA) and creatine (CC) metabolites are indicated by dark shading. The main ATP-consuming processes related to EC coupling are the myofibrillar ATPase (VAM) and SERCA (Jup), and in the sarcolemma, the Na^+, K^+ ATPase (INaK), and the Ca^{2+} ATPase (IpCa).
Source: Reproduced from Cortassa *et al.* (2006).

induces a rapid depolarization of the cell membrane. This facilitates voltage-dependent opening of L-type Ca^{2+}-channels, and the resulting Ca^{2+} influx triggers the opening of the ryanodine receptor (RyR2 subtype), eliciting a vast release of Ca^{2+} from the sarcoplasmic reticulum (Bers, 2001). Increased

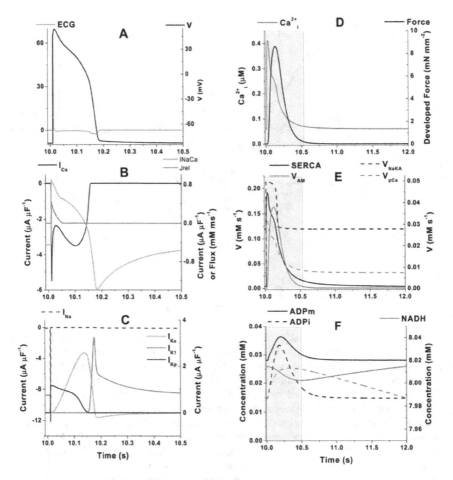

Figure 10.2.   Time course of state variables reflecting electrical, mechanical, and energetic processes during a myocyte beat as computed by the ECME model. Shown in the figures are the action potential (A, right axis), and underlying ionic currents and pump activities: Na^+ current, activated during the depolarization phase (C, left axis), L-type Ca^{2+} current (B, left) followed by Ca^{2+} release via ryanodine receptors, J_{rel} (B, right), SERCA activity (E, left), sarcolemmal Ca^{2+} pump (I_{pCa}) (E, right), Na^+ Ca^{2+} exchanger, I_{NaCa} (B right), and repolarizing K^+ currents I_{Ks}, I_{K1} and I_{KP} (C, right). Ca^{2+} transients in cytoplasm (D, left) trigger myofibrils contraction, developing force (D, right axis), reflected by ATP consumption via acto-myosyn ATPase activity, V_{AM} (E, left). Cytoplasm ADP (F, left), and in mitochondria, ADP_m (F left) reflect the cytoplasmic increase in ATP consumption during systole and uptake by mitochondria decreasing NADH levels (F, right). The shaded area in panels (D)–(F) corresponds to the same time range zoomed in panels (A)–(C). Parameters and initial conditions of the simulation as described elsewhere (Cortassa *et al.*, 2006). *Source*: Reproduced from Aon and Cortassa (2011) (in press).

binding of cytosolic Ca^{2+} to troponin C of the myofilaments induces contraction of the cardiomyocyte. Figure 10.2 also shows the temporal framework of mechanical and energetic processes according to which after the triggering of the AP, a Ca^{2+} transient and an ATP-fueled mechanical cycle follow; the latter increases to a maximum (systole, which coincides with the ejection of blood through aortic and pulmonary valves) and relaxes to a minimum (diastole, coinciding with passive refilling of the ventricles). During diastole, Ca^{2+} is either pumped back into the sarcoplasmic reticulum or transported out of the myocyte through the Na^+/Ca^{2+} exchanger.

Additionally, Fig. 10.2 highlights that overall cell function (e.g. AP in a cardiomyocyte) is spatially distributed, i.e. happens in several compartments. While the AP is mainly a sarcolemmal event, the Ca^{2+} transient takes place in the cytoplasm from Ca^{2+} released from the sarcoplasmic reticulum, the sequence of biochemical steps leading to cell contraction occur along the myofibrils in the cytoplasm, and the energy driving the whole process is delivered from the mitochondria. The latter involves exchange of ADP and Ca^{2+} signals between cytoplasmic and mitochondrial compartments, which prompt the energetic machinery to increase the energy supply (Cortassa *et al.*, 2006; Cortassa *et al.*, 2009a; Cortassa *et al.*, 2009b).

What do we learn from this almost instantaneously simultaneous orchestration of processes in different compartments in a specific, but crucial function for our lives? How can this knowledge affect our MCE strategies? First and foremost, we need an experimentally validated computational model in order to be able to predict the potential physiological impact of our MCE strategy. Second, the temporal and spatial interdependence between processes ensure that those occurring on a slow time scale (e.g. energetic) will determine the incidence of others happening faster (e.g. electric) due to their energetic dependence. The latter, in the case of heart, has shown that under acute energy crisis (e.g. ischemia/reperfusion) an escalation of failures initiated within the mitochondrial network may propagate to the whole organ provoking catastrophic arrhythmias (Akar *et al.*, 2005; Aon *et al.*, 2006b; Aon *et al.*, 2009; O'Rourke *et al.*, 2005). Third, the control and regulation of spatiotemporally nested networks is, in principle, exerted by all edges (processes: transport, biochemical reactions) on every node, and by each edge on every other edge and *vice versa*. This is also true for a whole system property (e.g. a flux), which both modulates and is modulated by the whole network of processes existing in a cell. We discuss this topic in the next section.

Signaling Networks: Connecting and Modulating the Mass–Energy–Information Networks

Signaling networks, distinct in composition, dynamics, and topology, connect layers of processes from the genome to the fluxome by modulating (activating or repressing) their function in space and time. The distinction between information (gene, mRNA, and protein circuits) and signaling networks is based on the following differences (Kiel *et al.*, 2010): (i) signaling systems operate rapidly (msec to min), whereas transcriptional responses are slow, ranging from minutes (prokaryotes) to hours (eukaryotes); (ii) subcellular localization plays an important role in signaling; and (iii) structure and folding of proteins in signaling is less predictable than in DNA. All three differences have consequences for engineering both kinds of network.

Conspicuous examples of network function exist in the literature whose activation has been described during varied environmental changes (Bhalla, 2003; Bhalla and Iyengar, 1999; Weng *et al.*, 1999). Other situations (some described below) are given by amplification of storage polysaccharides synthesis during light–dark transitions (Gomez-Casati *et al.*, 2003), and modulation of gene expression under oxidative stress mediated by reactive oxygen species (Giorgio *et al.*, 2007; Morel and Barouki, 1999). In all cases, different metabolites ratio such as AMP:ATP, 3PGA/Pi, and GSH:GSSG, activate different signaling responses, i.e. the AMP-activated protein kinase (AMPK) (Hardie *et al.*, 2003), the ultrasensitive switch of ADP glucose pyrophosphorylase, and the glycogen pathway (Gomez-Casati *et al.*, 2003), or transcription factors whose critical cysteine residues are oxidized resulting in decrease of gene promoter activity and subsequent gene expression (Haddad, 2004; Morel and Barouki, 1999).

Signaling networks are characterized by specific: (i) components and mechanisms; (ii) metabolic pathway targeted; (iii) conditions for signaling activation; and (iv) physiological response. Each one of these characteristics

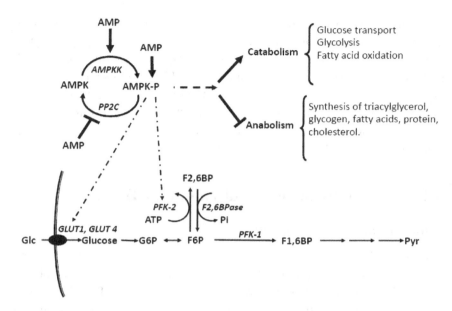

Figure 10.3.   AMP-activated protein kinase (AMPK) signaling network. Depicted are the main components of the network which include the kinase (AMPKK) and the phosphatise (PP2C) and the targets (glycolysis and other metabolic pathways). Rising AMP and falling ATP activate AMPKK which phosphorylates AMPK (AMPK-P). Dashed lines indicate activation through phosphorylation by AMPK-P of PFK-2 and glucose transport through increase of the levels of glucose transporters (*GLUT1* and *GLUT4*). The increase in PFK-2 activity augments the level of the allosteric regulator F2,6BP that in turn activates PFK-1, which is also activated by the decrease in ATP. Thus, activation of the glycolytic flux under ischemic conditions results from the concerted action of all these effects triggered by the AMPK pathway. Other downstream effects, positive and negative, are also indicated. Of note is that the metabolite AMP is produced by mass–energy networks and, at the same time, takes part of the AMPK signalling network, whose action will activate glycolysis thus down-modulating AMP while increasing ATP levels. This loop where AMP level is both cause and effect is one of the hallmarks of biological complexity (see text for further explanation).

can be identified in the AMPK signaling pathway, discussed here as a well-characterized example of transient behavior regulation (Fig. 10.3).

Time-dependent behavior regulation is of utmost importance for cells, particularly when challenged by sudden, transient changes in environmental conditions. The quick relaxation provided by molecular mechanisms involved in signaling networks is crucial for fast adaptation (Aon and Cortassa, 2002; Aon *et al.*, 2004b). The AMPK signaling network is a key example because the molecular components, and mechanisms involved (i.e. kinetic properties of AMPK toward the main effectors), physiological impact as well as conditions under which the signaling operates, are all well understood, and thus clearly identifiable (Fig. 10.3). Let us analyze one by one the components of this signaling network in the context of ischemia in the heart (Marsin *et al.*, 2000), and secondly as a response to a glucose pulse to chemostat cultures of *S. cerevisiae* under steady-state conditions (Vaseghi *et al.*, 1999; Vaseghi *et al.*, 2001).

Adaptation to Ischemic Conditions in the Heart

Components and mechanisms

Following the classical Monod/Wyman/Changeux model for allosteric enzymes, AMPK was proposed to be regulated by AMP and phosphorylation. The enzyme may exist in two conformations, R and T, each of which can also exist in phosphorylated and dephosphorylated forms, making four states in all (Fig. 10.4). Only the R state is a substrate for AMPKK, while

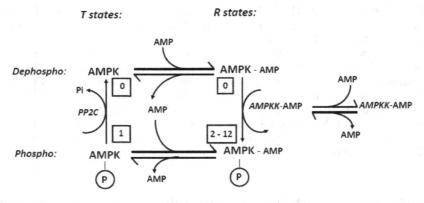

Figure 10.4. Model for the regulation of AMPK by AMP and phosphorylation. The enzyme may exist in two conformations, R and T. AMP binding promotes the T → R transitions by stabilizing the R states. Only the R state is a substrate for AMPKK, while only the T state is a substrate for the protein phosphatase. The figures in square boxes indicate the approximate kinase activity of that form, relative to that of the phosphorylated T state.
Source: Reproduced from *Bioessays* 23, Hardie DG, Hawley SA. AMP-activated protein kinase: The energy charge hypothesis revisited, 1112–1119. © (2001) with permission from John Wiley & Sons, Inc.

only the T state is a substrate for the protein phosphatase. Both, rising AMP or falling ATP activate the AMPK system, which therefore acts as an "energy charge sensor" (Hardie and Hawley, 2001).

Targets

In rats, hearts subjected to different periods of no-flow ischemia AMPK (which was almost completely inactive under aerobic conditions) was activated 10-fold after 10 min. The AMPK activation was followed, with a slight delay, by an increase in both 6-phosphofructo-2-kinase (PFK-2) activity and fructose 2,6-bisphosphate concentration (F2,6BP). AMPK activity returned toward basal levels after 30 min of ischemia (Marsin *et al.*, 2000). Thus, the activation of AMPK and PFK-2 by ischemia appears to be transient.

Many other downstream effects of the AMPK signaling network have been proposed and demonstrated; these effects would change the balance between anabolism and catabolism (Hardie and Hawley, 2001; Hardie *et al.*, 2003). The effect of nucleotides (AMP, ATP) is achieved not just by direct allosteric regulation of metabolic enzymes, but also by activation of complex signaling networks that regulate gene expression as well as the activity of pre-existing proteins (Hardie *et al.*, 2003).

Conditions for signaling activation

AMPK is activated by any stress treatment that interferes with ATP production (Hardie and Hawley, 2001). Such stresses include interruption of the blood supply (ischemia) (Marsin *et al.*, 2000), heat shock, glucose or oxygen deprivation (Hardie and Hawley, 2001, and refs. therein). Beyond abnormal, pathological events, exercise activates AMPK in skeletal muscle by increasing ATP consumption, both in animals and humans.

Physiological response

During ischemia in the heart, the AMPK-mediated activation of PFK-2 stimulates the flux through 6-phosphofructo-1-kinase (PFK-1) by increasing F2,6BP concentration leading to a stimulation of ATP production through glycolysis (Fig. 10.3). This phenomenon is superimposed on the well known and direct stimulation of PFK-1 by changes in adenine nucleotide concentrations (particularly a fall in ATP and a rise in AMP). Therefore, the increase in AMP:ATP ratio stimulates PFK-1 both through its direct allosteric stimulation and an indirect mechanism involving the phosphorylation of PFK-2 by AMPK. Besides PFK-1/PFK-2 activation, the stimulation of glucose transport and heart glycolysis under stress is an intrinsic part of a concerted mechanism mediated by AMPK (Hardie and Hawley, 2001; Marsin *et al.*, 2000).

Adaptation to a glucose pulse in S. cerevisiae

This example follows the studies of Vaseghi *et al.* (1999, 2001) on the dynamics of glycolysis and the pentose phosphate pathways after a glucose pulse. Addition of a glucose pulse to *S. cerevisiae* growing steadily in a chemostat culture activates glycolysis within a few seconds, causing a drastic decrease of the ATP pool and concurrently triggering the accumulation of F6P (Vaseghi *et al.*, 2001). Clearly, further downstream processing of F6P through glycolysis is transiently limited by PFK1 whose subsequent activation is mediated by the positive allosteric effector F2,6BP (Vaseghi *et al.*, 2001). The discharging of F6P occurs in parallel with the buildup of F2,6BP as a result of the higher activity of PFK2. Although Vaseghi *et al.* (2001) attributed the buildup of the glycolytic flux to the cyclic AMP-dependent protein kinase (PKA), it can also be explained by activation of the AMPK signaling pathway. The decrease in ATP levels and increase in cAMP and also likely that of AMP levels (Vaseghi *et al.*, 2001), will result in an increase of the AMP:ATP ratio that activates AMPK. This AMPK system has been shown to behave ultrasensitively with respect to the AMP:ATP ratio, i.e. a modest increase in AMP within a narrow concentration range is enough to take the AMPK kinase from 10% to 90% of its maximal activity (Hardie *et al.*, 1999).

Both examples illustrate several key features and concepts that need to be taken into account to engineer complex systems, and in particular signaling networks:

(1) A signalling network is defined by: (i) components and mechanisms; (ii) metabolic pathway targeted; (iii) conditions for signal activation; and (iv) physiological response.
(2) The time frame within which the signaling operates is transient.

(3) The downstream effects occur on both fast and slow time scales, the latter due to persistence of the effects even after the signaling is turned off.

(4) There are multiple targets.

Control and Regulation in Complex Networks

Both control and regulation are quantified by coefficients (see Chap. 5). *Flux control coefficients* are systemic properties of the network, whereas *elasticity coefficients* are local properties, i.e. they concern the behavior of individual enzymes namely with respect to their substrates, products, and effectors. Other measure is the *response coefficient*, which has major practical and conceptual implications for MCE purposes since it allows a correlation of high throughput proteomic information to mathematical models (see Chap. 5). The response coefficient quantifies a fractional change in flux in response to an effector P (e.g. Ca^{2+}, ADP, cAMP). Since the response of a pathway to an effector depends on (i) the sensitivity of the pathway to the activity of the enzyme targeted by the effector, and (ii) the strength of the effect of P on the enzyme (Fell, 1992; Fell, 1996), the response coefficient is a rigorous and practical measure of regulation (Cortassa *et al.*, 2009a; Cortassa *et al.*, 2009b). Since effectors can be both endogenous (e.g. metabolites) or exogenous (e.g. nutritional or growth factors) we may assess quantitatively the impact of an environmental challenge on a (patho)physiological cellular response. The response coefficient also underscores the importance of knowing the control exerted by other processes over the concentration of certain metabolites (e.g. AMP).

Using these tools we can demonstrate that the control is overall shared between processes in a way that transcends the specific compartment where a certain process is taking place. As an example, according to the ECME model when mitochondria are integrated to the ensemble of electromechanical processes, control of respiratory or ATP synthesis flux is not only localized within the mitochondrial compartment (e.g. ATP synthase, adenine nucleotide translocator (ANT)) but are also exerted by cytoplasmic and sarcolemmal membrane-linked processes (e.g. myofibrillar and Na/K ATPases). This is especially true under working conditions, when the interaction between cytoplasmic and mitochondrial processes is quantitatively more important. Thus the notion of "nested networks" (networks within networks) stresses that not all the control of a certain flux in an organelle resides within the organelle itself. In fact mitochondria, as the main energy suppliers of mechano-electrical processes in the heart, both modulate and are modulated by the overall network.

Heterarchical control in multilevel spatiotemporal networks of reactions is shown in Fig. 10.5, i.e. reciprocal control between edges, and edges over nodes. Placed at the inner mitochondrial membrane as one of the most abundant proteins, the ANT represents a key mechanistic link between cytoplasmic and mitochondrial compartments (Klingenberg, 2008). Comparing the first two panels, we can see that the ANT activity is mainly controlled by mitochondrial processes (top panel) but the ANT controls processes beyond the mitochondrial compartment (mid panel), e.g. SERCA, NaKATPase. Also evident is the difference between the degrees of control exerted under working or resting conditions. As an example, the F0F1 ATPase controls the ANT more under work than under rest (top panel), but the ANT controls the ATPase more under rest than under working conditions. This apparently counter intuitive result can be explained by the tighter commitment of ATP production to fulfilling the energy demand under work than when energy demand is low; in the latter case, flux of ATP synthesis can display larger variations (Cortassa *et al.*, 2009b).

Another counter-intuitive finding was that the myofibrillar ATPase controlled *negatively* the rate of ATP synthesis when one expects rather the contrary, i.e. that an ATP-consuming reaction controls positively ATP production — the higher the enzyme activity the higher the stimulus for production. The explanation of this result led to the discovery of *control by diffuse loops*. "Control by diffuse loops" is defined as the control that a process A exerts over process C or D without an apparent direct mechanistic link between them (Cortassa *et al.*, 2009b). In a diffuse control loop, there is at least one intermediary process between A and C. The uncovering of control by diffuse loops results from the existence of heterarchical control and regulation in complex networks (Fig. 5.1), throwing new light into the understanding of the secondary effects of pharmacological agents (Aon and Cortassa, 2011). The action of these agents on a complex network of reactions brings about changes in processes without direct mechanistic links between them. The

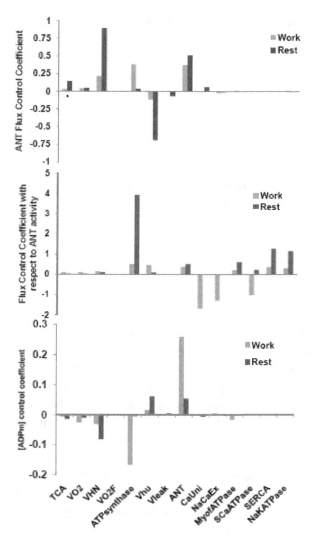

Figure 10.5. Heterarchical control in a complex network of metabolic and transport reactions. Top panel shows the control that the processes (edges) accounted for by the ECME model exert upon the ANT, whereas the mid panel shows the opposite, i.e. how the ANT controls other processes. Bottom panel depicts the control exerted by processes on the mitochondrial ADP concentration (node). These calculations were performed according to Cortassa *et al.* (2009b).

existence of diffuse loops is also evident in the control that the ANT has over processes outside the mitochondrion (Fig. 10.5, mid panel).

How can this knowledge affect our MCE strategies? First, when performing MCE, we need to look at the *overall* response of the system and not only the process that we are trying to improve. Second, we should consider the possibility of "unintended consequences" of our MCE strategy, in order to avoid them or, eventually, to use them to our advantage. Third, an incidental feature of our MCE may pass unnoticed under certain conditions but may turn out to be significant in another environment, similarly to the concept of a Darwinian preadaptation (Kauffman, 2008). A novel functionality may come into existence in this way

Systems Bioengineering

Whole system approaches are directed to identifying the overall control and regulation of networks distributed in different cellular compartments (Oliver, 2006). Related theoretical and practical frameworks are in place, such as the transdisciplinary approach to bioengineering (Chaps. 1 and 4), "physiological engineering" (Nielsen, 1997; Zhang *et al.*, 2009), and "industrial systems biology" (Otero and Nielsen, 2010). The frameworks mentioned share the common goal of understanding the function of important pathways in cells or microorganisms by using an integrated approach including several disciplines (microbial physiology, genetics, molecular biology, fermentation technology, biochemistry, computational modeling) and methodological and theoretical tools (MFA, MCA, kinetic and networks modeling). They are also aimed at generating fundamental knowledge as well as improving cellular and microbial performance of existing or novel functionalities introduced by genetic means (Lee *et al.*, 2007; Steen *et al.*, 2010). Another commonality among these approaches is that their main focus is the microorganism.

Physiological engineering stresses strain improvement to meet industrial demands such as robustness and adaptation to challenging conditions under industrial production such as temperature, acidity, high concentrations of products, and inhibitors. Specific examples are strain tolerance to ethanol and resistance to inhibitors in cellulosic ethanol production (Dien *et al.*,

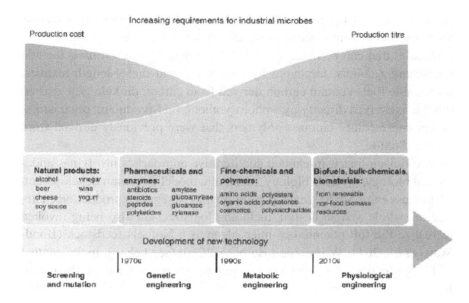

Figure 10.6.   Evolution in the aims of engineering microbes.
Source: Reprinted from *Trends Biotechnol* 27, Zhang Y, Zhu Y, Li Y. The importance of engineering physiological functionality into microbes, 664–672. © (2009) with permission from Elsevier.

2003; Zhang *et al.*, 2009), or additionally high temperature together with the ability to convert both hexose and pentose as in ethanol production from corn straw that has been hydrolyzed by sulfuric acid (Alper *et al.*, 2006; Shaw *et al.*, 2008).

Figure 10.6 stresses the evolution about the aims of engineering microbes and cells. From the traditional brewing and fermentation industries, such as the production of soy sauce, cheese, alcohol, antibiotics and other natural products (before 1970) to development in engineering microbes to meet desired requirements using genetic engineering, which enabled the introduction of novel microbial metabolic pathways. Introduced at the beginning of the 1990s, ME facilitated the broadening of substrates spectra, enabling improved titers and yields thus lowering costs, as well as the synthesis of new bioproducts, fine-chemicals, amino acids, and biodegradable polymers. Drastic increases in food demands, looming energy crises, and environmental pollution are driving the application of microbes for the production of biofuels, bulk-chemicals, and biomaterials from renewable nonfood biomass (Steen *et al.*, 2010), uncovering new candidate fuel molecules that can be made via ME. These next generation fuels include long-chain alcohols, terpenoid hydrocarbons, and diesel-length alkanes. Renewable fuels contain carbon derived from carbon dioxide. The carbon dioxide is derived directly by a photosynthetic fuel-producing organism(s) or via intermediary biomass polymers that were previously derived from carbon dioxide.

Multidisciplinary, integrated and iterative approaches involving fundamental biology, systems biology, metabolic modeling, strain development, bioprocess engineering, scaleup, biorefineries, integrated production chain, and the whole system design, including logistics, are also being invoked to realize the full potential of microalgae as a biofuels feedstock (Bond-Watts *et al.*, 2011; Wijffels and Barbosa, 2010) (see below). In the context of H_2 production by *C. reinhardtii*, we have shown in Chap. 4, the potential of a synergic approach between fundamental biology, high throughput metabolomics, MFA, strain development, and ME.

The Challenges Ahead

MCE started as an exciting new scientific endeavor dedicated to the purposeful modification of cells and organisms. Today, MCE is more than a new scientific enterprise since it has become crucially decisive for coping with the challenges ahead of mankind, which have increased dramatically in the past decades. The magnitude of the problems has augmented in parallel to our awareness of their existence and comprehension as a result of the ever increasing power of instantaneous and global communication across the planet. Anytime from almost anywhere we can have access to what is happening even in remote areas of the globe.

In this overall context, the pressure and the task upon the political and scientific communities appear both overwhelming and unavoidable. Due to the recognition that global temperature rise must be limited to less than 2°C to avoid dangerous climate change, it is now widely accepted that CO_2 emission reductions of 25%–40% by 2020 and up to 80% by 2050 are required to stay within this temperature range (Kruse and Hankamer, 2010).

The prediction that the global population will increase from 6.6 billion in 2008 to 9.2 billion by 2050 (United Nations, 2007) allows to certainly anticipate an increase in energy demand in parallel with global food demand, food security, and environmental pollution. This will be further exacerbated by the rapidly expanding economies of China and India (Stephens *et al.*, 2010b). In this context, the time for our fossil fuels-based economies is running out. Very recent calculations show that solely relying on fossil fuels to supply global energy demand then proven reserves (1P reserves: observed and marketable reserves of oil, gas, coal, and nuclear) are predicted to be completely depleted between 2069 and 2088 (Stephens *et al.*, 2010b). These calculations take into account the compounding effects of increasing the global population, economic growth, and a 1% increase in energy efficiency per year. Even accounting for the less certain 2P, 3P, and 1C reserves, and assuming the estimated reserves and a stable population of 9 billion people beyond 2050, a secure fossil fuel supply would only be possible until 2084–2112 at economic growth rates of 1.5%–3%. The authors conclude that "a major transition to renewable energy before 2030 should therefore be supported to ensure an orderly changeover from finite reserves and to address the more pressing constraints of climate change" (Stephens *et al.*, 2010b).

The Next Generation of Biofuels

The potentially explosive loop of increasing CO_2 emissions, climatic change, and energy crises, has turned the attention to the importance of developing CO_2-neutral fuel sources. This perception has been triggered by the detailed modeling of climate change effects, its global and national economic impacts, and the increasing competition for fossil fuel reserves.

The next generation of fuels will come from renewable resources sustained with energy from the sun (photons) or from biomass. Overall, a self-sustained cyclic interconversion of biomass from photosynthetic organisms (cyanobacteria, plants, algae) will be turned into biofuels (alcohols, aldehydes, esters, hydrocarbons, or hydrogen), by photoautotrophic organisms themselves or microbes which have been genetically engineered for increasing the overall efficiency (Wackett, 2010). In this self-sustainable cyclic energy loop, renewability as well cleanness (i.e. only H_2O as a product after combustion) are important. Moreover, given the role played in global warming by the accumulation of CO_2 in the atmosphere, the importance of developing a clean energy cycle has become obvious. In principle, solar-driven biohydrogen fulfils both criteria: is renewable and clean; the combustion of the evolved hydrogen, H_2, yields only H_2O, thereby completing a clean energy cycle.

A select group of photosynthetic organisms have evolved the ability to harness the huge solar energy resource to drive H_2 fuel production from H_2O. To date, the green microalgae *C. reinhardtii* remains the best photosynthetic eukaryotic H_2 producer with the cyanobacteria *Nostoc* and *Synechocystis* PCC 6803 also being promising H_2 production candidates (Kruse and Hankamer, 2010). Microalgae are efficient transducers of sunlight into chemical energy while able to grow in salt water year round under diverse conditions (Beer *et al.*, 2009). As a limitation, H_2 production by microalgae requires anaerobiosis, although very recent findings with the cyanobacterium *Cyanothece sp.* show that highly efficient H_2 generation is possible under natural aerobic environments (Jermy, 2011). Microalgal biofuel systems can produce fuels other than H_2 from single-celled microalgae (eukaryotes or cyanobacteria). Thus, they have the key potentiality of making a substantial contribution to global energy demand and CO_2 sequestration from the atmosphere, without increasing the pressure on arable land or important forest ecosystems (Stephens *et al.*, 2010b). Although the economic feasibility of microalgal biofuel systems is under debate (Larkum, 2010; Norsker *et al.*, 2010; Stephens *et al.*, 2010a, Stephens *et al.*, 2010b), the attractiveness of their advantages and sustainability as a fuel makes them worth of consideration.

Concerning H_2 production by microalgae (*C. reinhardtii*), Melis *et al.* (2000) described that sustainable H_2 production can happen under sulphur deprivation. This key discovery allowed a "two-stage photosynthesis and H_2 production" by temporally splitting O_2 and H_2 productions through bypassing the sensitivity of the reversible hydrogenase to O_2. Hydrogen production by *C. reinhardtii* constitutes one of the best studied examples because high throughput metabolomics in wild type and highly producing mutants in the presence or absence of sulphur deprivation, combined with ME, has been performed (see Chap. 4).

In order to appreciate the contribution that solar-powered biofuel production by microalgae can make to the global energy demand, we need a quantitative perspective. Solar energy, the largest source of renewable energy available, represents $\sim$5,700 times the global energy demand (Council, 2007; Stephens *et al.*, 2010b), which is 500 exajoules per year (EJ, or 5 10^{20} J yr^{-1}). About 80%–90% of the energy demand is supplied by fossil fuels (Larkum, 2010).

The surface of the Earth which is $\sim$510,000,000 km^2 with land representing 29.2% (3.9% arable; 25.3% nonarable), receive 2,200,000 EJ of sun energy per year. This energy amount accounts for the photosynthetically active radiation ($\sim$40% of the incident solar energy), thus ignoring the near infrared wavelengths that are not available to photosynthesis (Larkum, 2010; Stephens *et al.*, 2010b).

Given the earth's surface area and the value of annual solar radiation, 1,150,925 km^2 ($\sim$0.02% of the total surface area or 0.07% of the land surface) would be needed to satisfy the world's energy needs if 10% of the solar energy could be converted to useable energy (Larkum, 2010). The surface area needed would increase drastically ($\sim$4.3%) if one considers that most natural plant ecosystems have a solar energy-to-biomass conversion efficiency of $\sim$1% (Posten and Schaub, 2009; Stephens *et al.*, 2010b). Microalgae are already reported to have achieved light-to-biomass conversion efficiencies of 1%–4% in conventional open pond systems (Hase *et al.*, 2000). With 3% efficiency (and 350–650 Wm^{-2}, the irradiation in temperate and tropical regions), 0.38%–0.7% of the surface of the Earth would be needed using microalgae. Importantly, this surface comprises only $\sim$1.5%–2.7% of nonarable land for the production of the total global energy demand (Stephens *et al.*, 2010b).

From an economic perspective, Stephens *et al.* (2010a) presented a viable microalgal biofuel production base case scenario. The authors assumed the following: (i) production of microalgal biomass using 500 ha of microalgal production systems; (ii) the extraction of oil; (iii) co-production and extraction of a high value product (e.g. β-carotene at 0.1% of biomass, $600/kg); and (iv) the sale of the remaining biomass as feedstock (e.g. soy meal or fishmeal substitute) (Stephens *et al.*, 2010a). According to these authors, economically viable microalgal biofuel production systems have not yet been demonstrated because existing pilot and demonstration plants (at <5 ha) are well below the size threshold for economic viability (>200 ha), and also insufficient time has passed for the industry to evolve from recent capital injection (2006–2007) through to largescale commercial production (Stephens *et al.*, 2010a).

Another recent assessment compared three different microalgal production systems (open ponds, horizontal tubular, and flat panel photobioreactors) at currently operating conditions in commercial scale (Norsker *et al.*, 2010). Irradiation conditions, mixing, photosynthetic efficiency of systems, medium, and carbon dioxide costs were among the important biomass production cost factors. After optimization with respect to these factors, these authors concluded that the cost levels determined place microalgae as a promising feedstock for biodiesel and bulk chemicals (Norsker *et al.*, 2010).

Systems Bioengineering as Applied to Biomedicine

The regeneration of living tissues and organs is one of the most promising and demanding for the future of Biomedicine. Underlying several fields — cell transplantation, tissue engineering, stem cells, and nuclear transfer — regenerative medicine undertakes the endeavor for every type of tissue and organ in the human body (Atala, 2009).

Reprogramming of adult somatic cells to produce patient-specific pluripotent stem cells without the use of embryos are part of exciting ongoing studies (Takahashi and Yamanaka, 2006; Wong and Chiu, 2010). The groundbreaking discovery that somatic cells can be reprogrammed using specific transcription factor-encoding genes to form induced pluripotent stem cells (iPSCs) has allowed the derivation of pluripotent stem cells from easily attainable somatic cells (e.g. dermal fibroblasts) (Takahashi *et al.*, 2007; Yu *et al.*, 2009). These studies that have received worldwide confirmation demonstrate that terminally differentiated somatic cells can regain embryonic stem cell (ESC)-like pluripotency (Wong and Chiu, 2010). Recent results demonstrate that iPS cells can be obtained after reprogramming of human somatic cells without requiring genomic integration or the continued presence of exogenous reprogramming factors (Yu *et al.*, 2009). These cells are similar to human embryonic stem cells in proliferative and developmental potential.

There is great promise in the clinical application of these discoveries because they potentially offer a virtually inexhaustible source of autologous pluripotent stem cells, avoiding immune rejection problems and ethical issues associated with ESCs. However, safety issues linked to the long-term effects of the reprogramming process on cells and their progeny along with the risk of tumor formation after iPSC treatment remain to be solved before clinical applications (Wong and Chiu, 2010).

Bibliography

Abe K, Higuchi, T. (1998) Selective fermentation of xylose by a mutant of tetragenococcus halophila defective in phosphoenolpyruvate: Mannose phosphotransferase, phospho-fructokinase, and glucokinase. *Biosci Biotechnol Biochem* **62**, 2062–2064.

Abraham RH. (1987) Dynamics and self-organization. In: Yates EF, Garfinkel A, Walter DO, Yates GB (eds), *Self-Organizing Systems. The Emergence of Order*, pp. 599–613. Plenum Press, New York.

Abraham RH, Shaw CD. (1987) Dynamics: A visual introduction. In: Yates EF, Garfinkel A, Walter DO, Yates GB (eds), *Self-Organizing Systems. The Emergence of Order*, pp. 543–597. Plenum Press, New York.

Adams MD, Celniker SE, Holt RA, Evans CA, Gocayne JD, Amanatides PG, Scherer SE, Li PW, Hoskins RA, Galle RF, George RA, *et al.* (2000) The genome sequence of Drosophila melanogaster. *Science* **287**, 2185–2195.

Ainscow EK, Brand MD. (1999) Internal regulation of ATP turnover, glycolysis and oxida-tive phosphorylation in rat hepatocytes. *Eur J Biochem* **266**, 737–749.

Akar FG, Aon MA, Tomaselli GF, O'Rourke B. (2005) The mitochondrial origin of postis-chemic arrhythmias. *J Clin Invest* **115**, 3527–3535.

Akinterinwa O, Khankal R, Cirino PC. (2008) Metabolic engineering for bioproduction of sugar alcohols. *Curr Opin Biotechnol* **19**, 461–467.

Albert R. (2005) Scale-free networks in cell biology. *J Cell Sci* **118**, 4947–4957.

Albert R, Barabasi AL. (2002) Statistical mechanics of complex networks. *Reviews of Mod-ern Physics* **74**, 47–97.

Alberts B, Bray D, Lewis J, Raff M, Roberts K, Watson JD. (1989) *Molecular Biology of the Cell*. Garland Publishing, New York.

Alexander RD, Morris PC. (2006) A proteomic analysis of 14-3-3 binding proteins from developing barley grains. *Proteomics* **6**, 1886–1896.

Al Khateeb WM, Schroeder DF. (2009) Overexpression of Arabidopsis damaged DNA binding protein 1A (DDB1A) enhances UV tolerance. *Plant Mol Biol* **70**, 371–383.

Allen DK, Libourel IG, Shachar-Hill Y. (2009) Metabolic flux analysis in plants: Coping with complexity. *Plant Cell Environ* **32**, 1241–1257.

Alliance (2002) Overview of the alliance for cellular signaling. *Nature* **420**, 703–706.

Almaas E, Kovacs B, Vicsek T, Oltvai ZN, Barabasi AL. (2004) Global organization of metabolic fluxes in the bacterium *Escherichia coli*. *Nature* **427**, 839–843.

Alon U. (2003) Biological networks: The tinkerer as an engineer. *Science* **301**, 1866–1867.

Alper H, Moxley J, Nevoigt E, Fink GR, Stephanopoulos G. (2006) Engineering yeast transcription machinery for improved ethanol tolerance and production. *Science* **314**, 1565–1568.

Alsam S, Kim KS, Stins M, Rivas AO, Sissons J, Khan NA. (2003) Acanthamoeba interactions with human brain microvascular endothelial cells. *Microb Pathog* **35**, 235–241.

Altaras NE, Cameron DC. (1999) Metabolic engineering of a 1,2-propanediol pathway in *Escherichia coli*. *Appl Environ Microbiol* **65**, 1180–1185.

Andersen DC, Swartz J, Ryll T, Lin N, Snedecor B. (2001) Metabolic oscillations in an *E. coli* fermentation. *Biotechnol Bioeng* **75**, 212–218.

Anderson AJ, Dawes EA. (1990) Occurrence, metabolism, metabolic role, and industrial uses of bacterial polyhydroxyalkanoates. *Microbiol Rev* **54**, 450–472.

Andrews JF. (1968) A mathematical model for the continuous culture of microorganisms utilizing inhibitory substances. *Biotechnol Bioeng* **10**, 707–723.

Aon JC, Caimi RJ, Taylor AH, Lu Q, Oluboyede F, Dally J, Kessler MD, Kerrigan JJ, Lewis TS, Wysocki LA, Patel PS. (2008) Suppressing posttranslational gluconoylation of heterologous proteins by metabolic engineering of *Escherichia coli*. *Appl Environ Microbiol* **74**, 950–958.

Aon JC, Cortassa S. (2001) Involvement of nitrogen metabolism in the triggering of ethanol fermentation in aerobic chemostat cultures of *Saccharomyces cerevisiae*. *Metab Eng* **3**, 250–264.

Aon MA (2010) From isolated to networked: A paradigmatic shift in mitochondrial physiology. *Frontiers in Physiology* **1**, doi: 10.3389/fphys.2010.00020.

Aon MA, Cortassa S. (1991) Thermodynamic evaluation of energy metabolism in mixed substrate catabolism: Modeling studies of stationary and oscillatory states. *Biotechnol Bioeng* **37**, 197–204.

Aon MA, Cortassa S. (1993) An allometric interpretation of the spatio-temporal organization of molecular and cellular processes. *Mol Cell Biochem* **120**, 1–13.

Aon MA, Cortassa S. (1995) Cell growth and differentiation from the perspective of dynamical organization of cellular and subcellular processes. *Prog Biophys Mol Biol* **64**, 55–79.

Aon MA, Cortassa S. (1997) *Dynamic Biological Organization. Fundamentals as Applied to Cellular Systems*. Chapman & Hall, London.

Aon MA, Cortassa S. (1998) Catabolite repression mutants of *Saccharomyces cerevisiae* show altered fermentative metabolism as well as cell cycle behavior in glucose-limited chemostat cultures. *Biotechnol Bioeng* **59**, 203–213.

Aon MA, Cortassa S. (1999) Quantitation of the effects of disruption of catabolite (de)repression genes on the cell cycle behavior of *Saccharomyces cerevisiae*. *Curr Microbiol* **38**, 57–60.

Aon MA, Cortassa S. (2002) Coherent and robust modulation of a metabolic network by cytoskeletal organization and dynamics. *Biophys Chem* **97**, 213–231.

Aon MA, Cortassa S. (2005) Metabolic dynamics in cells viewed as multilayered, distributed, mass-energy-information networks. In: Jorde L, Little P, Dunn M, Subramaniam S (eds), *Encyclopedia of Genetics, Genomics, Proteomics and Bioinformatics*, Vol 3. Willey Interscience John Willey & Sons.

Aon MA, Cortassa S. (2009) Chaotic dynamics, noise and fractal space in biochemistry. In: Meyers R (ed), *Encyclopedia of Complexity and Systems Science*. Springer, New York.

Aon MA, Cortassa S. (2011) Mitochondrial network energetics in the heart. *Adv Exp Med Biol.* (in press)

Aon MA, Cortassa S, Akar FG, Brown DA, Zhou L, O'Rourke B. (2009) From mitochondrial dynamics to arrhythmias. *Int J Biochem Cell Biol* **41**, 1940–1948.

Aon MA, Cortassa S, Akar FG, O'Rourke, B. (2006b) Mitochondrial criticality: A new concept at the turning point of life or death. *Biochim Biophys Acta* **1762**, 232–240.

Aon MA, Cortassa S, Caceres A. (1996) Models of cytoplasmic structure and function. In: Cuthbertson R, Holcombe M, Paton R (eds), *Computation in Cellular and Molecular Biological Systems*, pp. 195–207. World Scientific, Singapore.

Aon MA, Cortassa S, Gomez Casati DF, Iglesias AA. (2000) Effects of stress on cellular infrastructure and metabolic organization in plant cells. *Int Rev Cytol* **194**, 239–273.

Aon MA, Cortassa S, Lemar KM, Hayes AJ, Lloyd D. (2007b) Single and cell population respiratory oscillations in yeast: A 2-photon scanning laser microscopy study. *FEBS Lett* **581**, 8–14.

Aon MA, Cortassa S, Lloyd D. (2000b) Chaotic dynamics and fractal space in biochemistry: Simplicity underlies complexity. *Cell Biol Int* **24**, 581–587.

Aon MA, Cortassa S, Lloyd D. (2011) Chaos in biochemistry and physiology. In: Meyers R (ed), *Encyclopedia of Molecular Cell Biology and Molecular Medicine. Systems Biology*. Wiley-VCH, Weinheim, Germany.

Aon MA, Cortassa S, Marban E, O'Rourke B. (2003) Synchronized whole cell oscillations in mitochondrial metabolism triggered by a local release of reactive oxygen species in cardiac myocytes. *J Biol Chem* **278**, 44735–44744.

Aon MA, Cortassa S, O'Rourke B. (2004a) Percolation and criticality in a mitochondrial network. *Proc Natl Acad Sci U S A* **101**, 4447–4452.

Aon MA, Cortassa S, O'Rourke B. (2006a) The fundamental organization of cardiac mitochondria as a network of coupled oscillators. *Biophys J* **91**, 4317–4327.

Aon MA, Cortassa S, O'Rourke B. (2007a) On the network properties of mitochondria. In: Saks V (ed), *Molecular System Bioenergetics: Energy for Life*, pp. 111–135. Wiley-VCH, Weinheim, Germany.

Aon MA, Cortassa S, O'Rourke B. (2008b) Mitochondrial oscillations in physiology and pathophysiology. *Adv Exp Med Biol* **641**, 98–117.

Aon MA, Cortassa S, O'Rourke B. (2008c) Is there a mitochondrial clock? In: Lloyd D, Rossi, EL (eds), *Ultradian Rhythms from Molecules to Mind. A New Vision of Life*, pp. 129–144. Springer-Verlag, New York.

Aon MA, Cortassa S, O'Rourke B. (2010) Redox-optimized ROS balance: A unifying hypothesis. *Biochim Biophys Acta* **1797**, 865–877.

Aon MA, Cortassa S, Westerhoff HV, Berden JA, Van Spronsen E, Van Dam K. (1991) Dynamic regulation of yeast glycolytic oscillations by mitochondrial functions. *J Cell Sci* **99**(Pt 2), 325–334.

Aon MA, Gomez-Casati DF, Iglesias AA, Cortassa, S. (2001) Ultrasensitivity in (supra)molecularly organized and crowded environments. *Cell Biol Int* **25**, 1091–1099.

Aon MA, Monaco ME, Cortassa S. (1995) Carbon and energetic uncoupling are associated with block of division at different stages of the cell cycle in several cdc mutants of *Saccharomyces cerevisiae*. *Exp Cell Res* **217**, 42–51.

Aon MA, O'Rourke B, Cortassa S. (2004b) The fractal architecture of cytoplasmic organization: Scaling, kinetics and emergence in metabolic networks. *Mol Cell Biochem* **256–257**, 169–184.

Aon MA, Roussel MR, Cortassa S, O'Rourke B, Murray DB, Beckmann M, Lloyd D. (2008a) The scale-free dynamics of eukaryotic cells. *PLoS ONE* **3**, e3624.

Apse MP, Blumwald E. (2002) Engineering salt tolerance in plants. *Curr Opin Biotechnol* **13**, 146–150.

Arabidopsis Genome Initiative T. (2000) Analysis of the genome sequence of the flowering plant *Arabidopsis thaliana*. *Nature* **408**, 796–815.

Araujo RP, Liotta LA. (2006) A control theoretic paradigm for cell signaling networks: A simple complexity for a sensitive robustness. *Curr Opin Chem Biol* **10**, 81–87.

Archibald JM. (2009) The puzzle of plastid evolution. *Curr Biol* **19**, R81–R88.

Arias DG, Piattoni CV, Guerrero SA, Iglesias AA. (2011) Biochemical mechanisms for the maintenance of oxidative stress under control in plants. In: Pessarakli M (ed), *Handbook of Plant and Crop Stress*, 3rd ed., pp 157–190. CRC Press, Taylor & Francis Group, Boca Raton, Florida.

Aristidou AA, San KY, Bennett GN. (1995) Metabolic engineering of *Escherichia coli* to enhance recombinant protein production through acetate reduction. *Biotechnol Prog* **11**, 475–478.

Atala A. (2009) Engineering organs. *Curr Opin Biotechnol* **20**, 575–592.

Atkinson B, Mavituna F. (1983) *Biochemical Engineering and Biotechnology Handbook*. Nature Press, Macmillan, London.

Atsumi S, Higashide W, Liao JC. (2009) Direct photosynthetic recycling of carbon dioxide to isobutyraldehyde. *Nat Biotechnol* **27**, 1177–1180.

Auberson LCM, Kanbier T., von Stockar U. (1993) Monitoring synchronized yeast cultures by calorimetry. *J Biotechnol* **29**, 205–215.

Bailey JE. (1991) Toward a science of metabolic engineering. *Science* **252**, 1668–1675.

Bailey JE, Axe DD, Doran PM, Galazzo JL, Reardon KF, Seressiotis A, Shanks JV. (1987) Redirection of cellular metabolism. Analysis and synthesis. *Ann N Y Acad Sci* **506**, 1–23.

Bailey JE, Ollis DF. (1977) *Biochemical Engineering Fundamentals*. McGraw-Hill, New York.

Bairoch A. (2000) The ENZYME database in 2000. *Nucleic Acids Res* **28**, 304–305.

Bak P. (1996) *How Nature Works: The Science of Self-Organized Criticality*. Copernicus, New York.

Balaban RS. (2002) Cardiac energy metabolism homeostasis: Role of cytosolic calcium. *J Mol Cell Cardiol* **34**, 1259–1271.

Ball P. (2004) *The Self-Made Tapestry. Pattern Formation in Nature*. Oxford University Press, Oxford, UK.

Ball S. (2002) The intricate pathway of starch biosynthesis and degradation in the monocellular alga *Chlamydomonas reinhardtii*. *Aust J Chem* **55**, 49–59.

Ballicora MA, Iglesias AA, Preiss J. (2003) ADP-glucose pyrophosphorylase, a regulatory enzyme for bacterial glycogen synthesis. *Microbiol Mol Biol Rev* **67**, 213–225, table of contents.

Ballicora MA, Iglesias AA, Preiss J. (2004) ADP-glucose pyrophosphorylase: A regulatory enzyme for plant starch synthesis. *Photosynth Res* **79**, 1–24.

Barabasi AL. (2003) *Linked*. Plume, Penguin Group, New York.

Barabasi AL, Oltvai ZN. (2004) Network biology: Understanding the cell's functional organization. *Nat Rev Genet* **5**, 101–113.

Barabasi AL. (2009) Scale-free networks: a decade and beyond. *Science* **325**: 412–3.

Barford JP, Pamment NB, Hall RJ. (1982) Lag phases and transients. In: Bazin MJ (ed), *Microbial Population Dynamics*. CRC Press, Boca Raton, Florida.

Barrett AJ, Rawlings ND, Woessner JF. (2003) *The Handbook of Proteolytic Enzymes*. Academic Press, London.

Barrow JD. (1999) *Impossibility. The Limits of Science and the Science of Limits*. Oxford University Press, New York.

Barthelmes J, Ebeling C, Chang A, Schomburg I, Schomburg D. (2007) BRENDA, AMENDA and FRENDA: The enzyme information system in 2007. *Nucleic Acids Res* **35**, D511–D514.

Basaran P, Rodriguez-Cerezo E. (2008) Plant molecular farming: Opportunities and challenges. *Crit Rev Biotechnol* **28**, 153–172.

Bascompte J. (2009) Disentangling the web of life. *Science* **325**: 416–9.

Bassingthwaighte JB, Liebovitch LS, West BJ. (1994) *Fractal Physiology*. Oxford University Press for the American Physiological Society, New York.

Bauchop T, Elsden SR. (1960) The growth of micro-organisms in relation to their energy supply. *J Gen Microbiol* **23**, 457–469.

Beck C, von Meyenburg HK. (1968) Enzyme pattern and aerobic growth of *Saccharomyces cerevisiae* under various degrees of glucose limitation. *J Bacteriol* **96**, 479–486.

Beckmann J, Lehr F, Finazzi G, Hankamer B, Posten C, Wobbe L, Kruse O. (2009) Improvement of light to biomass conversion by de-regulation of light-harvesting protein translation in *Chlamydomonas reinhardtii*. *J Biotechnol* **142**, 70–77.

Beer LL, Boyd ES, Peters JW, Posewitz MC. (2009) Engineering algae for biohydrogen and biofuel production. *Curr Opin Biotechnol* **20**, 264–271.

Bell SL, Bebbington C, Scott MF, Wardell JN, Spier RE, Bushell ME, Sanders PG. (1995) Genetic engineering of hybridoma glutamine metabolism. *Enzyme Microb Technol* **17**, 98–106.

Berger SI, Iyengar R. (2009) Network analyses in systems pharmacology. *Bioinformatics* **25**, 2466–2472.

Berridge MJ, Galione A. (1988) Cytosolic calcium oscillators. *Faseb J* **2**, 3074–3082.

Bers DM. (2001) *Excitation-Contraction Coupling and Cardiac Contractile Force*, 2nd ed. Vol. 237. Kluwer Academic Publishers, Dordrecht, Boston.

Bhalla US. (2003) Understanding complex signaling networks through models and metaphors. *Prog Biophys Mol Biol* **81**, 45–65.

Bhalla US, Iyengar R. (1999) Emergent properties of networks of biological signaling pathways. *Science* **283**, 381–387.

Birch RG. (1997) PLANT TRANSFORMATION: Problems and strategies for practical application. *Annu Rev Plant Physiol Plant Mol Biol* **48**, 297–326.

Boghigian BA, Shi H, Lee K, Pfeifer BA. (2010) Utilizing elementary mode analysis, pathway thermodynamics, and a genetic algorithm for metabolic flux determination and optimal metabolic network design. *BMC Syst Biol* **4**, 49.

Bonarius HP, Hatzimanikatis V, Meesters KP, de Gooijer CD, Schmid G, Tramper J. (1996) Metabolic flux analysis of hybridoma cells in different culture media using mass balances. *Biotechnol Bioeng* **50**, 299–318.

Bonaventura C, Myers J. (1969) Fluorescence and oxygen evolution from *Chlorella pyrenoidosa*. *Biochim Biophys Acta* **189**, 366–383.

Bond-Watts BB, Bellerose RJ, Chang MC. (2011) Enzyme mechanism as a kinetic control element for designing synthetic biofuel pathways. *Nat Chem Biol* **7**, 222–227.

Borland AM, Griffiths H, Hartwell J, Smith JA. (2009) Exploiting the potential of plants with crassulacean acid metabolism for bioenergy production on marginal lands. *J Exp Bot* **60**, 2879–96.

Bots M, Maughan S, Nieuwland J. (2006) RNAi Nobel ignores vital groundwork on plants. *Nature* **443**, 906.

Boyle NR, Morgan JA. (2009) Flux balance analysis of primary metabolism in *Chlamydomonas reinhardtii*. *BMC Syst Biol* **3**, 4.

Brady NR, Hamacher-Brady A, Westerhoff HV, Gottlieb RA. (2006) A wave of reactive oxygen species (ROS)-induced ROS release in a sea of excitable mitochondria. *Antioxid Redox Signal* **8**, 1651–1665.

Brandes R, Bers DM. (2002) Simultaneous measurements of mitochondrial NADH and Ca(2+) during increased work in intact rat heart trabeculae. *Biophys J* **83**, 587–604.

Branyik T, Vicente AA, Dostalek P, Teixeira JA. (2005) Continuous beer fermentation using immobilized yeast cell bioreactor systems. *Biotechnol Prog* **21**, 653–663.

Bray D. (1995) Protein molecules as computational elements in living cells. *Nature* **376**, 307–312.

Brazil GM, Kenefick L, Callanan M, Haro A, de Lorenzo V, Dowling DN, O'Gara F. (1995) Construction of a rhizosphere pseudomonad with potential to degrade polychlorinated biphenyls and detection of bph gene expression in the rhizosphere. *Appl Environ Microbiol* **61**, 1946–1952.

Brim H, McFarlan SC, Fredrickson JK, Minton KW, Zhai M, Wackett LP, Daly MJ. (2000) Engineering Deinococcus radiodurans for metal remediation in radioactive mixed waste environments. *Nat Biotechnol* **18**, 85–90.

Brodsky VY. (1998) On the nature of circahoralian (ultradian) intracellular rhythms: Similarity to fractals. *Biol Bull* **25**, 253-264

Brodsky VY. (2006) Direct cell–cell communication: A new approach derived from recent data on the nature and self-organisation of ultradian (circahoralian) intracellular rhythms. *Biol Rev Camb Philos Soc* **81**, 143–162.

Brown DA, Aon MA, Frasier CR, Sloan RC, Maloney AH, Anderson EJ, O'Rourke B. (2010) Cardiac arrhythmias induced by glutathione oxidation can be inhibited by preventing mitochondrial depolarization. *J Mol Cell Cardiol* **48**, 673–679.

Brown JH, West GB, Enquist BJ. (2000) *Scaling in Biology*. Oxford University Press, New York.Brown PH, Bellaloui N, Hu H, Dandekar A. (1999) Transgenically enhanced sorbitol synthesis facilitates phloem boron transport and increases tolerance of tobacco to boron deficiency. *Plant Physiol* **119**, 17–20.

Bruinenberg PM, van Dijken JP, Scheffers WA. (1983) A theoretical analysis of NADPH productin and consumption in yeast. *J Gen Microbiol* **129**, 953–964.

Burns N, Grimwade B, Ross-Macdonald PB, Choi EY, Finberg K, Roeder GS, Snyder M. (1994) Large-scale analysis of gene expression, protein localization, and gene disruption in *Saccharomyces cerevisiae*. *Genes Dev* **8**, 1087–1105.

Bustos DM, Bustamante CA, Iglesias AA. (2008) Involvement of non-phosphorylating glyceraldehyde-3-phosphate dehydrogenase in response to oxidative stress. *J Plant Physiol* **165**, 456–461.

Bustos DM, Iglesias AA. (2002) Non-phosphorylating glyceraldehyde-3-phosphate dehydrogenase is post-translationally phosphorylated in heterotrophic cells of wheat (Triticum aestivum). *FEBS Lett* **530**, 169–173.

Bustos DM, Iglesias AA. (2003) Phosphorylated non-phosphorylating glyceraldehyde-3-phosphate dehydrogenase from heterotrophic cells of wheat interacts with 14-3-3 proteins. *Plant Physiol* **133**, 2081–2088.

Caddick MX, Greenland AJ, Jepson I, Krause KP, Qu N, Riddell KV, Salter MG, Schuch W, Sonnewald U, Tomsett AB. (1998) An ethanol inducible gene switch for plants used to manipulate carbon metabolism. *Nat Biotechnol* **16**, 177–180.

Cameron DC, Altaras NE, Hoffman ML, Shaw AJ. (1998) Metabolic engineering of propane-diol pathways. *Biotechnol Prog* **14**, 116–125.

Cameron DC, Chaplen FWR. (1997) Developments in metabolic engineering. *Curr Opin Biotechnol* **8**, 175–180.

Cameron DC, Tong IT. (1993) Cellular and metabolic engineering. An overview. *Appl Biochem Biotechnol* **38**, 105–140.

Capell T, Christou P. (2004) Progress in plant metabolic engineering. *Curr Opin Biotechnol* **15**, 148–154.

Capra F. (1996) *The Web of Life*. Anchor books Doubleday, New York.

Carlsson AS. (2009) Plant oils as feedstock alternatives to petroleum — A short survey of potential oil crop platforms. *Biochimie* **91**, 665–670.

Carrari F, Urbanczyk-Wochniak E, Willmitzer L, Fernie AR. (2003) Engineering central metabolism in crop species: Learning the system. *Metab Eng* **5**, 191–200.

Carroll A, Somerville C. (2009) Cellulosic biofuels. *Annu Rev Plant Biol* **60**, 165–182.

Casati DF, Aon MA, Iglesias AA. (2000) Kinetic and structural analysis of the ultrasensitive behaviour of cyanobacterial ADP-glucose pyrophosphorylase. *Biochem J* **350**(Pt 1), 139–147.

Cascante M, Benito A, Zanuy M, Vizan P, Marin S, de Atauri P. (2010) Metabolic network adaptations in cancer as targets for novel therapies. *Biochem Soc Trans* **38**, 1302–1306.

Cascante M, Marin S. (2008) Metabolomics and fluxomics approaches. *Essays Biochem* **45**, 67–81.

Caspi R, Foerster H, Fulcher CA, Hopkinson R, Ingraham J, Kaipa P, Krummenacker M, Paley S, Pick J, Rhee SY, Tissier C, *et al.* (2006) MetaCyc: A multiorganism database of metabolic pathways and enzymes. *Nucleic Acids Res* **34**, D511–D516.

Celenza JL, Carlson M. (1986) A yeast gene that is essential for release from glucose repression encodes a protein kinase. *Science* **233**, 1175–1180.

Celinska E. (2010) Debottlenecking the 1,3-propanediol pathway by metabolic engineering. *Biotechnol Adv* **28**, 519–530.

Chakrabarti SK, Lutz KA, Lertwiriyawong B, Svab Z, Maliga P. (2006) Expression of the cry9Aa2 B.t. gene in tobacco chloroplasts confers resistance to potato tuber moth. *Transgenic Res* **15**, 481–488.

Chance B, Pye EK, Ghosh AD, Hess B. (1973) *Biological and Biochemical Oscillations*. Academic Press, New York.

Chance B, Pye K, Higgins J. (1967) Waveform generation by enzymatic oscillators. *IEEE Spectrum* **4**, 79–86.

Chandrashekaran MK. (2005) *Time in the Living World*. Universities Press (India), Hyderabad.

Chang C, Stewart RC. (1998) The two-component system. Regulation of diverse signaling pathways in prokaryotes and eukaryotes. *Plant Physiol* **117**, 723–731.

Chen H, Xiong L. (2009) Enhancement of vitamin B(6) levels in seeds through metabolic engineering. *Plant Biotechnol J* **7**, 673–681.

Chen S, Li H, Liu G. (2006) Progress of vitamin E metabolic engineering in plants. *Transgenic Res* **15**, 655–665.

Chisholm ST, Coaker G, Day B, Staskawicz BJ. (2006) Host-microbe interactions: Shaping the evolution of the plant immune response. *Cell* **124**, 803–814.

Christensen B, Nielsen J. (2000) Metabolic network analysis. A powerful tool in metabolic engineering. *Adv Biochem Eng Biotechnol* **66**, 209–231.

Christensen TS, Oliveira AP, Nielsen J. (2009) Reconstruction and logical modeling of glucose repression signaling pathways in *Saccharomyces cerevisiae*. *BMC Syst Biol* **3**,7.

Clegg JS. (1984) Properties and metabolism of the aqueous cytoplasm and its boundaries. *Am J Physiol* **246**, R133–R151.

Clegg JS. (1991) Metabolic organization and the ultrastructure of animal cells. *Biochem Soc Trans* **19**, 986–991.

Cleveland DW. (1988) Autoregulated instability of tubulin mRNAs: A novel eukaryotic regulatory mechanism. *Trends Biochem Sci* **13**, 339–343.

CMC, Biotech Working Group (2009) A-Mab: A case study in bioprocess development.

Colon AM, Sengupta N, Rhodes D, Dudareva N, Morgan J. (2010) A kinetic model describes metabolic response to perturbations and distribution of flux control in the benzenoid network of Petunia hybrida. *Plant J* **62**, 64–76.

Consortium, International Human Genome Sequencing (2004) Finishing the euchromatic sequence of the human genome. *Nature* **431**, 931–945.

Cooney CL, Wang HY, Wang DI. (1977) Computer-aided material balancing for prediction of fermentation parameters. *Biotechnol Bioeng* **19**, 55–67.

Cooper S. (1991) *Bacterial Growth and Division*. Academic Press, San Diego.

Cortassa S, Aon JC, Aon MA. (1995) Fluxes of carbon, phosphorylation, and redox intermediates during growth of *Saccharomyces cerevisiae* on different carbon sources. *Biotechnol Bioeng* **47**, 193–208.

Cortassa S, Aon JC, Aon MA, Spencer JF. (2000) Dynamics of metabolism and its interactions with gene expression during sporulation in *Saccharomyces cerevisiae*. *Adv Microb Physiol* **43**, 75–115.

Cortassa S, Aon MA. (1994a) Spatiotemporal regulation of glycolysis and oxidative-phosphorylation *in-vivo* in tumor and yeast-cells. *Cell Biol Int* **18**, 687–713.

Cortassa S, Aon MA. (1994b) Metabolic control analysis of glycolysis and branching to ethanol production in chemostat cultures of *Saccharomyces cerevisiae* under carbon, nitrogen, or phosphate limitations. *Enzyme Microb Technol* **16**, 761–770.

Cortassa S, Aon MA. (1996) Entrainment of enzymatic activity by the dynamics of cytoskeleton. In: Westerhoff HV, Snoep J (eds), *Biothermokinetics*, BioThermoKinetics Press, Amsterdam.

Cortassa S, Aon MA. (1997) Distributed control of the glycolytic flux in wild-type cells and catabolite repression mutants of *Saccharomyces cerevisiae* growing in carbon-limited chemostat cultures. *Enzyme Microb Technol* **21**, 596–602.

Cortassa S, Aon MA. (1998) The onset of fermentative metabolism in continuous cultures depends on the catabolite repression properties of *Saccharomyces cerevisiae*. *Enzyme Microb Technol* **22**, 705–712.

Cortassa S, Aon MA, Iglesias AA, Lloyd D. (2002) *An Introduction to Metabolic and Cellular Engineering*, 1st ed. World Scientific Publishers, Singapore.

Cortassa S, Aon MA, Marban E, Winslow RL, O'Rourke B. (2003) An integrated model of cardiac mitochondrial energy metabolism and calcium dynamics. *Biophys J* **84**, 2734–2755.

Cortassa S, Aon MA, O'Rourke B, Jacques R, Tseng HJ, Marban E, Winslow RL. (2006) A computational model integrating electrophysiology, contraction, and mitochondrial bioenergetics in the ventricular myocyte. *Biophys J* **91**, 1564–1589.

Cortassa S, Aon MA, Thomas D. (1990) Thermodynamic and kinetic studies in a stoichiometric model of energetic metabolism under starvation conditions. *FEMS Microbiol Lett* **66**, 249–256.

Cortassa S, Aon MA, Westerhoff HV. (1991) Linear nonequilibrium thermodynamics describes the dynamics of an autocatalytic system. *Biophys J* **60**, 794–803.

Cortassa S, Aon MA, Winslow RL, O'Rourke B. (2004) A mitochondrial oscillator dependent on reactive oxygen species. *Biophys J* **87**, 2060–2073.

Cortassa S, O'Rourke B, Winslow RL, Aon MA. (2009a) Control and regulation of integrated mitochondrial function in metabolic and transport networks. *Int J Mol Sci* **10**, 1500–1513.

Cortassa S, O'Rourke B, Winslow RL, Aon MA. (2009b) Control and regulation of mitochondrial energetics in an integrated model of cardiomyocyte function. *Biophys J* **96**, 2466–78.

Coschigano PW, Miller SM, Magasanik B. (1991) Physiological and genetic analysis of the carbon regulation of the NAD-dependent glutamate dehydrogenase of *Saccharomyces cerevisiae*. *Mol Cell Biol* **11**, 4455–4465.

Cotelle V, Meek SE, Provan F, Milne FC, Morrice N, MacKintosh C. (2000) 14-3-3s regulate global cleavage of their diverse binding partners in sugar-starved Arabidopsis cells. *EMBO J* **19**, 2869–2876.

Council, World Energy (2007) *Survey of Energy Resources, Executive Summary*. World Energy Council, London.

Covert MW, Schilling CH, Famili I, Edwards JS, Goryanin II, Selkov E, Palsson BO. (2001) Metabolic modeling of microbial strains in silico. *Trends Biochem Sci* **26**, 179–186.

Crank J. (1975) *The Mathematics of Diffusion*, 2nd ed. Clarendon Press, Oxford, UK.

Cromwell ME, Hilario E, Jacobson F. (2006) Protein aggregation and bioprocessing. *Aaps J* **8**, E572–E579.

Curds CR, Cockburn A. (1971) Continuous monoxenic culture of *Tetrahymena pyriformis*. *J Gen Microbiol* **66**, 95–108.

Dang VD, Bohn C, Bolotin-Fukuhara M., Daignan-Fornier B. (1996) The CCAAT box-binding factor stimulates ammonium assimilation in *Saccharomyces cerevisiae*, defining a new cross-pathway regulation between nitrogen and carbon metabolisms. *J Bacteriol* **178**, 1842–1849.

Daniell H, Datta R, Varma S, Gray S, Lee SB. (1998) Containment of herbicide resistance through genetic engineering of the chloroplast genome. *Nat Biotechnol* **16**, 345–348.

da Silva DA, Bicego MC. (2010) Polycyclic aromatic hydrocarbons and petroleum biomarkers in Sao Sebastiao Channel, Brazil: Assessment of petroleum contamination. *Mar Environ Res* **69**, 277–286.

Dave E, Guest JR, Attwood MM (1995) Metabolic engineering in *Escherichia coli*: Lowering the lipoyl domain content of the pyruvate dehydrogenase complex adversely affects the growth rate and yield. *Microbiol* **141**(Pt 8), 1839–1849.

Davey HM, Davey CL, Woodward AM, Edmonds AN, Lee AW, Kell DB. (1996) Oscillatory, stochastic and chaotic growth rate fluctuations in permittistatically controlled yeast cultures. *Biosystems* **39**, 43–61.

Davies J. (2009) *Animal Cell Culture*. Scion Publishing, Bloxham, UK.

Dawson PSS. (1985) Continuous cultivation of microorganisms. *CRC Crit Rev Biotechnol* **2**, 315–374.

de Groot MJ, Bundock P, Hooykaas PJ, Beijersbergen AG. (1998) Agrobacterium tumefaciens-mediated transformation of filamentous fungi. *Nat Biotechnol* **16**, 839–842.

de Jong-Gubbels P, Vanrolleghem P, Heijnen S, van Dijken JP, Pronk JT. (1995) Regulation of carbon metabolism in chemostat cultures of *Saccharomyces cerevisiae* grown on mixtures of glucose and ethanol. *Yeast* **11**, 407–418.

Dean AC, Moss DA. (1970) Interaction of nalidixic acid with Klebsiella (Aerobacter) aerogenes growing in continuous culture. *Chem Biol Interact* **2**, 281–296.

Deanda K, Zhang M, Eddy C, Picataggio S. (1996) Development of an arabinose-fermenting Zymomonas mobilis strain by metabolic pathway engineering. *Appl Environ Microbiol* **62**, 4465–4470.

Degn H, Harrison DE. (1969) Theory of oscillations of respiration rate in continuous culture of Klebsiella aerogenes. *J Theor Biol* **22**, 238–248.

DellaPenna D. (2001) Plant metabolic engineering. *Plant Physiol* **125**, 160–163.

Delmer DP, Haigler CH. (2002) The regulation of metabolic flux to cellulose, a major sink for carbon in plants. *Metab Eng* **4**, 22–28.

Dhooge A, Govaerts W, Kuznetsov YA, Meijer HGE, Sautois B. (2008) New features of the software MatCont for bifurcation analysis of dynamical systems. *Math Comput Model Dyn Syst* **14**, 147–175.

Dien BS, Cotta MA, Jeffries TW. (2003) Bacteria engineered for fuel ethanol production: Current status. *Appl Microbiol Biotechnol* **63**, 258–266.

DiFrancesco D, Noble D. (1985) A model of cardiac electrical activity incorporating ionic pumps and concentration changes. *Philos Trans R Soc Lond B Biol Sci* **307**, 353–398.

Doblin MS, Kurek I, Jacob-Wilk D, Delmer DP. (2002) Cellulose biosynthesis in plants: From genes to rosettes. *Plant Cell Physiol* **43**, 1407–1420.

Doebbe A, Keck M, La Russa M, Mussgnug JH, Hankamer B, Tekce E, Niehaus K, Kruse O. (2010) The interplay of proton, electron, and metabolite supply for photosynthetic H2 production in *Chlamydomonas reinhardtii. J Biol Chem* **285**, 30247–30260.

Doebbe A, Rupprecht J, Beckmann J, Mussgnug JH, Hallmann A, Hankamer B, Kruse, O. (2007) Functional integration of the HUP1 hexose symporter gene into the genome of *C. reinhardtii*: Impacts on biological H(2) production. *J Biotechnol* **131**, 27–33.

Domach MM, Leung SK, Cahn RE, Cocks GG, Shuler ML. (1984) Computer model for glucose-limited growth of a single cell of *Escherichia coli* B/r-A. *Biotechnol Bioeng* **26**, 203–216.

Domach MM, Shuler ML. (1984) A finite representation model for an asynchronous culture of *E. coli*. *Biotechnol Bioeng* **26**, 877–884.

Dong X, Quinn PJ, Wang X. (2011) Metabolic engineering of *Escherichia coli* and *Corynebacterium glutamicum* for the production of L-threonine. *Biotechnol Adv* **29**, 11–23.

Doran PM. (1995) *Bioprocess Engineering Principles*. Academic Press, London.

dos Santos MM, Gombert AK, Christensen B, Olsson L, Nielsen J. (2003) Identification of *in vivo* enzyme activities in the cometabolism of glucose and acetate by *Saccharomyces cerevisiae* by using 13C-labeled substrates. *Eukaryot Cell* **2**, 599–608.

Drake PM, Thangaraj H. (2010) Molecular farming, patents and access to medicines. *Expert Rev Vaccines* **9**, 811–819.

Duboc P, Marison I, von Stockar U. (1996) Physiology of *Saccharomyces cerevisiae* during cell cycle oscillations. *J Biotechnol* **51**, 57–72.

Dudareva N, Pichersky E. (2008) Metabolic engineering of plant volatiles. *Curr Opin Biotechnol* **19**, 181–189.

Duke SO. (2010) Glyphosate degradation in glyphosate-resistant and -susceptible crops and weeds (dagger). *J Agric Food Chem*

Dunbar J, Campbell SL, Banks DJ, Warren DR. (1998) Metabolic aspects of a commercial continuous fermentation system. *Aust. N. Z. Sect. Proc. 20th Conv. Brisbane* 151–158.

Duport C, Spagnoli R, Degryse E, Pompon D. (1998) Self-sufficient biosynthesis of pregnenolone and progesterone in engineered yeast. *Nat Biotechnol* **16**, 186–189.

Durrett TP, Benning C, Ohlrogge J. (2008) Plant triacylglycerols as feedstocks for the production of biofuels. *Plant J* **54**, 593–607.

Dykhuizen DE, Dean AM, Hartl DL. (1987) Metabolic flux and fitness. *Genetics* **115**, 25–31.

Edmunds LNJ. (1988) *Cellular and Molecular Bases of Biological Clocks. Models and Mechanisms for Circadian Timekeeping*. Springer, New York.

Edwards, JS, Ibarra RU, Palsson BO. (2001) In silico predictions of *Escherichia coli* metabolic capabilities are consistent with experimental data. *Nat Biotechnol* **19**, 125–130.

Edwards SW, Lloyd D. (1978) Properties of mitochondria isolated from cyanide-sensitive and cyanide-stimulated cultures of *Acanthamoeba castellanii*. *Biochem J* **174**, 203–211.

Edwards SW, Lloyd D. (1980) Oscillations in protein and RNA content during synchronous growth of *Acanthamoeba castellanii*. Evidence for periodic turnover of macromolecules during the cell cycle. *FEBS Lett* **109**, 21–26.

Eggeling L, Sahm H. (1999) Amino acid production: Principles of metabolic engineering. In: Lee SY, Papoutsakis ET (eds), *Metabolic Engineering*, pp. 153–176. Marcel Dekker, New York.

Ehrenberg M, Elf J, Hohmann S. (2009) Systems biology: Nobel symposium 146. *FEBS Lett* **583**, 3881.

Emes MJ, Bowcher CG, Debnam PM, Dennis DT, Hanke G, Rawsthorne S, Tetlow IJ. (1999) Implications of inter- and intracellular compartmentation for the movement of metabolites in plant cells. In: Bryant JA, Burnell MM, Kruger NT (eds), *Plant Carbohydrate Biochemistry*, pp. 231–244. BIOS Scientific Publishers, Oxford, UK.

Emmerling M, Bailey JE, Sauer U. (1999) Glucose catabolism of *Escherichia coli* strains with increased activity and altered regulation of key glycolytic enzymes. *Metab Eng* **1**, 117–127.

Enquist BJ, West GB, Brown JH. (2000) *Quarter-Power Allometric Scaling in Vascular Plants: Functional Basis and Ecological Consequences.* Oxford University Press, New York.

Entian KD, Barnett JA. (1992) Regulation of sugar utilization by *Saccharomyces cerevisiae. Trends Biochem Sci* **17**, 506–510.

Entian KD, Zimmermann FK. (1982) New genes involved in carbon catabolite repression and derepression in the yeast *Saccharomyces cerevisiae. J Bacteriol* **151**, 1123–1128.

Erdös P, Rényi A. (1960) On the evolution of random graphs. *Publ Math Inst Hung Acad Sci* **5**, 17–61.

Erickson LE, Minkevich IG, Eroshin VK. (2000) Application of mass and energy balance regularities in fermentation. *Biotechnol Bioeng* **67**, 748–774. Reprinted from *Biotechnol Bioeng* **XX**(10), 1595–1621 (1978).

Esener AA, Roels JA, Kossen NW. (1983) Theory and applications of unstructured growth models: Kinetic and energetic aspects. *Biotechnol Bioeng* **25**, 2803–2841.

Farinati S, DalCorso G, Bona E, Corbella M, Lampis S, Cecconi D, Polati R, Berta G, Vallini G, Furini A. (2009) Proteomic analysis of *Arabidopsis halleri* shoots in response to the heavy metals cadmium and zinc and rhizosphere microorganisms. *Proteomics* **9**, 4837–4850.

Farmer WR, Liao JC. (1996) Progress in metabolic engineering. *Curr Opin Biotechnol* **7**, 198–204.

Farmer WR, Liao JC. (2000) Improving lycopene production in *Escherichia coli* by engineering metabolic control. *Nat Biotechnol* **18**, 533–537.

Farmer WR, Liao JC. (2001) Precursor balancing for metabolic engineering of lycopene production in *Escherichia coli. Biotechnol Prog* **17**, 57–61.

FDA (2006a) *ICH Q8 FDA Guidance for Industry.* Rockville, MD.

FDA (2006b) *Q9 Quality Risk Management.* Rockville, MD.

FDA (2006c) *Q10 Quality Systems Approach to Pharmaceutical GMP Regulations.* Rockville, MD.

Feder J. (1988) *Fractals.* Plenum Press, New York.

Feder J, Tolbert WR. (1985) *Large Scale Mammalian Culture.* Academic Press, New York.

Feist AM, Herrgard MJ, Thiele I, Reed JL, Palsson BO. (2009) Reconstruction of biochemical networks in microorganisms. *Nat Rev Microbiol* **7**, 129–143.

Fell DA. (1992) Metabolic control analysis: A survey of its theoretical and experimental development. *Biochem J* **286**(Pt 2), 313–330.

Fell DA. (1996) *Understanding the Control of Metabolism.* Portland Press, London.

Fell DA. (1998) Increasing the flux in metabolic pathways: A metabolic control analysis perspective. *Biotechnol Bioeng* **58**, 121–124.

Ferrell JE Jr., Machleder EM. (1998) The biochemical basis of an all-or-none cell fate switch in *Xenopus oocytes. Science* **280**, 895–898.

Feuillet C, Leach JE, Rogers J, Schnable PS, Eversole K. (2011) Crop genome sequencing: Lessons and rationales. *Trends Plant Sci* **16**, 77–88.

Fiehn O, Kopka J, Dormann P, Altmann T, Trethewey RN, Willmitzer L. (2000) Metabolite profiling for plant functional genomics. *Nat Biotechnol* **18**, 1157–1161.

Field CB, Behrenfeld MJ, Randerson JT, Falkowski P. (1998) Primary production of the biosphere: Integrating terrestrial and oceanic components. *Science* **281**, 237–240.

Figueroa CM, Iglesias AA. (2010) Aldose-6-phosphate reductase from apple leaves: Importance of the quaternary structure for enzyme activity. *Biochimie* **92**, 81–88.

Figueroa CM, Iglesias AA, Podestá FE. (2011) Carbon metabolism and plant stress. In: Pessarakli M (ed), *Handbook of Plant and Crop Stress*, 3rd. ed., pp. 447–463. CRC Press, Taylor & Francis Group, Boca Raton, Florida.

Flores-Samaniego B, Olivera H, Gonzalez A. (1993) Glutamine synthesis is a regulatory signal controlling glucose catabolism in *Saccharomyces cerevisiae*. *J Bacteriol* **175**, 7705–7706.

Flores N, Xiao J, Berry A, Bolivar F, Valle F. (1996) Pathway engineering for the production of aromatic compounds in *Escherichia coli*. *Nat Biotechnol* **14**, 620–623.

Forsburg SL, Nurse P. (1991) Cell cycle regulation in the yeasts *Saccharomyces cerevisiae* and *Schizosaccharomyces pombe*. *Annu Rev Cell Biol* **7**, 227–256.

Fraenkel DG. (1992) Genetics and intermediary metabolism. *Annu Rev Genet* **26**, 159–177.

French CE, Rosser SJ, Davies GJ, Nicklin S, Bruce NC. (1999) Biodegradation of explosives by transgenic plants expressing pentaerythritol tetranitrate reductase. *Nat Biotechnol* **17**, 491–494.

Fridman E, Carrari F, Liu YS, Fernie AR, Zamir D. (2004) Zooming in on a quantitative trait for tomato yield using interspecific introgressions. *Science* **305**, 1786–1789.

Fryer JL, Lannan CN. (1994) Three decades of fish cell culture; a current listing of cell lines derived from fishes. *Methods Cell Sci* **16**, 87–94.

Fu Z, Leighton JM, Aili CA, Appelbaum E, Patel PS, Aon JC. (submitted) Optimization of a *Saccharomyces cerevisae* fermentation process for production of a therapeutic recombinant protein using a multivariate Bayesian approach.

Fuhrer T, Sauer U. (2009) Different biochemical mechanisms ensure network-wide balancing of reducing equivalents in microbial metabolism. *J Bacteriol* **191**, 2112–2121.

Fukushima A, Kusano M, Redestig H, Arita M, Saito K. (2009) Integrated omics approaches in plant systems biology. *Curr Opin Chem Biol* **13**, 532–538.

Fussenegger M, Sburlati A, Bailey JE. (1999) Metabolic engineering of mammalian cells. In: Lee SY, Papoutsakis ET (eds), *Metabolic Engineering*, pp. 353–389. Marcel Dekker, New York.

Gancedo JM. (1992) Carbon catabolite repression in yeast. *Eur J Biochem* **206**, 297–313.

Garcia-Ochoa F, Gomez E. (2009) Bioreactor scale-up and oxygen transfer rate in microbial processes: An overview. *Biotechnol Adv* **27**, 153–176.

Gargalovic PS, Imura M, Zhang B, Gharavi NM, Clark MJ, Pagnon J, Yang WP, He A, Truong A, Patel S, Nelson SF., *et al.* (2006) Identification of inflammatory gene modules based on variations of human endothelial cell responses to oxidized lipids. *Proc Natl Acad Sci U S A* **103**, 12741–12746.

Gasteiger E, Gattiker A, Hoogland C, Ivanyi I, Appel RD, Bairoch, A. (2003) ExPASy: The proteomics server for in-depth protein knowledge and analysis. *Nucleic Acids Res* **31**, 3784–3788.

Genta HD, Monaco ME, Aon MA. (1995) Decreased mitochondrial biogenesis in temperature-sensitive cell division cycle mutants of *Saccharomyces cerevisiae*. *Curr Microbiol* **31**, 327–331.

Gerasimenko OV, Gerasimenko JV, Belan PV, Petersen OH. (1996) Inositol trisphosphate and cyclic ADP-ribose-mediated release of Ca2+ from single isolated pancreatic zymogen granules. *Cell* **84**, 473–480.

Ghirardi ML, Posewitz MC, Maness PC, Dubini A, Yu J, Seibert M. (2007) Hydrogenases and hydrogen photoproduction in oxygenic photosynthetic organisms. *Annu Rev Plant Biol* **58**, 71–91.

Ghirardi ML, Zhang L, Lee JW, Flynn T, Seibert M, Greenbaum E, Melis A. (2000) Microalgae: A green source of renewable H(2). *Trends Biotechnol* **18**, 506–511.

Ghosh A, Chance B. (1964) Oscillations of glycolytic intermediates in yeast cells. *Biochem Biophys Res Commun* **16**, 174–181.

Gibson DG, Glass JI, Lartigue C, Noskov VN, Chuang RY, Algire MA, Benders GA, Montague MG, Ma L, Moodie MM, Merryman C, *et al.* (2010) Creation of a bacterial cell controlled by a chemically synthesized genome. *Science* **329**, 52–56.

Ginger ML, Fritz-Laylin LK, Fulton C, Cande WZ, Dawson SC. (2010) Intermediary metabolism in protists: A sequence-based view of facultative anaerobic metabolism in evolutionarily diverse eukaryotes. *Protist* **161**, 642–671.

Giordano PC, Martinez HD, Iglesias AA, Beccaria AJ, Goicoechea HC. (2010) Application of response surface methodology and artificial neural networks for optimization of recombinant *Oryza sativa* non-symbiotic hemoglobin 1 production by *Escherichia coli* in medium containing byproduct glycerol. *Bioresour Technol* **101**, 7537–7544.

Giorgio M, Trinei M, Migliaccio E, Pelicci PG. (2007) Hydrogen peroxide: A metabolic by-product or a common mediator of ageing signals? *Nat Rev Mol Cell Biol* **8**, 722–728.

Giroux MJ, Shaw J, Barry G, Cobb BG, Greene T, Okita T, Hannah LC. (1996) A single mutation that increases maize seed weight. *Proc Natl Acad Sci U S A* **93**, 5824–5829.

Gisiger T. (2001) Scale invariance in biology: Coincidence or footprint of a universal mechanism? *Biol Rev Camb Philos Soc* **76**, 161–209.

Giuseppin ML, Almkerk JW, Heistek JC, Verrips CT. (1993) Comparative study on the production of guar alpha-galactosidase by *Saccharomyces cerevisiae* SU50B and *Hansenula polymorpha* 8/2 in continuous cultures. *Appl Environ Microbiol* **59**, 52–59.

Giuseppin ML, van Riel NA. (2000) Metabolic modeling of *Saccharomyces cerevisiae* using the optimal control of homeostasis: A cybernetic model definition. *Metab Eng* **2**, 14–33.

Gleick J. (1988) *Chaos: Making a New Science.* Penguin Books, New York.

Godge MR, Purkayastha A, Dasgupta I, Kumar PP. (2008) Virus-induced gene silencing for functional analysis of selected genes. *Plant Cell Rep* **27**, 209–219.

Goel A, Lee J, Domach MM, Ataai MM. (1995) Suppressed acid formation by cofeeding of glucose and citrate in Bacillus cultures: Emergence of pyruvate kinase as a potential metabolic engineering site. *Biotechnol Prog* **11**, 380–385.

Goldberg B. (2004) Seymour Hutner's soup kitchen for protozoa. *J Eukaryot Microbiol* **51**, 590–593.

Goldbeter A. (1996) *Biochemical Oscillations and Cellular Rhythms.* Cambridge University Press, Cambridge.

Goldbeter A, Koshland DE Jr. (1984) Ultrasensitivity in biochemical systems controlled by covalent modification. Interplay between zero-order and multistep effects. *J Biol Chem* **259**, 14441–14447.

Goldemberg J. (2007) Ethanol for a sustainable energy future. *Science* **315**, 808–810.

Gomez-Casati D, Aon MA, Iglesias AA. (1999) Ultrasensitive glycogen synthesis in Cyanobacteria. *FEBS Lett* **446**, 117–121.

Gomez-Casati DF, Cortassa S, Aon MA, Iglesias AA. (2003) Ultrasensitive behavior in the synthesis of storage polysaccharides in cyanobacteria. *Planta* **216**, 969–975.

Gomez C, Aon MA, Iglesias AA. (1999) Ultrasensitive glycogen synthesis in Cyanobacteria. *FEBS Lett* **446**, 117–121.

Gomez Casati DF, Aon MA, Cortassa S, Iglesias AA. (2001) Measurement of the glycogen synthetic pathway in permeabilized cells of cyanobacteria. *FEMS Microbiol Lett* **194**, 7–11.

Gomez Casati DF, Aon MA, Iglesias AA. (2000) Kinetic and structural analysis of the ultrasensitive behaviour of cyanobacterial ADP-glucose pyrophosphorylase. *Biochem J* **350**(Pt 1), 139–147.

Gommers PJ, van Schie BJ, van Dijken JP, Kuenen JG. (1988) Biochemical limits to microbial growth yields: An analysis of mixed substrate utilization. *Biotechnol Bioeng* **32**, 86–94.

Goodenough UW. (1992) Green yeast. *Cell* **70**, 533–538.

Goosen MF. (1991) Large-scale insect cell culture: Methods, applications and products. *Curr Opin Biotechnol* **2**, 365–369.

Gosset G, Yong-Xiao J, Berry A. (1996) A direct comparison of approaches for increasing carbon flow to aromatic biosynthesis in *Escherichia coli*. *J Ind Microbiol* **17**, 47–52.

Gould SB, Waller RF, McFadden GI. (2008) Plastid evolution. *Annu Rev Plant Biol* **59**, 491–517.

Gray VL, Muller CT, Watkins ID, Lloyd D. (2008) Peptones from diverse sources: Pivotal determinants of bacterial growth dynamics. *J Appl Microbiol* **104**, 554–565.

Gray VL, O'Reilly M, Muller CT, Watkins ID, Lloyd D. (2006) Low tyrosine content of growth media yields aflagellate *Salmonella enterica* serovar *Typhimurium*. *Microbiol* **152**, 23–28.

Gubb D. (1993) Genes controlling cellular polarity in *Drosophila*. *Dev Suppl*, 269–277.

Gundersen GG, Cook TA. (1999) Microtubules and signal transduction. *Curr Opin Cell Biol* **11**, 81–94.

Gunn RB, Curran PF. (1971) Membrane potentials and ion permeability in a cation exchange membrane. *Biophys J* **11**, 559–571.

Gurramkonda C, Polez S, Skoko N, Adnan A, Gabel T, Chugh D, Swaminathan S, Khanna N, Tisminetzky S, Rinas U. (2010) Application of simple fed-batch technique to high-level secretory production of insulin precursor using *Pichia pastoris* with subsequent purification and conversion to human insulin. *Microb Cell Fact* **9**, 31.

Gutteridge A, Pir P, Castrillo JI, Charles PD, Lilley KS, Oliver SG. (2010) Nutrient control of eukaryote cell growth: A systems biology study in yeast. *BMC Biol* **8**, 68.

Haddad JJ. (2004) Oxygen sensing and oxidant/redox-related pathways. *Biochem Biophys Res Commun* **316**, 969–977.

Haken H. (1983) *Synergetics. An Introduction*. Springer-Verlag, Berlin, Heidelberg.

Hannah LC, James M. (2008) The complexities of starch biosynthesis in cereal endosperms. *Curr Opin Biotechnol* **19**, 160–165.

Hardie DG, Hawley SA. (2001) AMP-activated protein kinase: The energy charge hypothesis revisited. *Bioessays* **23**, 1112–1119.

Hardie DG, Salt IP, Hawley SA, Davies SP. (1999) AMP-activated protein kinase: An ultrasensitive system for monitoring cellular energy charge. *Biochem J* **338**(Pt 3), 717–722.

Hardie DG, Scott JW, Pan DA, Hudson ER. (2003) Management of cellular energy by the AMP-activated protein kinase system. *FEBS Lett* **546**, 113–120.

Harold FM. (1990) To shape a cell: An inquiry into the causes of morphogenesis of microorganisms. *Microbiol Rev* **54**, 381–431.

Harrison DE. (1970) Undamped oscillations of pyridine nucleotide and oxygen tension in chemostat cultures of *Klebsiella aerogenes*. *J Cell Biol* **45**, 514–521.

Harrison DE, Loveless JE. (1971) Transient responses of facultatively anaerobic bacteria growing in chemostat culture to a change from anaerobic to aerobic conditions. *J Gen Microbiol* **68**, 45–52.

Harrison DE, Pirt SJ. (1967) The influence of dissolved oxygen concentration on the respiration and glucose metabolism of *Klebsiella aerogenes* during growth. *J Gen Microbiol* **46**, 193–211.

Harrison DEF, MacLennan DG, Pirt SJ. (1969) *Responses of Bacteria to Dissolved Oxygen Tension. Fermentation Advances*. Academic Press, New York.

Harrison DEF, Topiwala HH. (1974) Transient and oscillatory states of continuous culture. In: Ghose TK, Fiechter A, Blakebrough N (eds), *Advances in Biochemical Engineering*, Vol 3. Springer-Verlag, Berlin.

Hartl DL, Jones EW. (2009) *Genetics: Analysis of Genes and Genomes*. Jones and Bartlett Publishers, Sudbury, MA.

Hartwell LH. (1991) Twenty-five years of cell cycle genetics. *Genetics* **129**, 975–980.

Hase R, Oikawa H, Sasao C, Morita M, Watanabe Y. (2000) Photosynthetic production of microalgal biomass in a raceway system under greenhouse conditions in Sendai city. *J Biosci Bioeng* **89**, 157–163.

Hatzimanikatis V, Bailey JE. (1997) Effects of spatiotemporal variations on metabolic control: Approximate analysis using (log)linear kinetic models. *Biotechnol Bioeng* **54**, 91–104.

Hatzimanikatis V, Floudas CA, Bailey JE. (1996) Optimization of regulatory architectures in metabolic reaction networks. *Biotechnol Bioeng* **52**, 485–500.

Hebert CG, Valdes JJ, Bentley WE. (2008) Beyond silencing — engineering applications of RNA interference and antisense technology for altering cellular phenotype. *Curr Opin Biotechnol* **19**, 500–505.

Heijnen JJ, Van Dijken JP. (1992) In search of a thermodynamic description of biomass yields for the chemotrophic growth of microorganisms. *Biotechnol Bioeng* **39**, 833–858.

Heinrich R, Rapoport SM, Rapoport TA. (1977) Metabolic regulation and mathematical models. *Prog Biophys Mol Biol* **32**, 1–82.

Heinrich R, Rapoport TA. (1974) A linear steady-state treatment of enzymatic chains. General properties, control and effector strength. *Eur J Biochem* **42**, 89–95.

Hejazi M, Fettke J, Paris O, Steup M. (2009) The two plastidial starch-related dikinases sequentially phosphorylate glucosyl residues at the surface of both the A- and B-type allomorphs of crystallized maltodextrins but the mode of action differs. *Plant Physiol* **150**, 962–976.

Hemmat S, Lieberman DM, Most SP. (2010) An introduction to stem cell biology. *Facial Plast Surg* **26**, 343–349.

Hensing M. (1995) *Production of Extracellular Proteins by Kluyveromyces Yeasts.* Delft University of Technology, Delft, Netherlands.

Herbers K, Sonnewald U. (1996) Manipulating metabolic partitioning in transgenic plants. *Trends Biotechnol* **14**, 198–205.

Herrmann KM, Weaver LM. (1999) The Shikimate pathway. *Annu Rev Plant Physiol Plant Mol Biol* **50**, 473–503.

Heyer AG, Lloyd JR, Kossmann J. (1999) Production of modified polymeric carbohydrates. *Curr Opin Biotechnol* **10**, 169–174.

Higgins J. (1967) The theory of oscillating reactions. *Ind Engin Chem* **59**, 19–62.

Hildebrandt G. (1982) *The Time Structure of Adaptive Processes.* George Thieme Verlag, Stuttgart, Germany.

Hill TL, Chay TR. (1979) Theoretical methods for study of kinetics of models of the mitochondrial respiratory chain. *Proc Natl Acad Sci U S A* **76**, 3203–3207.

Hillel AT, Elisseeff JH. (2010) Embryonic progenitor cells in adipose tissue engineering. *Facial Plast Surg* **26**, 405–412.

Himmel ME, Ding SY, Johnson DK, Adney WS, Nimlos MR, Brady JW, Foust TD. (2007) Biomass recalcitrance: Engineering plants and enzymes for biofuels production. *Science* **315**, 804–807.

Hodgkin AL, Huxley AF. (1952) A quantitative description of membrane current and its applicatin to conduction and excitation in nerve. *J Physiol* **117**, 500–544.

Hofmeyr JH, Cornish-Bowden A. (1991) Quantitative assessment of regulation in metabolic systems. *Eur J Biochem* **200**, 223–236.

Holmberg N, Lilius G, Bailey JE, Bulow L. (1997) Transgenic tobacco expressing *Vitreoscilla hemoglobin* exhibits enhanced growth and altered metabolite production. *Nat Biotechnol* **15**, 244–247.

Holms WH. (1986) The central metabolic pathways of *Escherichia coli*: Relationship between flux and control at a branch point, efficiency of conversion to biomass, and excretion of acetate. *Curr Top Cell Regul* **28**, 69–105.

Hols P, Kleerebezem M, Schanck AN, Ferain T, Hugenholtz J, Delcour J, de Vos WM. (1999) Conversion of *Lactococcus lactis* from homolactic to homoalanine fermentation through metabolic engineering. *Nat Biotechnol* **17**, 588–592.

Holtz WJ, Keasling JD. (2010) Engineering static and dynamic control of synthetic pathways. *Cell* **140**, 19–23.

Hu WJ, Harding SA, Lung J, Popko JL, Ralph J, Stokke DD, Tsai CJ, Chiang VL. (1999) Repression of lignin biosynthesis promotes cellulose accumulation and growth in transgenic trees. *Nat Biotechnol* **17**, 808–812.

Hua Q, Araki M, Koide Y, Shimizu K. (2001) Effects of glucose, vitamins, and DO concentrations on pyruvate fermentation using *Torulopsis glabrata* IFO 0005 with metabolic flux analysis. *Biotechnol Prog* **17**, 62–68.

Hughes EH, Shanks JV. (2002) Metabolic engineering of plants for alkaloid production. *Metab Eng* **4**, 41–48.

Iglesias AA. (1990) On the metabolism of triose-phosphate in photosynthetic cells. Their involvement on the traffic of ATP and NADPH. *Biochem Educ* **18**, 2–5.

Iglesias AA, Podestá FE. (2005) Photosynthate formation and partitioning in crop plants. In: Pessarakli M (ed), *Handbook of Photosynthesis*, 2nd ed., pp. 525–545. CRC Press, Taylor & Francis Group, Boca Raton, Florida.

Iglesias AA, Podestá FE. (2008) Carbon metabolism in turfgrasses. In: Pessarakli M (ed), *Handbook of Turfgrass Management and Physiology*, pp. 29–45. CRC Press. Taylor & Francis Group, Boca Raton, Florida.

Ikonomon L, Schneider YJ, Agathos SN. (2003) Insect cell culture for industrial production of recombinant proteins. *Appl Microbiol Biotechnol* **62**, 1–20.

Ingalls BP, Sauro HM. (2003) Sensitivity analysis of stoichiometric networks: An extension of metabolic control analysis to non-steady state trajectories. *J Theor Biol* **222**, 23–36.

Ishizaki-Nishizawa O, Fujii T, Azuma M, Sekiguchi K, Murata N, Ohtani T, Toguri, T. (1996) Low-temperature resistance of higher plants is significantly enhanced by a nonspecific cyanobacterial desaturase. *Nat Biotechnol* **14**, 1003–1006.

Jafri MS, Rice JJ, Winslow RL. (1998) Cardiac Ca2+ dynamics: The roles of ryanodine receptor adaptation and sarcoplasmic reticulum load. *Biophys J* **74**, 1149–1168.

Jamshidi N, Palsson BO. (2008) Formulating genome-scale kinetic models in the post-genome era. *Mol Syst Biol* **4**, 171.

Jarman TR, Pace GW. (1984) Energy requirements for microbial exopolysaccharide synthesis. *Arch Microbiol* **137**, 231–235.

Jensen KF, Pedersen S. (1990) Metabolic growth rate control in *Escherichia coli* may be a consequence of subsaturation of the macromolecular biosynthetic apparatus with substrates and catalytic components. *Microbiol Rev* **54**, 89–100.

Jeong H, Tombor B, Albert R, Oltvai ZN, Barabasi AL. (2000) The large-scale organization of metabolic networks. *Nature* **407**, 651–654.

Jermy A. (2011) Soil fungi helped ancient plants to make land. *Nat Rev Microbiol* **9**, 6.

Jones JG. (1979) *A Guide to Methods for Estimating Microbial Numbers and Biomass*. Freshwater Biological Association, Windermere, UK.

Junker BH. (2010) Networks in biology. In: Junker BH, Schreiber F (eds), *Analysis of Biological Networks*, pp. 3–14. Wiley-Interscience, Hoboken, New Jersey.

Kacser H, Burns JA. (1973) The control of flux. *Symp Soc Exp Biol* **27**, 65–104.

Kacser H, Burns JA. (1981) The molecular basis of dominance. *Genetics* **97**, 639–666.

Kacser H, Burns JA. (1995) The control of flux. *Biochem Soc Trans* **23**, 341–366.

Kacser H, Porteous JW. (1987) Control of metabolism — what do we have to measure. *Trends in Biochemical Sciences* **12**, 5.

Kahn D, Westerhoff HV. (1991) Control theory of regulatory cascades. *J Theor Biol* **153**, 255–285.

Kajiwara S, Fraser PD, Kondo K, Misawa N. (1997) Expression of an exogenous isopentenyl diphosphate isomerase gene enhances isoprenoid biosynthesis in *Escherichia coli. Biochem J* **324**(Pt 2), 421–426.

Kanehisa M, Goto S. (2000) KEGG: Kyoto encyclopedia of genes and genomes. *Nucleic Acids Res* **28**, 27–30.

Kanehisa M, Goto S, Furumichi M, Tanabe M, Hirakawa M. (2010) KEGG for representation and analysis of molecular networks involving diseases and drugs. *Nucleic Acids Res* **38**, D355–D360.

Kanehisa M, Goto S, Hattori M, Aoki-Kinoshita KF, Itoh M, Kawashima S, Katayama T, Araki M, Hirakawa M. (2006) From genomics to chemical genomics: New developments in KEGG. *Nucleic Acids Res* **34**, D354–D357.

Kantz H, Schreiber T. (2005) *Nonlinear Time Series Analysis*. Cambridge University Press, New York.

Karp PD. (1998) Metabolic databases. *Trends Biochem Sci* **23**, 114–116.

Karp PD, Paley SM, Krummenacker M, Latendresse M, Dale JM, Lee TJ, Kaipa P, Gilham F, Spaulding A, Popescu L, Altman T, *et al.* (2009) Pathway tools version 13.0: Integrated software for pathway/genome informatics and systems biology. *Brief Bioinform* **11**, 40–79.

Kauffman SA. (1989) Adaptation on rugged fitness landscapes. In: Stein DL (ed), *Lectures in the Sciences of Complexity*, Vol. 1, pp. 527–712. Addison-Wesley Publishing, New Mexico.

Kauffman SA. (1995) *At Home in the Universe. The Search for the Laws of Self-Organization and Complexity*. Oxford University Press, New York.

Kauffman SA. (2008) *Reinventing the Sacred. A New View of Science, Reason, and Religion*. Basic Books, Perseus Book Group, New York.

Keasling JD. (1999) Gene-expression tools for the metabolic engineering of bacteria. *Trends Biotechnol* **17**, 452–460.

Keen N, Staskawicz B, Mekalanos J, Ausubel F, Cook RJ. (2000) Pathogens and hosts: The dance is the same, the couples are different. *Proc Natl Acad Sci U S A* **97**, 8752–8753.

Kell DB. (2006) Theodor Bucher Lecture. Metabolomics, modelling and machine learning in systems biology — towards an understanding of the languages of cells. Delivered on 3 July 2005 at the 30th FEBS Congress and the 9th IUBMB conference in Budapest. *Febs J* **273**, 873–894.

Kell DB, Van Dam K, Westerhoff HV. (1989) Control Analysis of Microbial Growth and Productivity. Cambridge University Press, London.

Kensy F, Engelbrecht C, Buchs J. (2009) Scale-up from microtiter plate to laboratory fermenter: Evaluation by online monitoring techniques of growth and protein expression in *Escherichia coli* and *Hansenula polymorpha* fermentations. *Microb Cell Fact* **8**, 68.

Keseler IM, Bonavides-Martinez C, Collado-Vides J, Gama-Castro S, Gunsalus RP, Johnson DA, Krummenacker M, Nolan LM, Paley S, Paulsen IT, Peralta-Gil M, *et al.* (2009) EcoCyc: A comprehensive view of *Escherichia coli* biology. *Nucleic Acids Res* **37**, D464–D470.

Keseler IM, Collado-Vides J, Gama-Castro S, Ingraham J, Paley S, Paulsen IT, Peralta-Gil M, Karp PD. (2005) EcoCyc: A comprehensive database resource for *Escherichia coli*. *Nucleic Acids Res* **33**, D334–D337.

Keulers M, Satroutdinov AD, Suzuki T, Kuriyama H. (1996a) Synchronization affector of autonomous short-period-sustained oscillation of *Saccharomyces cerevisiae*. *Yeast* **12**, 673–682.

Keulers M, Suzuki T, Satroutdinov AD, Kuriyama H. (1996b) Autonomous metabolic oscillation in continuous culture of *Saccharomyces cerevisiae* grown on ethanol. *FEMS Microbiol Lett* **142**, 253–258.

Khan NA. (2009) Novel in vitro and in vivo models to study central nervous system infections due to *Acanthamoeba* spp. *Exp Parasitol* **126**, 69–72.

Khetan A, Hu WS. (1999) Metabolic engineering of antibiotic biosynthesis for process improvement. In: Lee SY, Papoutsakis ET (eds), *Metabolic Engineering*, pp. 177–202. Marcel Dekker, New York.

Khetan A, Malmberg LH, Sherman DH, Hu WS. (1996) Metabolic engineering of cephalosporin biosynthesis in *Streptomyces clavuligerus*. *Ann N Y Acad Sci* **782**, 17–24.

Kholodenko BN. (2006) Cell-signalling dynamics in time and space. *Nat Rev Mol Cell Biol* **7**, 165–176.

Kiel C, Yus E, Serrano L. (2010) Engineering signal transduction pathways. *Cell* **140**, 33–47.

Kim HD, Shay T, O'Shea EK, Regev A. (2009) Transcriptional regulatory circuits: Predicting numbers from alphabets. *Science* **325**, 429–432.

Kirschner M, Mitchison T. (1986) Beyond self-assembly: From microtubules to morphogenesis. *Cell* **45**, 329–342.

Kitano H. (2001) *Foundations of Systems Biology*. The MIT Press, Cambridge, MA.

Kitano H. (2004) Biological robustness. *Nat Rev Genet* **5**, 826–837.

Kjeldsen T, Pettersson AF. (2003) Relationship between self-association of insulin and its secretion efficiency in yeast. *Protein Expr Purif* **27**, 331–337.

Klamt S, Gagneur J, von Kamp A. (2005) Algorithmic approaches for computing elementary modes in large biochemical reaction networks. *Syst Biol (Stevenage)* **152**, 249–255.

Klapa MI, Aon JC, Stephanopoulos G. (2003a) Systematic quantification of complex metabolic flux networks using stable isotopes and mass spectrometry. *Eur J Biochem* **270**, 3525–3542.

Klapa MI, Aon JC, Stephanopoulos G. (2003b) Ion-trap mass spectrometry used in combination with gas chromatography for high-resolution metabolic flux determination. *Biotechniques* **34**, 832–836, 838, 840 passim.

Klevecz RR, Bolen J, Forrest G, Murray DB. (2004) A genomewide oscillation in transcription gates DNA replication and cell cycle. *Proc Natl Acad Sci USA* **101**, 1200–1205.

Klingenberg M. (2008) The ADP and ATP transport in mitochondria and its carrier. *Biochim Biophys Acta* **1778**, 1978–2021.

Kloas W, Lutz I, Einspanier R. (1999) Amphibians as a model to study endocrine disruptors: II. Estrogenic activity of environmental chemicals *in vitro* and *in vivo*. *Sci Total Environ* **225**, 59–68.

Ko CH, Liang H, Gaber RF. (1993) Roles of multiple glucose transporters in *Saccharomyces cerevisiae*. *Mol Cell Biol* **13**, 638–648.

Koffas MA, Jung GY, Aon JC, Stephanopoulos G. (2002) Effect of pyruvate carboxylase overexpression on the physiology of *Corynebacterium glutamicum*. *Appl Environ Microbiol* **68**, 5422–5428.

Koizumi S, Endo T, Tabata K, Ozaki A. (1998) Large-scale production of UDP-galactose and globotriose by coupling metabolically engineered bacteria. *Nat Biotechnol* **16**, 847–850.

Kopka J, Schauer N, Krueger S, Birkemeyer C, Usadel B, Bergmuller E, Dormann P, Weckwerth W, Gibon Y, Stitt M, Willmitzer L, *et al.* (2005) GMD@CSB.DB: The Golm Metabolome Database. *Bioinformatics* **21**, 1635–1638.

Koshland DE Jr. (1998) The era of pathway quantification. *Science* **280**, 852–853.

Kotting O, Kossmann J, Zeeman SC, Lloyd JR. (2010) Regulation of starch metabolism: The age of enlightenment? *Curr Opin Plant Biol* **13**, 321–329.

Kruckeberg AL. (1996) The hexose transporter family of *Saccharomyces cerevisiae. Arch Microbiol* **166**, 283–292.

Kruse O, Hankamer B. (2010) Microalgal hydrogen production. *Curr Opin Biotechnol* **21**, 238–243.

Kruse O, Rupprecht J, Bader KP, Thomas-Hall S, Schenk PM, Finazzi G, Hankamer, B. (2005) Improved photobiological H2 production in engineered green algal cells. *J Biol Chem* **280**, 34170–34177.

Kruse O, Rupprecht J, Mussgnug JH, Dismukes GC, Hankamer B. (2005) Photosynthesis: A blueprint for solar energy capture and biohydrogen production technologies. *Photochem Photobiol Sci* **4**, 957–970.

Ku MS, Agarie S, Nomura M, Fukayama H, Tsuchida H, Ono K, Hirose S, Toki S, Miyao M, Matsuoka M. (1999) High-level expression of maize phosphoenolpyruvate carboxylase in transgenic rice plants. *Nat Biotechnol* **17**, 76–80.

Kuenzi MT, Fiechter A. (1969) Changes in carbohydrate composition and trehalase-activity during the budding cycle of *Saccharomyces cerevisiae. Arch Mikrobiol* **64**, 396–407.

Lakra WS, Swaminathan TR, Joy KP. (2010) Development, characterization, conservation and storage of fish cell lines: A review. *Fish Physiol Biochem*

Lamprecht I. (1980) Growth and metabolism in yeasts. In: Beezer AE (ed), *Biological Microcalorimetry*, pp. 43–112. Academic Press, London.

Lange CC, Wackett LP, Minton KW, Daly MJ. (1998) Engineering a recombinant *Deinococcus radiodurans* for organopollutant degradation in radioactive mixed waste environments. *Nat Biotechnol* **16**, 929–933.

Lara AR, Galindo E, Ramirez OT, Palomares LA. (2006) Living with heterogeneities in bioreactors: Understanding the effects of environmental gradients on cells. *Mol Biotechnol* **34**, 355–381.

Lara AR, Taymaz-Nikerel H, Mashego MR, van Gulik WM, Heijnen JJ, Ramirez OT and van Winden WA. (2009) Fast dynamic response of the fermentative metabolism of *Escherichia coli* to aerobic and anaerobic glucose pulses. *Biotechnol Bioeng* **104**, 1153–1161.

Larkum AW. (2010) Limitations and prospects of natural photosynthesis for bioenergy production. *Curr Opin Biotechnol* **21**, 271–276.

Larsson C, von Stockar U, Marison I, Gustafsson L. (1993) Growth and metabolism of *Saccharomyces cerevisiae* in chemostat cultures under carbon-, nitrogen-, or carbon- and nitrogen-limiting conditions. *J Bacteriol* **175**, 4809–4816.

Laskey RA. (1970) The use of antibiotics in the preparation of amphibian cell cultures from highly contaminated material. *J Cell Sci* **7**, 653–659.

Le Corre D, Bras J, Dufresne A. (2010) Starch nanoparticles: A review. *Biomacromolecules* **11**, 1139–1153.

Lee KH, Park JH, Kim TY, Kim HU, Lee SY. (2007) Systems metabolic engineering of *Escherichia coli* for L-threonine production. *Mol Syst Biol* **3**, 149.

Lee LE, Dayeh VR, Schirmer K, Bols NC. (2009) Applications and potential uses of fish gill cell lines: Examples with RTgill-W1. *In Vitro Cell Dev Biol Anim* **45**, 127–134.

Lee SB, Bailey JE. (1984) A mathematical model for lambda dv plasmid replication: Analysis of copy number mutants. *Plasmid* **11**, 166–177.

Lee SY, Papoutsakis ET. (1999) The challenges and promise of metabolic engineering. In: Lee SY, Papoutsakis ET (eds), *Metabolic Engineering*, pp. 1–12. Marcel Dekker Inc., New York.

Lee SY, Woo HM, Lee DY, Choi HS, Kim TY, Yun H. (2005) Systems-level analysis of genome-scale in silico metabolic models using MetaFluxNet. *Biotechnol Bioprocess Eng* **10**, 425–431.

Lee WN, Go VL. (2005) Nutrient-gene interaction: Tracer-based metabolomics. *J Nutr* **135**, 3027S–3032S.

Lein W, Bornke F, Reindl A, Ehrhardt T, Stitt M, Sonnewald U. (2004) Target-based discovery of novel herbicides. *Curr Opin Plant Biol* **7**, 219–225.

Lemaux PG. (2008) Genetically engineered plants and foods: A scientist's analysis of the issues (part I). *Annu Rev Plant Biol* **59**, 771–812.

Lemaux PG. (2009) Genetically engineered plants and foods: A scientist's analysis of the issues (part II). *Annu Rev Plant Biol* **60**, 511–559.

Lenbury Y, Neamvong A, Amornsamankul S, Puttapiban P. (1999) Modelling effects of high product and substrate inhibition on oscillatory behavior in continuous bioreactors. *Biosystems* **49**, 191–203.

Lerouxel O, Cavalier DM, Liepman AH, Keegstra K. (2006) Biosynthesis of plant cell wall polysaccharides — a complex process. *Curr Opin Plant Biol* **9**, 621–630.

Liao JC, Delgado J. (1992) Dynamic metabolic control theory. A methodology for investigating metabolic regulation using transient metabolic data. *Ann N Y Acad Sci* **665**, 27–38.

Liao JC, Hou SY, Chao YP. (1996) Pathway analysis, engineering, and physiological considerations for redirecting central metabolism. *Biotechnol Bioeng* **52**, 129–140.

Liebovitch LS, Todorov AT. (1996) Using fractals and nonlinear dynamics to determine the physical properties of ion channel proteins. *Crit Rev Neurobiol* **10**, 169–187.

Lienard D, Sourrouille C, Gomord V, Faye L. (2007) Pharming and transgenic plants. *Biotechnol Annu Rev* **13**, 115–147.

Linton JD, Stephenson RJ. (1978) A preliminary study on growth yields in relation to the carbon and energy content of various organic growth substrates. *FEMS Microbiol Lett* **3**, 95–98.

Liu CW, Lin CC, Yiu JC, Chen JJ, Tseng MJ. (2008) Expression of a *Bacillus thuringiensis* toxin (cry1Ab) gene in cabbage (*Brassica oleracea* L. var. capitata L.) chloroplasts confers high insecticidal efficacy against *Plutella xylostella*. *Theor Appl Genet* **117**, 75–88.

Lligadas G, Ronda JC, Galia M, Cadiz V. (2010) Plant oils as platform chemicals for polyurethane synthesis: Current state-of-the-art. *Biomacromolecules*

Lloyd AL, Lloyd D. (1993) Hypothesis: The central oscillator of the circadian clock is a controlled chaotic attractor. *Biosystems* **29**, 77–85.

Lloyd AL, Lloyd D. (1995) Chaos: Its significance and detection in biology. *Biol Rhythm Res* **26**, 233–252.

Lloyd AL, May RM. (2001) Epidemiology. How viruses spread among computers and people. *Science* **292**, 1316–1317.

Lloyd D. (1974) *The Mitochondria of Microorganisms*. Academic Press, London.

Lloyd D. (1992) Intracellular time keeping: Epigenetic oscillations reveal the functions of an ultradian clock. In: Lloyd D, Rossi ER (eds), *Ultradian Rhythms in Life Processes*, pp. 5–22. Springer-Verlag, London.

Lloyd D. (1993) *Flow Cytometry in Microbiology*. Springer-Verlag, London.

Lloyd D. (1998) Circadian and ultradian clock-controlled rhythms in unicellular microorganisms. *Adv Microb Physiol* **39**, 291–338.

Lloyd D. (2002) Noninvasive methods for the investigation of organisms at low oxygen levels. *Adv Appl Microbiol* **51**, 155–183.

Lloyd D. (2003) Effects of uncoupling of mitochondrial energy conservation on the ultradian clock-driven oscillations in *Saccharomyces cerevisiae* continuous culture. *Mitochondrion* **3**, 139–146.

Lloyd D. (2008) Respiratory oscillations in yeasts. *Adv Exp Med Biol* **641**, 118–140.

Lloyd D. (2009) Oscillations, synchrony and deterministic chaos. *Progress in Botany* **70**, 70–91.

Lloyd D, Aon MA, Cortassa S. (2001) Why homeodynamics, not homeostasis? *Scientific-WorldJournal* **1**, 133–145.

Lloyd D, Lloyd AL. (1994) Hypothesis: A controlled chaotic attractor constitutes the central oscillator of the circadian clock. *Biochem Soc Trans* **22**, 322S.

Lloyd D, Lloyd AL, Olsen LF. (1992) The cell division cycle: A physiologically plausible dynamic model can exhibit chaotic solutions. *Biosystems* **27**, 17–24.

Lloyd D, Murray DB. (2000) Redox cycling of intracellular thiols: State variables for ultradian, cell cycle division and circadian cycles? In: Vanden Driessche T, Guisset JL, Petiau-de Vries GM (eds), *The Redox State and Circadian Rhythms*, pp. 85–94. Kluwer Academic Publishers, Amsterdam.

Lloyd D, Murray DB. (2005) Ultradian metronome: Timekeeper for orchestration of cellular coherence. *Trends Biochem Sci* **30**, 373–377.

Lloyd D, Murray DB. (2006) The temporal architecture of eukaryotic growth. *FEBS Lett* **580**, 2830–2835.

Lloyd D, Murray DB. (2007) Redox rhythmicity: Clocks at the core of temporal coherence. *Bioessays* **29**, 465–473.

Lloyd D, Murray DB, Klevecz RR, Wolf J, Kuriyama H. (2008) The ultradian clock (40min) in yeast. In: Lloyd D, Rossi EL (eds), *Ultradian Rhythms from Molecules to Mind*, pp 11–42. Springer Science+Business Media B.V., New York.

Lloyd D, Poole RK, Edwards SW. (1982) *The Cell Division Cycle: Temporal Organization and Control of Cellular Growth and Reproduction*. Academic Press, London.

Lloyd D, Rossi EL. (2008) Epilogue: A new vision of life. In: Lloyd D, Rossi EL (eds), *Ultradian Rhythms from Molecules to Mind*, pp. 431–439. Springer Science+Business Media B.V., New York.

Lloyd D, Salgado LE, Turner MP, Suller MT, Murray D. (2002) Cycles of mitochondrial energization driven by the ultradian clock in a continuous culture of *Saccharomyces cerevisiae*. *Microbiol* **148**, 3715–3724.

Lopez de Felipe F, Kleerebezem M, de Vos WM, Hugenholtz, J. (1998) Cofactor engineering: A novel approach to metabolic engineering in *Lactococcus lactis* by controlled expression of NADH oxidase. *J Bacteriol* **180**, 3804–3808.

Lorenz EN. (1963) Deterministic nonperiodic flow. *J Atmos Sci* **20**, 130–141.

Lu X, Vora H, Khosla C. (2008) Overproduction of free fatty acids in *E. coli*: Implications for biodiesel production. *Metab Eng* **10**, 333–339.

Lubiniecki A. (1990) *Large-Scale Mammalian Cell Culture Technology*, Vol 10. Marcel Dekker Inc., New York and Basel.

Luisi PL. (2006) *The Emergence of Life. From Chemical Origins to Synthetic Biology.* Cambridge University Press, Cambridge, UK.

Lunn JE. (2007) Compartmentation in plant metabolism. *J Exp Bot* **58**, 35–47.

Luo CH, Rudy Y. (1991) A model of the ventricular cardiac action potential. Depolarization, repolarization, and their interaction. *Circ Res* **68**, 1501–1526.

Luo CH, Rudy Y. (1994) A dynamic model of the cardiac ventricular action potential. I. Simulations of ionic currents and concentration changes. *Circ Res* **74**, 1071–1096.

Lusis AJ, Weiss JN. (2010) Cardiovascular networks: Systems-based approaches to cardiovascular disease. *Circulation* **121**, 157–170.

Lutke-Eversloh T, Stephanopoulos G. (2008) Combinatorial pathway analysis for improved L-tyrosine production in *Escherichia coli*: Identification of enzymatic bottlenecks by systematic gene overexpression. *Metab Eng* **10**, 69–77.

Lyon AR, Joudrey PJ, Jin D, Nass RD, Aon MA, O'Rourke B, Akar FG. (2010) Optical imaging of mitochondrial function uncovers actively propagating waves of mitochondrial membrane potential collapse across intact heart. *J Mol Cell Cardiol* **49**, 565–575.

Lytovchenko A, Sonnewald U, Fernie AR. (2007) The complex network of non-cellulosic carbohydrate metabolism. *Curr Opin Plant Biol* **10**, 227–235.

Maack C, Cortassa S, Aon MA, Ganesan AN, Liu T, O'Rourke B. (2006) Elevated cytosolic Na+ decreases mitochondrial Ca2+ uptake during excitation–contraction coupling and impairs energetic adaptation in cardiac myocytes. *Circ Res* **99**, 172–182.

MacFarlane AGJ. (1973) *Dynamical Systems Models*. Harrap, London.

Magnus G, Keizer J. (1997) Minimal model of beta-cell mitochondrial Ca2+ handling. *Am J Physiol* **273**, C717–C733.

Mandelbrot BB. (1983) *The Fractal Geometry of Nature*. W.H. Freeman, New York.

Mandelbrot BB, Hudson RL. (2004) *The (Mis)behavior of Markets. A Fractal View of Risk, Ruin, and Reward*. Basic Books, New York.

Mansergh FC, Daly CS, Hurley AL, Wride MA, Hunter SM, Evans MJ. (2009) Gene expression profiles during early differentiation of mouse embryonic stem cells. *BMC Dev Biol* **9**, 5.

Marsin AS, Bertrand L, Rider MH, Deprez J, Beauloye C, Vincent MF, Van den Berghe, G, Carling D, Hue L. (2000) Phosphorylation and activation of heart PFK-2 by AMPK has a role in the stimulation of glycolysis during ischaemia. *Curr Biol* **10**, 1247–1255.

Marsin AS, Bouzin C, Bertrand L, Hue L. (2002) The stimulation of glycolysis by hypoxia in activated monocytes is mediated by AMP-activated protein kinase and inducible 6-phosphofructo-2-kinase. *J Biol Chem* **277**, 30778–30783.

Martegani E, Porro D, Ranzi BM, Alberghina L. (1990) Involvement of a cell size control mechanism in the induction and maintenance of oscillations in continuous cultures of budding yeast. *Biotechnol Bioeng* **36**, 453–459.

Martone PT, Estevez JM, Lu F, Ruel K, Denny MW, Somerville C, Ralph J. (2009) Discovery of lignin in seaweed reveals convergent evolution of cell-wall architecture. *Curr Biol* **19**, 169–175.

Mathews CK, van Holde KE, Ahern KG. (2000) *Biochemistry*. Addison Wesley Longman, Inc., San Francisco, CA.

Matthew T, Zhou W, Rupprecht J, Lim L, Thomas-Hall SR, Doebbe A, Kruse O, Hankamer B, Marx UC, Smith SM, Schenk PM. (2009) The metabolome of *Chlamydomonas reinhardtii* following induction of anaerobic H2 production by sulfur depletion. *J Biol Chem* **284**, 23415–23425.

McChesney JD, Venkataraman SK, Henri JT. (2007) Plant natural products: Back to the future or into extinction? *Phytochemistry* **68**, 2015–2022.

McPherson RM, MacRae TC. (2009) Evaluation of transgenic soybean exhibiting high expression of a synthetic *Bacillus thuringiensis* cry1A transgene for suppressing lepidopteran population densities and crop injury. *J Econ Entomol* **102**, 1640–1648.

Melendez-Hevia E. (1990) The game of the pentose phosphate cycle: A mathematical approach to study the optimization in design of metabolic pathways during evolution. *Biomed Biochim Acta* **49**, 903–916.

Melis A, Happe T. (2001) Hydrogen production. Green algae as a source of energy. *Plant Physiol* **127**, 740–748.

Melis A, Zhang L, Forestier M, Ghirardi M, Seibert, M. (2000) Sustained photobiological hydrogen gas production upon reversible inactivation of oxygen evolution in the green alga *Chlamydomonas reinhardtii*. *Plant Physiol* **122**, 127–136.

Mikkelsen R, Baunsgaard L, Blennow A. (2004) Functional characterization of alpha-glucan,water dikinase, the starch phosphorylating enzyme. *Biochem J* **377**, 525–532.

Miklos, GL, Maleszka R. (2000) Deus ex genomix. *Nat Neurosci* **3**, 424–425.

Miklos GL, Rubin GM. (1996) The role of the genome project in determining gene function: Insights from model organisms. *Cell* **86**, 521–529.

Mitchison T, Kirschner M. (1984) Dynamic instability of microtubule growth. *Nature* **312**, 237–242.

Mittendorf V, Robertson EJ, Leech RM, Kruger N, Steinbuchel A, Poirier Y. (1998) Synthesis of medium-chain-length polyhydroxyalkanoates in arabidopsis thaliana using intermediates of peroxisomal fatty acid beta-oxidation. *Proc Natl Acad Sci U S A* **95**, 13397–13402.

Mol JNM, Holton TS, Koes RE. (1995) Floriculture: Genetic engineering of commercial traits. *Trends Biotechnol* **13**, 350–355.

Monaco ME. (1996) *Crecimiento y proliferacion de* Saccharomyces cerevisiae. PhD thesis, Universidad Nacional de Tucuman, Tucuman, Argentina.

Monaco ME, Valdecantos PA, Aon MA. (1995) Carbon and energy uncoupling associated with cell cycle arrest of cdc mutants of *Saccharomyces cerevisiae* may be linked to glucose-induced catabolite repression. *Exp Cell Res* **217**, 52–56.

Monaghan SR, Kent ML, Watral VG, Kaufman RJ, Lee LE, Bols NC. (2009) Animal cell cultures in microsporidial research: Their general roles and their specific use for fish microsporidia. *In Vitro Cell Dev Biol Anim* **45**, 135–147.

Mooney BP. (2009) The second green revolution? Production of plant-based biodegradable plastics. *Biochem J* **418**, 219–232.

Morandini P. (2009) Rethinking metabolic control. *Plant Science* **176**, 441–451.

Morel Y, Barouki R. (1999) Repression of gene expression by oxidative stress. *Biochem J* **342**(Pt 3), 481–496.

Morell MK, Myers AM. (2005) Towards the rational design of cereal starches. *Curr Opin Plant Biol* **8**, 204–210.

Morgan JA, Rijhwani SK, Shanks JV. (1999) Metabolic engineering for the production of plant secondary metabolites. In: Lee SY, Papoutsakis ET (eds), *Metabolic Engineering*, pp. 325–351. Marcel Dekker, New York.

Mukhopadhyay A, Redding AM, Rutherford BJ, Keasling JD. (2008) Importance of systems biology in engineering microbes for biofuel production. *Curr Opin Biotechnol* **19**, 228–234.

Munch T, Sonnleitner B, Fiechter A. (1992) New insights into the synchronization mechanism with forced synchronous cultures of *Saccharomyces cerevisiae*. *J Biotechnol* **24**, 299–314.

Munns R, Tester M. (2008) Mechanisms of salinity tolerance. *Annu Rev Plant Biol* **59**, 651–681.

Murai T, Ueda M, Yamamura M, Atomi H, Shibasaki Y, Kamasawa N, Osumi M, Amachi T, Tanaka, A. (1997) Construction of a starch-utilizing yeast by cell surface engineering. *Appl Environ Microbiol* **63**, 1362–1366.

Murata N. (1969) Control of excitation transfer in photosynthesis. I. Light-induced change of chlorophyll a fluorescence in *Porphyridium cruentum*. *Biochim Biophys Acta* **172**, 242–251.

Murray DB, Beckmann M, Kitano H. (2007a) Regulation of yeast oscillatory dynamics. *Proc Natl Acad Sci U S A* **104**, 2241–2246.

Murray DB, Engelen F, Lloyd D, Kuriyama H. (1999) Involvement of glutathione in the regulation of respiratory oscillation during a continuous culture of *Saccharomyces cerevisiae*. *Microbiol* **145**(Pt 10), 2739–2745.

Murray DB, Engelen FA, Keulers M, Kuriyama H, Lloyd D. (1998) NO+, but not NO., inhibits respiratory oscillations in ethanol-grown chemostat cultures of *Saccharomyces cerevisiae*. *FEBS Lett* **431**, 297–299.

Murray DB, Lloyd D. (2007a) A tuneable attractor underlies yeast respiratory dynamics. *Biosystems* **90**, 287–294.

Murray DB, Lloyd D, Kitano H. (2007b) Frequency modulation of the yeast reaction network. *FEBS J* **274**, 240.

Murray DB, Roller S, Kuriyama H, Lloyd D. (2001) Clock control of ultradian respiratory oscillation found during yeast continuous culture. *J Bacteriol* **183**, 7253–7259.

Nadeau JH, Burrage LC, Restivo J, Pao YH, Churchill G, Hoit BD. (2003) Pleiotropy, homeostasis, and functional networks based on assays of cardiovascular traits in genetically randomized populations. *Genome Res* **13**, 2082–2091.

Nelson DL, Cox MM. (2005) *Lehninger Principles of Biochemistry*, 4th. ed. W.H. Freeman, New York.

Nicolis G, Prigogine I. (1977) *Self-organization in Nonequilibrium Systems: From Dissipative Structures to Order through Fluctuations*. Wiley, New York.

Niederberger P, Prasad R, Miozzari G, Kacser H. (1992) A strategy for increasing an *in vivo* flux by genetic manipulations. The tryptophan system of yeast. *Biochem J* **287**(Pt 2), 473–479.

Nielsen J. (1997) Metabolic control analysis of biochemical pathways based on a thermokinetic description of reaction rates. *Biochem J* **321**(Pt 1), 133–138.

Noor E, Eden E, Milo R, Alon U. (2010) Central carbon metabolism as a minimal biochemical walk between precursors for biomass and energy. *Mol Cell* **39**, 809–820.

Norsker NH, Barbosa MJ, Vermue MH, Wijffels RH. (2010) Microalgal production — A close look at the economics. *Biotechnol Adv* **29**, 24–27.

Novikova LA, Faletrov YV, Kovaleva IE, Mauersberger S, Luzikov VN, Shkumatov VM. (2009) From structure and functions of steroidogenic enzymes to new technologies of gene engineering. *Biochemistry (Mosc)* **74**, 1482–1504.

Nurse P. (2003a) The great ideas of biology. *Clin Med* **3**, 560–568.

Nurse P. (2003b) Systems biology: Understanding cells. *Nature* **424**, 883.

Nyiri LK. (1972) Applications of computers in biochemical engineering. In: *Advances in Biochemical Engineering*. Springer Verlag, New York.

O'Grady J, Morgan JA. (2010) Heterotrophic growth and lipid production of *Chlorella protothecoides* on glycerol. *Bioprocess Biosyst Eng* **34**, 121–125.

O'Rourke B, Cortassa S, Aon MA. (2005) Mitochondrial ion channels: Gatekeepers of life and death. *Physiology (Bethesda)* **20**, 303–315.

Obembe OO, Popoola JO, Leelavathi S, Reddy SV. (2011) Advances in plant molecular farming. *Biotechnol Adv* **29**, 210–222.

Ohlrogge J. (1999) Plant metabolic engineering: Are we ready for phase two? *Curr Opin Plant Biol* **2**, 121–122.

Oliver S. (2000) Guilt-by-association goes global. *Nature* **403**, 601–603.

Oliver SG. (1996) From DNA sequence to biological function. *Nature* **379**, 597–600.

Oliver SG. (2006) From genomes to systems: The path with yeast. *Philos Trans R Soc Lond B Biol Sci* **361**, 477–482.

Oltvai ZN, Barabasi AL. (2002) Systems biology. Life's complexity pyramid. *Science* **298**, 763–764.

Orth JD, Palsson BO. (2010) Systematizing the generation of missing metabolic knowledge. *Biotechnol Bioeng* **107**, 403–412.

Orth JD, Thiele I, Palsson BO. (2010) What is flux balance analysis? *Nat Biotechnol* **28**, 245–248.

Otero JM, Nielsen J. (2010) Industrial systems biology. *Biotechnol Bioeng* **105**, 439–460.

Ott E, Sauer T, Yorke JA. (1994) *Coping with Chaos*. Wiley and Sane, New York.

Otterstedt K, Larsson C, Bill RM, Stahlberg A, Boles E, Hohmann S, Gustafsson L. (2004) Switching the mode of metabolism in the yeast *Saccharomyces cerevisiae*. *EMBO Rep* **5**, 532–537.

Ovadi J, Srere PA. (2000) Macromolecular compartmentation and channeling. *Int Rev Cytol* **192**, 255–280.

Overbeek R, Larsen N, Pusch GD, D'Souza M, Selkov E, Jr, Kyrpides N, Fonstein M, Maltsev N, Selkov E. (2000) WIT: Integrated system for high-throughput genome sequence analysis and metabolic reconstruction. *Nucleic Acids Res* **28**, 123–125.

Ozcan S, Johnston M. (1999) Function and regulation of yeast hexose transporters. *Microbiol Mol Biol Rev* **63**, 554–569.

Page N, Kluepfel D, Shareck F, Morosoli R. (1996) Increased xylanase yield in *Streptomyces lividans*: Dependence on number of ribosome-binding sites. *Nat Biotechnol* **14**, 756–759.

Palma M, Seret ML, Baret PV. (2009) Combined phylogenetic and neighbourhood analysis of the hexose transporters and glucose sensors in yeasts. *FEMS Yeast Res* **9**, 526–534.

Palsson B. (2009) Metabolic systems biology. *FEBS Lett* **583**, 3900–3904.

Panikov NS. (1995) *Microbial Growth kinetics*, 1st ed. Springer Verlag, Berlin.

Papin JA, Hunter T, Palsson BO, Subramaniam S. (2005) Reconstruction of cellular signalling networks and analysis of their properties. *Nat Rev Mol Cell Biol* **6**, 99–111.

Papoutsakis ET, Bennett GN. (1999) Molecular regulation and metabolic engineering of solvent production by *Clostridium acetobutylicum*. In: Lee SY, Papoutsakis ET (eds), *Metabolic Engineering*, pp. 253–279. Marcel Dekker, New York.

Park SJ, Cotter PA, Gunsalus RP. (1995) Regulation of malate dehydrogenase (mdh) gene expression in *Escherichia coli* in response to oxygen, carbon, and heme availability. *J Bacteriol* **177**, 6652–6656.

Park SJ, McCabe J, Turna J, Gunsalus RP. (1994) Regulation of the citrate synthase (gltA) gene of *Escherichia coli* in response to anaerobiosis and carbon supply: Role of the arcA gene product. *J Bacteriol* **176**, 5086–5092.

Parulekar SJ, Semones GB, Rolf MJ, Lievense JC, Lim HC. (1986) Induction and elimination of oscillations in continuous cultures of *Saccharomyces cerevisiae*. *Biotechnol Bioeng* **28**, 700–710.

Paul MJ. (2008) Trehalose 6-phosphate: A signal of sucrose status. *Biochem J* **412**, e1–2.

Paul MJ, Foyer CH. (2001) Sink regulation of photosynthesis. *J Exp Bot* **52**, 1383–1400.

Paul MJ, Primavesi LF, Jhurreea D, Zhang Y. (2008) Trehalose metabolism and signaling. *Annu Rev Plant Biol* **59**, 417–441.

Peng B, Petrov V, Showalter K. (1991) Controlling chemical chaos. *J Phys Chem* **95**, 4957–4959.

Penman S, Fulton A, Capco D, Ben Ze'ev A, Wittelsberger S, Tse CF. (1982) Cytoplasmic and nuclear architecture in cells and tissue: Form, functions, and mode of assembly. *Cold Spring Harb Symp Quant Biol* **46**(Pt 2), 1013–1028.

Perez-Garcia O, Escalante FM, de-Bashan LE, Bashan Y. (2011) Heterotrophic cultures of microalgae: Metabolism and potential products. *Water Res* **45**, 11–36.

Petersen OH, Findlay I. (1987) Electrophysiology of the pancreas. *Physiol Rev* **67**, 1054–1116.

Petersen OH, Maruyama Y. (1984) Calcium-activated potassium channels and their role in secretion. *Nature* **307**, 693–696.

Peterson JJ. (2004) A posterior predictive approach to multiple response surface optimization. *J Quality Technol* **36**, 139–153.

Pfeifer TA. (1998) Expression of heterologous proteins in stable insect cell culture. *Curr Opin Biotechnol* **9**, 518–521.

Pichersky E, Gang DR. (2000) Genetics and biochemistry of secondary metabolites in plants: An evolutionary perspective. *Trends Plant Sci* **5**, 439–445.

Pierik RLM. (1987) *In vitro Culture of Higher Plants*. Martinus Nijhoff Publishers, Dordrecht, The Netherlands.

Pietrobon D, Zoratti M, Azzone GF, Caplan SR. (1986) Intrinsic uncoupling of mitochondrial proton pumps. 2. Modeling studies. *Biochem* **25**, 767–775.

Pines O, Shemesh S, Battat E, Goldberg I. (1997) Overexpression of cytosolic malate dehydrogenase (MDH2) causes overproduction of specific organic acids in *Saccharomyces cerevisiae*. *Appl Microbiol Biotechnol* **48**, 248–255.

Pirt SJ. (1975) *Principles of Microbe and Cell Cultivation*. Blackwell Scientific, Oxford, UK.

Pittman JK, Dean AP, Osundeko O. (2011) The potential of sustainable algal biofuel production using wastewater resources. *Bioresour Technol* **102**, 17–25.

Pitzschke A, Hirt H. (2010) Bioinformatic and systems biology tools to generate testable models of signaling pathways and their targets. *Plant Physiol* **152**, 460–469.

Platteeuw C, Hugenholtz J, Starrenburg M, van Alen-Boerrigter I, de Vos WM. (1995) Metabolic engineering of *Lactococcus lactis*: Influence of the overproduction of alpha-acetolactate synthase in strains deficient in lactate dehydrogenase as a function of culture conditions. *Appl Environ Microbiol* **61**, 3967–3971.

Plaxton WC. (1996) The organization and regulation of plant glycolysis. *Annu Rev Plant Physiol Plant Mol Biol* **47**, 185–214.

Plaxton WC, Podestá FE. (2006) The functional organization and control of plant respiration. *Crit Rev Plant Sci* **25**, 159–198.

Poirier Y. (1999) Production of new polymeric compounds in plants. *Curr Opin Biotechnol* **10**, 181–185.

Polevoda B, Sherman F. (2003) N-terminal acetyltransferases and sequence requirements for N-terminal acetylation of eukaryotic proteins. *J Mol Biol* **325**, 595–622.

Poole RK, Lloyd D, Kemp RB. (1973) Respiratory oscillations and heat evolution in synchronously dividing cultures of fission yeast *Schizosaccharomyces-Pombe* 972h-. *J Gen Microbiol* **77**, 209–220.

Porro D, Martegani E, Ranzi BM, Alberghina L. (1988) Oscillations in continuous cultures of budding yeast: A segregated parameter analysis. *Biotechnol Bioeng* **32**, 411–417.

Posten C, Schaub G. (2009) Microalgae and terrestrial biomass as source for fuels — a process view. *J Biotechnol* **142**, 64–69.

Price ND, Schellenberger J, Palsson BO. (2004) Uniform sampling of steady-state flux spaces: Means to design experiments and to interpret enzymopathies. *Biophys J* **87**, 2172–2186.

Raamsdonk LM, Teusink B, Broadhurst D, Zhang N, Hayes A, Walsh MC, Berden JA, Brindle KM, Kell DB, Rowland JJ, Westerhoff HV, *et al.* (2001) A functional genomics strategy that uses metabolome data to reveal the phenotype of silent mutations. *Nat Biotechnol* **19**, 45–50.

Rakoczy-Trojanowska M. (2002) Alternative methods of plant transformation — a short review. *Cell Mol Biol Lett* **7**, 849–858.

Rathore AS, Winkle H. (2009) Quality by design for biopharmaceuticals. *Nat Biotechnol* **27**, 26–34.

Reaves ML, Rabinowitz JD. (2010) Metabolomics in systems microbiology. *Curr Opin Biotechnol*

Reddy AS, Thomas TL. (1996) Expression of a cyanobacterial delta 6-desaturase gene results in gamma-linolenic acid production in transgenic plants. *Nat Biotechnol* **14**, 639–642.

Reder C. (1988) Metabolic control theory: A structural approach. *J Theor Biol* **135**, 175–201.

Reed JL, Vo TD, Schilling CH, Palsson BO. (2003) An expanded genome-scale model of *Escherichia coli* K-12 (iJR904 GSM/GPR). *Genome Biol* **4**, R54.

Reed SI. (1992) The role of p34 kinases in the G1 to S-phase transition. *Annu Rev Cell Biol* **8**, 529–561.

Reich JG, Sel'kov EE. (1981) *Energy Metabolism of the Cell*. Academic Press, London.

Reifenberger E, Freidel K, Ciriacy M. (1995) Identification of novel HXT genes in *Saccharomyces cerevisiae* reveals the impact of individual hexose transporters on glycolytic flux. *Mol Microbiol* **16**, 157–167.

Reitz M, Sacher O, Tarkhov A, Trumbach D, Gasteiger J. (2004) Enabling the exploration of biochemical pathways. *Org Biomol Chem* **2**, 3226–3237.

Rice JJ, Jafri MS, Winslow RL. (2000) Modeling short-term interval-force relations in cardiac muscle. *Am J Physiol Heart Circ Physiol* **278**, H913–H931.

Rice JJ, Winslow RL, Hunter WC. (1999) Comparison of putative cooperative mechanisms in cardiac muscle: Length dependence and dynamic responses. *Am J Physiol* **276**, H1734–H1754.

Ritte G, Heydenreich M, Mahlow S, Haebel S, Kotting O, Steup M. (2006) Phosphorylation of C6- and C3-positions of glucosyl residues in starch is catalysed by distinct dikinases. *FEBS Lett* **580**, 4872–4876.

Rius SP, Casati P, Iglesias AA, Gomez-Casati DF. (2006) Characterization of an *Arabidopsis thaliana* mutant lacking a cytosolic non-phosphorylating glyceraldehyde-3-phosphate dehydrogenase. *Plant Mol Biol* **61**, 945–957.

Rius SP, Casati P, Iglesias AA, Gomez-Casati DF. (2008) Characterization of *Arabidopsis* lines deficient in GAPC-1, a cytosolic NAD-dependent glyceraldehyde-3-phosphate dehydrogenase. *Plant Physiol* **148**, 1655–1667.

Robson PR, McCormac AC, Irvine AS, Smith H. (1996) Genetic engineering of harvest index in tobacco through overexpression of a phytochrome gene. *Nat Biotechnol* **14**, 995–998.

Rocha I, Maia P, Evangelista P, Vilaca P, Soares S, Pinto JP, Nielsen J, Patil KR, Ferreira EC, Rocha M. (2010) OptFlux: An open-source software platform for in silico metabolic engineering. *BMC Syst Biol* **4**, 45.

Roels JA. (1983) *Energetics and Kinetics in Biotechnology*. Elsevier Biomedical Press, Amsterdam.

Roessner U, Luedemann A, Brust D, Fiehn O, Linke T, Willmitzer L, Fernie A. (2001) Metabolic profiling allows comprehensive phenotyping of genetically or environmentally modified plant systems. *Plant Cell* **13**, 11–29.

Roitsch T. (1999) Source-sink regulation by sugar and stress. *Curr Opin Plant Biol* **2**, 198–206.

Ronne H. (1995) Glucose repression in fungi. *Trends Genet* **11**, 12–17.

Rosa da Silva M, Sun J, Hongwu MA, He F, Zeng AP. (2008) Metabolic networks. In: Junker BH, Schreiber F (eds), *Analysis of Biological Networks*, pp. 233–253. Wiley-Interscience, Hoboken, New Jersey.

Rosati C, Aquilani R, Dharmapuri S, Pallara P, Marusic C, Tavazza R, Bouvier F, Camara B, Giuliano G. (2000) Metabolic engineering of beta-carotene and lycopene content in tomato fruit. *Plant J* **24**, 413–419.

Rosen R. (1967) *Optimality Principles in Biology*. Butterworths, London.

Rosen R. (1970) *Dynamical System Theory in Biology*. John Wiley & Sons, New York.

Roussel MR, Ivlev AA, Igamberdiev AU. (2006) Oscillations of the internal $CO(2)$ concentration in tobacco leaves transferred to low $CO(2)$. *J Plant Physiol* **34**, 1188–1196.

Roussel MR, Lloyd D. (2007) Observation of a chaotic multioscillatory metabolic attractor by real-time monitoring of a yeast continuous culture. *FEBS J* **274**, 1011–1018.

Rouwenhorst RJ, Visser LE, Van Der Baan AA, Scheffers WA, Van Dijken JP. (1988) Production, distribution, and kinetic properties of inulinase in continuous cultures of *Kluyveromyces marxianus* CBS 6556. *Appl Environ Microbiol* **54**, 1131–1137.

Rubin GM. (2001) The draft sequences. Comparing species. *Nature* **409**, 820–821.

Rubin GM, Yandell MD, Wortman JR, Gabor Miklos GL, Nelson CR, Hariharan IK, Fortini ME, Li PW, Apweiler R, Fleischmann W, Cherry JM, *et al.* (2000) Comparative genomics of the eukaryotes. *Science* **287**, 2204–2215.

Rugh CL, Senecoff JF, Meagher RB, Merkle SA. (1998) Development of transgenic yellow poplar for mercury phytoremediation. *Nat Biotechnol* **16**, 925–928.

Ruoppolo M, Vinci F, Klink TA, Raines RT, Marino G. (2000) Contribution of individual disulfide bonds to the oxidative folding of ribonuclease A. *Biochem* **39**, 12033–12042.

Ruppin E, Papin JA, de Figueiredo LF, Schuster S. (2010) Metabolic reconstruction, constraint-based analysis and game theory to probe genome-scale metabolic networks. *Curr Opin Biotechnol* **21**, 502–510.

Rupprecht J, Hankamer B, Mussgnug JH, Ananyev G, Dismukes C, Kruse O. (2006) Perspectives and advances of biological H2 production in microorganisms. *Appl Microbiol Biotechnol* **72**, 442–449.

Saito K, Matsuda F. (2010) Metabolomics for functional genomics, systems biology, and biotechnology. *Annu Rev Plant Biol* **61**, 463–489.

Saks V, Dzeja P, Schlattner U, Vendelin M, Terzic A, Wallimann T. (2006) Cardiac system bioenergetics: Metabolic basis of the Frank–Starling law. *J Physiol* **571**, 253–273.

Saks V, Monge C, Anmann T, Dzeja PP. (2007) Integrated and organized cellular energetic systems: Theories of cell energetics, compartmentation, and metabolic channeling. In: Saks V (ed), *Molecular System Bioenergetics: Energy for Life*, pp. 59–109. Wiley-VCH Verlag GmbH & Co. KGaA, Weinheim.

Saks VA, Kaambre T, Sikk P, Eimre M, Orlova E, Paju K, Piirsoo A, Appaix F, Kay L, Regitz-Zagrosek V, Fleck E, *et al.* (2001) Intracellular energetic units in red muscle cells. *Biochem J* **356**, 643–657.

Satroutdinov AD, Kuriyama H, Kobayashi H. (1992) Oscillatory metabolism of *Saccharomyces cerevisiae* in continuous culture. *FEMS Microbiol Lett* **77**, 261–267.

Sauer U. (2006) Metabolic networks in motion: 13C-based flux analysis. *Mol Syst Biol* **2**, 62.

Sauer U, Cameron DC, Bailey JE. (1998) Metabolic capacity of *Bacillus subtilis* for the production of purine nucleosides, riboflavin, and folic acid. *Biotechnol Bioeng* **59**, 227–238.

Sauer U, Hatzimanikatis V, Bailey JE, Hochuli M, Szyperski T, Wuthrich K. (1997) Metabolic fluxes in riboflavin-producing *Bacillus subtilis*. *Nat Biotechnol* **15**, 448–452.

Sauro HM, Small JR, Fell DA. (1987) Metabolic control and its analysis. Extensions to the theory and matrix method. *Eur J Biochem* **165**, 215–221.

Savageau MA. (1991) Biochemical systems theory: Operational differences among variant representations and their significance. *J Theor Biol* **151**, 509–530.

Savageau MA. (1995) Michaelis–Menten mechanism reconsidered: Implications of fractal kinetics. *J Theor Biol* **176**, 115–124.

Savinell JM, Palsson BO. (1992) Network analysis of intermediary metabolism using linear optimization. I. Development of mathematical formalism. *J Theor Biol* **154**, 421–454.

Savinell JM, Palsson BO. (1992) Network analysis of intermediary metabolism using linear optimization. II. Interpretation of hybridoma cell metabolism. *J Theor Biol* **154**, 455–473.

Savinell JM, Palsson BO. (1992) Optimal selection of metabolic fluxes for *in vivo* measurement. I. Development of mathematical methods. *J Theor Biol* **155**, 201–214.

Schneider ED, Sagan D. (2005) *Into the Cool. Energy Flow, Thermodynamics and Life.* The University of Chicago Press, Chicago.

Schomburg I, Chang A, Schomburg D. (2002) BRENDA, enzyme data and metabolic information. *Nucleic Acids Res* **30**, 47–49.

Schreiber F. (2010) Graph theory. In: Junker BH, Schreiber F (eds), *Analysis of Biological Networks*, pp. 15–28. Wiley-Interscience, Hoboken, New Jersey.

Schroeder M. (1991) *Fractals, Chaos, Power Laws. Minutes from an Infinite Paradise.* W.H. Freeman and Company, New York.

Schuller HJ, Entian KD. (1987) Isolation and expression analysis of two yeast regulatory genes involved in the derepression of glucose-repressible enzymes. *Mol Gen Genet* **209**, 366–373.

Schuster HG. (1988) *Deterministic Chaos: An Introduction.* VCH, Zurich, Germany.

Schuster S, Dandekar T, Fell DA. (1999) Detection of elementary flux modes in biochemical networks: A promising tool for pathway analysis and metabolic engineering. *Trends Biotechnol* **17**, 53–60.

Schuster S, Fell DA, Dandekar T. (2000) A general definition of metabolic pathways useful for systematic organization and analysis of complex metabolic networks. *Nat Biotechnol* **18**, 326–332.

Scott SA, Davey MP, Dennis JS, Horst I, Howe CJ, Lea-Smith DJ, Smith AG. (2010) Biodiesel from algae: Challenges and prospects. *Curr Opin Biotechnol* **21**, 277–286.

Seely RJ, Haury J. (2005) *Validation in Manufacturing of Pharmaceuticals.* Taylor & Francis, Boca Raton, Florida.

Segel IH. (1975) *Enzyme Kinetics : Behavior and Analysis of Rapid Equilibrium and Steady State Enzyme Systems.* Wiley, New York.

Segel LA. (1980) *Mathematical Models in Molecular and Cellular Biology.* Cambridge University Press, Cambridge [UK]; New York.

Sel'kov EE. (1968) Self-oscillations in glycolysis. 1. A simple kinetic model. *Eur J Biochem* **4**, 79–86.

Senior PJ, Beech GA, Ritchie GA, Dawes EA. (1972) The role of oxygen limitation in the formation of poly-β-hydroxybutyrate during batch and continuous culture of *Azotobacter beijerinckii. Biochem J* **128**, 1193–1201.

Sevenier R, Hall RD, van der Meer IM, Hakkert HJ, van Tunen AJ, Koops AJ. (1998) High level fructan accumulation in a transgenic sugar beet. *Nat Biotechnol* **16**, 843–846.

Shao H, Burrage LC, Sinasac DS, Hill AE, Ernest SR, O'Brien W, Courtland HW, Jepsen KJ, Kirby A, Kulbokas EJ, Daly MJ, *et al.* (2008) Genetic architecture of complex traits: Large phenotypic effects and pervasive epistasis. *Proc Natl Acad Sci U S A* **105**, 19910–19914.

Sharma AK, Sharma MK. (2009) Plants as bioreactors: Recent developments and emerging opportunities. *Biotechnol Adv.*

Shaw AJ, Podkaminer KK, Desai SG, Bardsley JS, Rogers SR, Thorne PG, Hogsett DA, Lynd LR. (2008) Metabolic engineering of a thermophilic bacterium to produce ethanol at high yield. *Proc Natl Acad Sci U S A* **105**, 13769–13774.

Shayne GC. (2007) *Textbook of Pharmaceutical Biotechnology.* John Wiley & Sons, Inc., New Jersey.

Shen B, Jensen RG, Bohnert HJ. (1997) Increased resistance to oxidative stress in transgenic plants by targeting mannitol biosynthesis to chloroplasts. *Plant Physiol* **113**, 1177–1183.

Shewry PR, Jones HD, Halford NG. (2008) Plant biotechnology: Transgenic crops. *Adv Biochem Eng Biotechnol* **111**, 149–186.

Shi NQ, Jeffries TW. (1998) Anaerobic growth and improved fermentation of *Pichia stipitis* bearing a URA1 gene from *Saccharomyces cerevisiae*. *Appl Microbiol Biotechnol* **50**, 339–345.

Shiloach J, Fass R. (2005) Growing *E. coli* to high cell density — a historical perspective on method development. *Biotechnol Adv* **23**, 345–357.

Shimada H, Kondo K, Fraser PD, Miura Y, Saito T, Misawa N. (1998) Increased carotenoid production by the food yeast *Candida utilis* through metabolic engineering of the isoprenoid pathway. *Appl Environ Microbiol* **64**, 2676–2680.

Shinbrot T, Ditto W, Grebogi C, Ott E, Spano M, Yorke JA. (1992) Using the sensitive dependence of chaos (the "butterfly effect") to direct trajectories in an experimental chaotic system. *Phys Rev Lett* **68**, 2863–2866.

Shu J, Shuler ML. (1989) A mathematical model for the growth of a single cell of *E. coli* on a glucose/glutamine/ammonium medium. *Biotechnol Bioeng* **33**, 1117–1126.

Shu J, Shuler ML. (1991) Prediction of effects of amino acid supplementation on growth of *E. coli* B/r. *Biotechnol Bioeng* **37**, 708–715.

Sierkstra LN, Verbakel JM, Verrips CT. (1992) Analysis of transcription and translation of glycolytic enzymes in glucose-limited continuous cultures of *Saccharomyces cerevisiae*. *J Gen Microbiol* **138**, 2559–2566.

Skyttner L. (2005) *General Systems Theory*. World Scientific, Singapore.

Slater JH. (1985) Microbial growth dynamics. In: Moo-Young M, Bull AT, Dalton H (eds), *Comprehensive Biotechnology. The Principles, Applications and Regulations of Biotechnology in Industry, Agriculture and Medicine*, Vol. 1, pp. 189–213. Pergamon Press, Oxford.

Slater S, Mitsky TA, Houmiel KL, Hao M, Reiser SE, Taylor NB, Tran M, Valentin HE, Rodriguez DJ, Stone DA, Padgette SR, *et al.* (1999) Metabolic engineering of Arabidopsis and Brassica for poly(3-hydroxybutyrate-co-3-hydroxyvalerate) copolymer production. *Nat Biotechnol* **17**, 1011–1016.

Slatyer RO, Tolbert NE. (1971) Photosynthesis and photorespiration. *Science* **173**, 1162–1167.

Slodzinski MK, Aon MA, O'Rourke B. (2008) Glutathione oxidation as a trigger of mitochondrial depolarization and oscillation in intact hearts. *J Mol Cell Cardiol* **45**, 650–660.

Smagghe G, Goodman CL, Stanley D. (2009) Insect cell culture and applications to research and pest management. *In Vitro Cell Dev Biol Anim* **45**, 93–105.

Smirnoff N. (1998) Plant resistance to environmental stress. *Curr Opin Biotechnol* **9**, 214–219.

Smith AM, Stitt M. (2007) Coordination of carbon supply and plant growth. *Plant Cell Environ* **30**, 1126–1149.

Smith TGJ, Lange GD. (1996) *Fractal Studies of Neuronal and Glial Cellular Morphology*. CRC Press, Boca Raton, Florida.

Smits HP, Hauf J, Muller S, Hobley TJ, Zimmermann FK, Hahn-Hagerdal B. (2000) Simultaneous overexpression of enzymes of the lower part of glycolysis can enhance the fermentative capacity of *Saccharomyces cerevisiae*. *Yeast* **16**, 1325–1334.

Sode K, Sugimoto S, Watanabe M, Tsugawa W. (1995) Effect of PQQ glucose dehydrogenase overexpression in *Escherichia coli* on sugar-dependent respiration. *J Biotechnol* **43**, 41–44.

Somerville C, Briscoe J. (2001) Genetic engineering and water. *Science* **292**, 2217.

Somerville CR, Bonetta D. (2001) Plants as factories for technical materials. *Plant Physiol* **125**, 168–171.

Sonnewald U, Hajirezaei MR, Kossmann J, Heyer A, Trethewey RN, Willmitzer L. (1997) Increased potato tuber size resulting from apoplastic expression of a yeast invertase. *Nat Biotechnol* **15**, 794–797.

Sornette D. (2000) *Critical Phenomena in Natural Sciences. Chaos, fractals, Selforganization and Disorder: Concepts and Tools*. Springer-Verlag, Berlin, Heidelberg.

Sornette D, Deschatres F, Gilbert T, Ageon Y. (2004) Endogenous versus exogenous shocks in complex networks: An empirical test using book sale rankings. *Phys Rev Lett* **93**, 228701.

Stam CJ. (2005) Nonlinear dynamical analysis of EEG and MEG: Review of an emerging field. *Clin Neurophysiol* **116**, 2266–2301.

Stark DM, Timmerman KP, Barry GF, Preiss J, Kishore GM. (1992) Regulation of the amount of starch in plant tissues by ADP glucose pyrophosphorylase. *Science* **258**, 287–292.

Staskawicz B. (2009) First insights into the genes that control plant-bacterial interactions. *Mol Plant Pathol* **10**, 719–720.

Staskawicz BJ, Ausubel FM, Baker BJ, Ellis JG, Jones JD. (1995) Molecular genetics of plant disease resistance. *Science* **268**, 661–667.

Stauffer D, Aharony A. (1994) *Introduction to Percolation Theory*. Taylor & Francis, London.

Steen EJ, Chan R, Prasad N, Myers S, Petzold CJ, Redding A, Ouellet M, Keasling JD. (2008) Metabolic engineering of *Saccharomyces cerevisiae* for the production of n-butanol. *Microb Cell Fact* **7**, 36.

Steen EJ, Kang Y, Bokinsky G, Hu Z, Schirmer A, McClure A, Del Cardayre SB, Keasling JD. (2010) Microbial production of fatty-acid-derived fuels and chemicals from plant biomass. *Nature* **463**, 559–562.

Stephanopoulos G. (2000) Bioinformatics and metabolic engineering. *Metab Eng* **2**, 157–158.

Stephanopoulos G. (2008) Metabolic engineering: Enabling technology for biofuels production. *Metab Eng* **10**, 293–294.

Stephanopoulos G, Aristidou AA, Nielsen J. (1998) *Metabolic Engineering. Principles and Methodologies*. Academic Press, San Diego, CA.

Stephanopoulos G, Vallino JJ. (1991) Network rigidity and metabolic engineering in metabolite overproduction. *Science* **252**, 1675–1681.

Stephens E, Ross IL, King Z, Mussgnug JH, Kruse O, Posten C, Borowitzka MA, Hankamer B. (2010a) An economic and technical evaluation of microalgal biofuels. *Nat Biotechnol* **28**, 126–128.

Stephens E, Ross IL, Mussgnug JH, Wagner LD, Borowitzka MA, Posten C, Kruse O, Hankamer B. (2010b) Future prospects of microalgal biofuel production systems. *Trends Plant Sci.* **15**, 554–564.

Stephenson RC, Clarke S. (1989) Succinimide formation from aspartyl and asparaginyl peptides as a model for the spontaneous degradation of proteins. *J Biol Chem* **264**, 6164–6170.

Steuer R, Zamora Lopez G. (2010) Global network properties. In: Junker BH, Schreiber F (eds), *Analysis of Biological Networks*, pp. 31–63. Wiley-Interscience, Hoboken, New Jersey.

Stitt M. (1999) The first will be the last and the last will be the first: Non-regulated enzymes call the tune? In: Bryant JA, Burnell MM, Kruger NJ (eds), *Plant Carbohydrate Biochemistry*, pp. 1–16. BIOS Scientific Publishers, Oxford, UK.

Stitt M, Fernie AR. (2003) From measurements of metabolites to metabolomics: An "on the fly" perspective illustrated by recent studies of carbon-nitrogen interactions. *Curr Opin Biotechnol* **14**, 136–144.

Stitt M, Sulpice R, Keurentjes J. (2010) Metabolic networks: How to identify key components in the regulation of metabolism and growth. *Plant Physiol* **152**, 428–444.

Stockdale GW, Cheng A. (2009) Finding design space and a reliable operating region using a multivariate Bayesian approach with experimental design. *Qual Technol Quant Manag* **6**, 391–408.

Stouthamer AH. (1979) The search for correlation between theoretical and experimental growth yields. *Int Rev Biochem* **21**, 1–47.

Stouthamer AH, van Verseveld HW. (1985) Stoichiometry of microbial growth. In: Moo-Young M, Bull AT, Dalton H (eds), *Comprehensive Biotechnology. The Principles, Applications and Regulations of Biotechnology in Industry, Agriculture and Medicine*, Vol 1, pp. 215–238. Pergamon Press, Oxford, UK.

Stouthamer AH, Van Verseveld HW. (1987) Microbial energetics should be considered in manipulating metabolism for biotechnological purposes. *TIBTECH* **5**, 149–155.

Stowers CC, Robertson JB, Ban H, Tanner RD, Boczko EM. (2009) Periodic fermentor yield and enhanced product enrichment from autonomous oscillations. *Appl Biochem Biotechnol* **156**, 59–75.

Strassle C, Sonnleitner B, Fiechter A. (1988/1989) A predictive model for the spontaneous synchronization of *Saccharomyces cerevisiae*. *J Biotechnol* **7**, 299–318.

Strogatz SH. (2001) Exploring complex networks. *Nature* **410**, 268–76.

Strom AR. (1998) Osmoregulation in the model organism *Escherichia coli*: Genes governing the synthesis of glycine betaine and trehalose and their use in metabolic engineering of stress tolerance. *J Biosci* **23**, 437–445.

Stucki JW. (1980) The optimal efficiency and the economic degrees of coupling of oxidative phosphorylation. *Eur J Biochem* **109**, 269–283.

Sudeep AB, Mourya DT, Mishra AC. (2005) Insect cell culture in research: Indian scenario. *Indian J Med Res* **121**, 725–738.

Surana U, Robitsch H, Price C, Schuster T, Fitch I, Futcher AB, Nasmyth K. (1991) The role of CDC28 and cyclins during mitosis in the budding yeast *S. cerevisiae*. *Cell* **65**, 145–161.

Sweetlove LJ, Muller-Rober B, Willmitzer L, Hill SA. (1999) The contribution of adenosine 5'-diphosphoglucose pyrophosphorylase to the control of starch synthesis in potato tubers. *Planta* **209**, 330–337.

Takahashi K, Tanabe K, Ohnuki M, Narita M, Ichisaka T, Tomoda K, Yamanaka S. (2007) Induction of pluripotent stem cells from adult human fibroblasts by defined factors. *Cell* **131**, 861–872.

Takahashi K, Yamanaka S. (2006) Induction of pluripotent stem cells from mouse embryonic and adult fibroblast cultures by defined factors. *Cell* **126**, 663–676.

Taticek RA, Choi C, Phan SE, Palomares LA, Shuler ML. (2001) Comparison of growth and recombinant protein expression in two different insect cell lines in attached and suspension culture. *Biotechnol Prog* **17**, 676–684.

Tcherkez GG, Farquhar GD, Andrews TJ. (2006) Despite slow catalysis and confused substrate specificity, all ribulose bisphosphate carboxylases may be nearly perfectly optimized. *Proc Natl Acad Sci U S A* **103**, 7246–7251.

Tempest DW, Hunter JR, Sykes J. (1965) Magnesium-limited growth of *Aerobacter aerogenes* in a chemostat. *J Gen Microbiol* **39**, 355–366.

Tempest DW, Neijssel OM. (1984) The status of YATP and maintenance energy as biologically interpretable phenomena. *Annu Rev Microbiol* **38**, 459–486.

Tharanathan RN. (2005) Starch — value addition by modification. *Crit Rev Food Sci Nutr* **45**, 371–384.

Thelen JJ, Ohlrogge JB. (2002) Metabolic engineering of fatty acid biosynthesis in plants. *Metab Eng* **4**, 12–21.

Theologis A. (1994) Control of ripening. *Curr Opin Biotechnol* **5**, 152–157.

Toivari MH, Aristidou A, Ruohonen L, Penttila M. (2001) Conversion of xylose to ethanol by recombinant *Saccharomyces cerevisiae*: Importance of xylulokinase (XKS1) and oxygen availability. *Metab Eng* **3**, 236–249.

Toussaint O, Weemaels G, Debacq-Chainiaux F, Scharffetter-Kochanek K, Wlaschek M. (2011) Artefactual effects of oxygen on cell culture models of cellular senescence and stem cell biology. *J Cell Physiol* **226**, 315–321.

Trager, W. (1964) Cultivation and physiology of erythrocytic stages of plasmodia. *Am J Trop Med Hyg* **13**(SUPPL), 162–166.

Tran LM, Rizk ML, Liao JC. (2008) Ensemble modeling of metabolic networks. *Biophys J* **95**, 5606–5617.

Trinh CT, Srienc F. (2009) Metabolic engineering of *Escherichia coli* for efficient conversion of glycerol to ethanol. *Appl Environ Microbiol* **75**, 6696–6705.

Tsuchiya HM. (1983) The holding time in pure and mixed culture fermentations. *Ann N Y Acad Sci* **413**, 184–192.

Tzfira T, Citovsky V. (2006) Agrobacterium-mediated genetic transformation of plants: Biology and biotechnology. *Curr Opin Biotechnol* **17**, 147–154.

United Nations PD. (2007) *World Population Prospects: The 2006 Revision*. United Nations, Department of Economic and Social Affairs, New York.

Vallino JJ, Stephanopoulos G. (1993) Metabolic flux distributions in *Corynebacterium glutamicum* during growth and lysine overproduction. *Biotechnol Bioeng* **41**, 633–646.

van Dam K. (1996) Role of glucose signaling in yeast metabolism. *Biotechnol Bioeng* **52**, 161–165.

Vanrolleghem PA, de Jong-Gubbels P, van Gulik WM, Pronk JT, van Dijken JP, Heijnen S. (1996) Validation of a metabolic network for *Saccharomyces cerevisiae* using mixed substrate studies. *Biotechnol Prog* **12**, 434–448.

Varela F, Maturana H, Uribe R. (1974) Autopoiesis: The organization of living sytems, its characterization and a model. *Biosystems* **5**, 187–196.

Varma A, Boesch BW, Palsson BO. (1993) Stoichiometric interpretation of *Escherichia coli* glucose catabolism under various oxygenation rates. *Appl Environ Microbiol* **59**, 2465–2473.

Varma A, Palsson BO. (1994) Metabolic flux balancing: Basic concepts, scientific and practical use. *Biotechnol* **12**, 994–998.

Varner J, Ramkrishna D. (1999) Metabolic engineering from a cybernetic perspective. 1. Theoretical preliminaries. *Biotechnol Prog* **15**, 407–425.

Vaseghi S, Baumeister A, Rizzi M, Reuss M. (1999) *In vivo* dynamics of the pentose phosphate pathway in *Saccharomyces cerevisiae*. *Metab Eng* **1**, 128–140.

Vaseghi S, Macherhammer F, Zibek S, Reuss M. (2001) Signal transduction dynamics of the protein kinase-A/phosphofructokinase-2 system in *Saccharomyces cerevisiae*. *Metab Eng* **3**, 163–172.

Vasil IK. (2008) A history of plant biotechnology: From the cell theory of Schleiden and Schwann to biotech crops. *Plant Cell Rep* **27**, 1423–1440.

Venter JC. (2007) *A Life Decoded. My Genome: My Life.* Penguin Books, New York.

Verdoni N, Aon MA, Lebeault JM. (1992) Metabolic and energetic control of *Pseudomonas mendocina* growth during transitions from aerobic to oxygen-limited conditions in chemostat cultures. *Appl Environ Microbiol* **58**, 3150–3156.

Verdoni N, Aon MA, Lebeault JM, Thomas D. (1990) Proton motive force, energy recycling by end product excretion, and metabolic uncoupling during anaerobic growth of *Pseudomonas mendocina*. *J Bacteriol* **172**, 6673–6681.

Verduyn C, Postma E, Scheffers WA, Van Dijken JP. (1992) Effect of benzoic acid on metabolic fluxes in yeasts: A continuous-culture study on the regulation of respiration and alcoholic fermentation. *Yeast* **8**, 501–517.

Vespignani, A. (2009) Predicting the behavior of techno-social systems. *Science* **325**: 425–8.

Vicsek T. (2001) *Fluctuations and Scaling in Biology.* Oxford University Press, New York.

Vidal M. (2009) A unifying view of 21st century systems biology. *FEBS Lett* **583**, 3891–3894.

Vlassara H. (2005) Advanced glycation in health and disease: Role of the modern environment. *Ann N Y Acad Sci* **1043**, 452–460.

Von Bertalanffy L. (1950) The theory of open systems in physics and biology. *Science* **111**, 23–29.

Von Bertalanffy L. (1955) General systems theory. *Main currents in modern thought* **71**.

Von Klitzing L, Betz A. (1970) Metabolic control in flow systems. 1. Sustained glycolytic oscillations in yeast suspensioins under continuous substrate infusion. *Arch Mikrobiol* **71**, 220–225.

von Meyenburg K. (1973) Stable synchrony oscillations in continuous cultures of Saccharomyces cerevisiae under glucose limitations. In: Chance B, Pye EK, Ghosh AK, Hess B (eds), *Biological and Biochemical Oscillators*, pp. 411–417. Academic Press, New York.

Voznesenskaya EV, Franceschi VR, Kiirats O, Freitag H, Edwards GE. (2001) Kranz anatomy is not essential for terrestrial C4 plant photosynthesis. *Nature* **414**, 543–546.

Vriezen N, van Dijken JP. (1998) Fluxes and enzyme activities in central metabolism of myeloma cells grown in chemostat culture. *Biotechnol Bioeng* **59**, 28–39.

Wackett LP. (2010) Engineering microbes to produce biofuels. *Curr Opin Biotechnol* **22**, 1–6.

Wagner A, Fell DA. (2001) The small world inside large metabolic networks. *Proc Biol Sci* **268**, 1803–1810.

Walsh MC, Scholte M, Valkier J, Smits HP, van Dam K. (1996) Glucose sensing and signalling properties in *Saccharomyces cerevisiae* require the presence of at least two members of the glucose transporter family. *J Bacteriol* **178**, 2593–2597.

Wang CL, Cooney AL, Demain P, Dunnil A, Ehumphrey M, Lilly D. (1979) *Fermentation and Enzyme Technology*. John Wiley, New York.

Wang X, Da Silva NA. (1996) Site-specific integration of heterologous genes in yeast via Ty3 retrotransposition. *Biotechnol Bioeng* **51**, 703–712.

Watts DJ, Strogatz SH. (1998) Collective dynamics of "small-world" networks. *Nature* **393**, 440–442.

Weber W, Weber E, Geisse S, Memmert K. (2002) Optimisation of protein expression and establishment of the Wave Bioreactor for Baculovirus/insect cell culture. *Cytotechnol* **38**, 77–85.

Wei ML, Webster DA, Stark BC. (1998) Genetic engineering of *Serratia marcescens* with bacterial hemoglobin gene: Effects on growth, oxygen utilization, and cell size. *Biotechnol Bioeng* **57**, 477–483.

Wei, AC, Aon MA, O'Rourke B, Winslow RL and Cortassa S (2011) Mitochondrial energetics, pH regulation, and ion dynamics: a computational-experimental approach. *Biophys J* **100**: 2894–903.

Weiss JN, Yang L, Qu Z. (2006) Systems biology approaches to metabolic and cardiovascular disorders: Network perspectives of cardiovascular metabolism. *J Lipid Res* **47**, 2355–2366.

Welch GR, Clegg JS. (2010) From protoplasmic theory to cellular systems biology: A 150-year reflection. *Am J Physiol Cell Physiol* **298**, C1280–C1290.

Wenefrida I, Utomo HS, Blanche SB, Linscombe SD. (2009) Enhancing essential amino acids and health benefit components in grain crops for improved nutritional values. *Recent Pat DNA Gene Seq* **3**, 219–225.

Weng G, Bhalla US, Iyengar R. (1999) Complexity in biological signaling systems. *Science* **284**, 92–96.

Weng JK, Li X, Bonawitz ND, Chapple C. (2008) Emerging strategies of lignin engineering and degradation for cellulosic biofuel production. *Curr Opin Biotechnol* **19**, 166–172.

Weselake RJ, Taylor DC, Rahman MH, Shah S, Laroche A, McVetty PB, Harwood JL. (2009) Increasing the flow of carbon into seed oil. *Biotechnol Adv* **27**, 866–878.

West, B. J. (1990) *Fractal Physiology and Chaos in Medicine*, Vol. 1. World Scientific, Singapore.

West BJ. (1999) *Physiology, Promiscuity and Prophecy at the Millennium: A Tale of Tails*, Vol. 7. World Scientific, Singapore.

Westerhoff HV, Kell DB. (1987) Matrix method for determining steps most rate-limiting to metabolic fluxes in biotechnological processes. *Biotechnol Bioeng* **30**, 101–107.

Westerhoff HV, Lolkema JS, Otto R, Hellingwerf KJ. (1982) Thermodynamics of growth. Non-equilibrium thermodynamics of bacterial growth. The phenomenological and the mosaic approach. *Biochim Biophys Acta* **683**, 181–220.

Westerhoff HV, Van Dam K. (1987) *Thermodynamics and Control of Biological Free-Energy Transduction*. Elsevier, Amsterdam.

Wijffels RH, Barbosa MJ. (2010) An outlook on microalgal biofuels. *Science* **329**, 796–799.

Wiley HS, Shvartsman SY, Lauffenburger DA. (2003) Computational modeling of the EGF-receptor system: A paradigm for systems biology. *Trends Cell Biol* **13**, 43–50.

Williams GP. (2003) *Chaos Theory Tamed*. Joseph Henry Press, Washington D.C.

Willis AT. (1977) *Anaerobic Bacteriology: Clinical and Laboratory Practice*. Butterworths, London.

Willmitzer L. (1999) Plant biotechnology: Output traits — the second generation of plant biotechnology products is gaining momentum. *Curr Opin Biotechnol* **10**, 161–162.

Wilson KG. (1979) Problems in physics with many scales of length. *Scientific American* **241**, 158–179.

Wilson LP, Bouwer EJ. (1997) Biodegradation of aromatic compounds under mixed oxygen/denitrifying conditions: A review. *J Industrial Microbiol Biotechnol* **18**, 116–130.

Winslow RL, Cortassa S, Greenstein JL. (2005) Using models of the myocyte for functional interpretation of cardiac proteomic data. *J Physiol* **563**, 73–81.

Winslow RL, Rice J, Jafri S, Marban E, O'Rourke B. (1999) Mechanisms of altered excitation-contraction coupling in canine tachycardia-induced heart failure, II: Model studies. *Circ Res* **84**, 571–586.

Wist AD, Berger SI, Iyengar R. (2009) Systems pharmacology and genome medicine: A future perspective. *Genome Med* **1**, 11.

Wodicka L, Dong H, Mittmann M, Ho MH, Lockhart DJ. (1997) Genome-wide expression monitoring in *Saccharomyces cerevisiae*. *Nat Biotechnol* **15**, 1359–1367.

Wolf K, Quimby MC. (1964) Amphibian cell culture: Permanent cell line from the Bullfrog (*Rana Catesbeiana*). *Science* **144**, 1578–1580.

Wong GK, Chiu AT. (2010) Gene therapy, gene targeting and induced pluripotent stem cells: Applications in monogenic disease treatment. *Biotechnol Adv* **28**, 715–724.

Wykoff DD, Davies JP, Melis A, Grossman AR. (1998) The regulation of photosynthetic electron transport during nutrient deprivation in *Chlamydomonas reinhardtii*. *Plant Physiol* **117**, 129–139.

Xia Y, Yu H, Jansen R, Seringhaus M, Baxter S, Greenbaum D, Zhao H, Gerstein M. (2004) Analyzing cellular biochemistry in terms of molecular networks. *Annu Rev Biochem* **73**, 1051–1087.

Xie JL, Zhang L, Ye Q, Zhou QW, Xin L, Du P, Gan RB. (2003) Angiostatin production in cultivation of recombinant *Pichia pastoris* fed with mixed carbon sources. *Biotechnol Lett* **25**, 173–177.

Xie L, Li J, Xie L, Bourne PE. (2009) Drug discovery using chemical systems biology: Identification of the protein-ligand binding network to explain the side effects of CETP inhibitors. *PLoS Comput Biol* **5**, e1000387.

Yang JH, Yang L, Qu Z, Weiss JN. (2008) Glycolytic oscillations in isolated rabbit ventricular myocytes. *J Biol Chem* **283**, 36321–36327.

Yang L, Korge P, Weiss JN, Qu Z. (2010) Mitochondrial oscillations and waves in cardiac myocytes: Insights from computational models. *Biophys J* **98**, 1428–1438.

Yarmush ML, Berthiaume F. (1997) Metabolic engineering and human disease. *Nat Biotechnol* **15**, 525–528.

Yates FE. (1987) *Self-Organizing Systems. The Emergence of Order.* Plenum Press, New York.

Yates FE. (1992) Fractal applications in biology: Scaling time in biochemical networks. *Methods Enzymol* **210**, 636–675.

Yates FE. (1993) Self-organizing systems. *In* Boyd CAR, Noble D (eds), *The Logic of Life. The Challenge of Integrative Physiology*, pp. 189–218. Oxford University Press, New York.

Ye X, Al-Babili S, Kloti A, Zhang J, Lucca P, Beyer P, Potrykus I. (2000) Engineering the provitamin A (beta-carotene) biosynthetic pathway into (carotenoid-free) rice endosperm. *Science* **287**, 303–305.

Yu J, Hu K, Smuga-Otto K, Tian S, Stewart R, Slukvin II, Thomson JA. (2009) Human induced pluripotent stem cells free of vector and transgene sequences. *Science* **324**, 797–801.

Zaslavskaia LA, Lippmeier JC, Shih C, Ehrhardt D, Grossman AR, Apt KE. (2001) Trophic conversion of an obligate photoautotrophic organism through metabolic engineering. *Science* **292**, 2073–2075.

Zeeman SC, Kossmann J, Smith AM. (2010) Starch: Its metabolism, evolution, and biotechnological modification in plants. *Annu Rev Plant Biol* **61**, 209–234.

Zhang L, Li Y, Wang Z, Xia Y, Chen W, Tang K. (2007) Recent developments and future prospects of *Vitreoscilla hemoglobin* application in metabolic engineering. *Biotechnol Adv* **25**, 123–136.

Zhang L, Melis A. (2002) Probing green algal hydrogen production. *Philos Trans R Soc Lond B Biol Sci* **357**, 1499–1507; discussion 1507–1511.

Zhang M, Eddy C, Deanda K, Finkelstein M, Picataggio S. (1995) Metabolic engineering of a pentose metabolism pathway in ethanologenic *Zymomonas mobilis*. *Science* **267**, 240–243.

Zhang P, Dreher K, Karthikeyan A, Chi A, Pujar A, Caspi R, Karp P, Kirkup V, Latendresse M, Lee C, Mueller LA, *et al.* (2010) Creation of a genome-wide metabolic pathway database for *Populus trichocarpa* using a new approach for reconstruction and curation of metabolic pathways for plants. *Plant Physiol* **153**, 1479–1491.

Zhang W, Czupryn MJ. (2002) Free sulfhydryl in recombinant monoclonal antibodies. *Biotechnol Prog* **18**, 509–513.

Zhang Y, Zhu Y, Zhu Y, Li Y. (2009) The importance of engineering physiological functionality into microbes. *Trends Biotechnol* **27**, 664–672.

Zhengan W. (1978) Cultivation of tissues and cells from amphibian. *Acta Zoologica Sinica.*

Zhou L, Aon MA, Almas T, Cortassa S, Winslow RL, O'Rourke B. (2010) A reaction-diffusion model of ROS-induced ROS release in a mitochondrial network. *PLoS Comput Biol* **6**, e1000657.

Zhou L, Cortassa S, Wei AC, Aon MA, Winslow RL, O'Rourke B. (2009) Modeling cardiac action potential shortening driven by oxidative stress-induced mitochondrial oscillations in guinea pig cardiomyocytes. *Biophys J* **97**, 1843–1852.

Zhou T, Carlson JM, Doyle J. (2005) Evolutionary dynamics and highly optimized tolerance. *J Theor Biol* **236**, 438–447.

Zimmerman WB. (2005) Metabolic pathways reconstruction by frequency and amplitude response to forced glycolytic oscillations in yeast. *Biotechnol Bioeng* **92**, 91–116.

Zimmermann FK, Kauffman I, Rasenberg H, Haussmann P. (1977) Genetics of carbon catabolite repression in *Saccharomyces cerevisiae*: Genes involved in the derepression process. *Mol Gen Genet* **151**, 95–103.

Zorov DB, Filburn CR, Klotz LO, Zweier JL, Sollott SJ. (2000) Reactive oxygen species (ROS)-induced ROS release: A new phenomenon accompanying induction of the mitochondrial permeability transition in cardiac myocytes. *J Exp Med* **192**, 1001–1014.

Zupan J, Muth TR, Draper O, Zambryski P. (2000) The transfer of DNA from agrobacterium tumefaciens into plants: A feast of fundamental insights. *Plant J* **23**, 11–28.

Zupke C, Sinskey AJ, Stephanopoulos G. (1995) Intracellular flux analysis applied to the effect of dissolved oxygen on hybridomas. *Appl Microbiol Biotechnol* **44**, 27–36.

Zupke C, Stephanopoulos G. (1995) Intracellular flux analysis in hybridomas using mass balances and *in vitro* (13)C NMR. *Biotechnol Bioeng* **45**, 292–303.

Index